Prebiotics:
Development & Application

Prebiotics: Development & Application

G.R. Gibson and R.A. Rastall

John Wiley & Sons, Ltd

John Wiley & Sons Ltd, The Atrium, Southern Gate, Chichester,
West Sussex PO19 8SQ, England

Telephone (+44) 1243 779777

Email (for orders and customer service enquiries): cs-books@wiley.co.uk
Visit our Home Page on www.wileyeurope.com or www.wiley.com

Other Wiley Editorial Offices

John Wiley & Sons Inc., 111 River Street, Hoboken, NJ 07030, USA

Jossey-Bass, 989 Market Street, San Francisco, CA 94103-1741, USA

Wiley-VCH Verlag GmbH, Boschstr. 12, D-69469 Weinheim, Germany

John Wiley & Sons Australia Ltd, 42 McDougall Street, Milton, Queensland 4064, Australia

John Wiley & Sons (Asia) Pte Ltd, 2 Clementi Loop #02-01, Jin Xing Distripark, Singapore 129809

John Wiley & Sons Canada Ltd, 22 Worcester Road, Etobicoke, Ontario, Canada M9W 1L1

Wiley also publishes its books in a variety of electronic formats. Some content that appears in print may not be available in electronic books.

Library of Congress Cataloging-in-Publication Data

Gibson, G.R.
Prebiotics : development and application/G.R. Gibson & R.A. Rastall.
p. cm.
ISBN-13: 978-0-470-02313-6 (cloth : alk. paper)
ISBN-10: 0-470-02313-9 (cloth : alk. paper)
1. Functional foods. 2. Bacteria–Health aspects. 3. Colon (Anatomy)–Microbiology. I. Rastall, R. A. II. Title.

QP144.F85G53 2006
616.9′201–dc22 2006000953

British Library Cataloguing in Publication Data

A catalogue record for this book is available from the British Library
ISBN-13 978-0-470-02313-6 (HB)
ISBN-10 0-470-02313-9 (HB)

Typeset in 10/12 pt Times by Thomson Press (India) Ltd., New Delhi, India
Printed and bound in Great Britain by Antony Rowe Ltd, Chippenham, Wiltshire
This book is printed on acid-free paper responsibly manufactured from sustainable forestry
in which at least two trees are planted for each one used for paper production.

Contents

Contributors

Wolfram M. Brück Postdoctoral Fellow, Division of Biomedical Marine Research, Harbor Branch Oceangraphic Institution, 5600 US 1 North for Pierce, FL 34946, USA

Natalie R. Bullock University of Sheffield, Human Nutrition Unit, Divison of Clinical Sciences (North), Coleridge House, Northern General Hospital, Herries Road, Sheffield S5 7AU, UK

Tamara Casci School of Food Biosciences, University of Reading, Whiteknights, P O Box 226, Reading RG6 6AP, UK

Ross Crittenden Food Science Australia, Private Bag 16, Werribee, Vic 3030, Australia

George C. Fahey Jr Professor of Animal Sciences and Nutritional Sciences, University of Illinois, Department of Animal Sciences, 132 Animal Sciences Laboratory, 1207 West Gregory Drive, Urbana, IL 61801, USA

Glenn R. Gibson Food Microbial Sciences Unit, School of Food Biosciences, University of Reading, Whiteknights, PO Box 226, Reading RG6 6AP, UK

Mark Jones School of Food Biosciences, University of Reading, Whiteknights, PO Box 226, Reading, RG6 6AP, UK

Jeff Leach Paleobiotics Lab, 144 Arenas Valley, Silver City, NM 88061, USA

Anne L. McCartney Food Microbial Sciences Unit, School of Food Biosciences, University of Reading, Whiteknights, PO Box 226, Reading RG6 6AP, UK

Bodun A. Rabiu Food Microbial Sciences Unit, School of Food Biosciences, University of Reading, Whiteknights, PO Box 226, Reading RG6 6AP, UK

Robert A. Rastall Professor of Biotechnology, School of Food Biosciences, University of Reading, Whiteknights, PO Box 226, Reading RG6 6AP, UK

Gregor Reid Canadian Research and Development Centre for Probiotics, Lawson Health Research Institute, 268 Grosvenor Street, London, Ontario, Canada N6A 4V2

Kelly S. Swanson Professor of Animal Sciences and Nutritional Sciences, University of Illinois, Department of Animal Sciences, 132 Animal Sciences Laboratory, 1207 West Gregory Drive, Urbana, IL 61801, USA

Kieran M. Tuohy Food Microbial Sciences Unit, School of Food Biosciences, University of Reading, Whiteknights, PO Box 226, Reading RG6 6AP, UK

Jan Van Loo ORAFTI, Aandorenstraat 1, B-33000 Tienen, Belgium

Claire L. Vernazza Food Microbial Sciences Unit, School of Food Biosciences, University of Reading, Whiteknights, PO Box 226, Reading RG6 6AP, UK

1

Human Colonic Microbiology and the Role of Dietary Intervention: Introduction to Prebiotics

Claire L. Vernazza, Bodun A. Rabiu and Glenn R. Gibson

1.1 Acquisition and Development of the Human Gut Flora

The human embryo is virtually sterile, but at birth microbial colonisation of the gastrointestinal tract occurs, with the neonate receiving an inoculum from the birth canal (Fuller, 1991; Zetterström *et al.*, 1994). The microbial pattern that ensues depends on the method of delivery (Beritzoglou, 1997; Salminen *et al.*, 1998a) and hygiene precautions associated with parturition (Lundequist *et al.*, 1985). In addition to characteristic vaginal flora such as lactobacilli, yeast, streptococci, staphylococci and *Escherichia coli*, the neonate is also likely to come into contact with faecal microorganisms and skin bacteria during birth (Fuller, 1991). Furthermore, inoculation from the general environment and other external contacts may also be significant, especially during Caesarean delivery (Beritzoglou, 1997; Gronlund *et al.*, 1999). During the acquisition period, some bacteria transiently colonise the gut whilst others survive and grow to form the indigenous microflora. Consequently, the neonatal gut experiences a rapid succession of microfloral components in the first days to months of development, selected for, initially, by luminal redox potential (Eh) but more frequently reported as being due to the feeding regime that follows birth (Zetterström *et al.*, 1994). Initial colonisers utilise any available oxygen, usually by 48 h, creating an environment sufficiently reduced to allow succession by obligate anaerobes, mainly those belonging to the bifidobacteria, bacteroides and clostridia groups. At this stage, it appears that

Prebiotics: Development and Application Edited by G. R. Gibson and R. A. Rastall

feeding methods have a significant influence on the relative proportions of bacteria that establish in the infant gut. Historically, breast-fed infants are thought to have relatively higher proportions of bifidobacteria than formula-fed babies of the same age, who possess a more complex composition (Fuller, 1991). Such purported differences have been linked with a lower risk of gastrointestinal, respiratory and urinary tract infections in breast-fed infants (Kunz and Rudloff, 1993). However, as the nature of commercial feeds has altered in recent times, the bifidobacterial predominance seen during breast feeding is less definitive. Nevertheless, such observations demonstrate the ability of diet to influence the gut microbiota composition and the possibilities for influencing health as a result. This has formed the basis for dietary intervention procedures that are extremely popular today (see later).

By the end of weaning there is a drop in the frequency of bifidobacteria. With the introduction of solid foods and by about 2 years after birth, infants start to adopt microflora profiles in proportions that approximate to those seen in adults (Fuller, 1991). The populations then seem to be relatively stable (>99 % anaerobic), aside for perturbations by diet and habit, until advanced ages when a significant decline in bifidobacteria, plus increases in clostridia and enterobacteriaceae are reported (Mitsuoka, 1990).

1.2 The Human Gastrointestinal Tract and its Microflora

Microorganisms occur along the whole length of the human alimentary tract with population numbers and species distribution characteristic of particular regions of the gut (Macfarlane *et al.*, 1997). After the mouth, colonisation is markedly influenced, in part by luminal pH, and by the progressively slower transit of food materials towards the colon. The movement of digesta through the stomach and small intestine is rapid (ca. 4–6 h), when compared with a typical colonic transit time of around 48–70 h for adults (Macfarlane and Gibson, 1994). This allows the establishment of a complex and relatively stable bacterial community in the large intestine (Table 1.1). The near neutral pH and the relatively low absorptive state of the colon further encourages extensive microbial colonisation and growth (Macfarlane *et al.*, 1997; O'Sullivan, 1996).

The human large intestine consists of the caecum, ascending colon, transverse colon, descending colon, sigmoid colon and rectum (Macfarlane and Macfarlane, 1997) (Figure 1.1). Through the microflora, the colon is capable of complex hydrolytic-digestive functions (Cummings and Macfarlane, 1991). This involves the breakdown of dietary components, principally complex carbohydrates, but also some proteins, that are not hydrolysed nor absorbed in the upper digestive tract (Macfarlane *et al.*, 1992). Carbohydrate availability subsequently diminishes as dietary residues pass from the proximal colon to the transverse and distal bowel.

For persons living on Western-style diets, the microbial biomass makes up over 50 % of colonic contents. There are more than 500 different culturable species of indigenous bacteria present in the adult large intestine comprising around 10^{12} bacteria per gram dry weight (Moore *et al.*, 1978; Simon and Gorbach, 1984). A summary of the principal bacterial groups present is shown in Table 1.2.

In very general terms, intestinal bacteria can be divided on the basis of whether they can exert health promoting, benign or potentially harmful activities in their host (Gibson

Table 1.1 *Host factors that may determine population levels of microbiota in various regions of the human gastrointestinal tract (Data sourced from Rowland and Mallett, 1990; Macfarlane and McBain, 1999)*

Anatomical regions	Mouth	Stomach	Duodenum	Ileum	Colon
pH	Alkaline	1.0–3.0	Acidic–Neutral	Neutral–alkaline	5.5–7.2
Stasis	Periodic	Periodic	Propulsions	Prolonged	Prolonged/retentive
Microenvironment	Teeth/tongue	Mucus	Mucus	Mucus	Mucus/food/crypts/epithelium
Function	Mastication/partial digestion	Digestion	Digestion/absorption	Digestion/absorption	Digestion/fluid and salts reabsorption
Approximate number of cells per ml or g content	10^8	10–100	10–1000	10^4–10^6	10^{12}

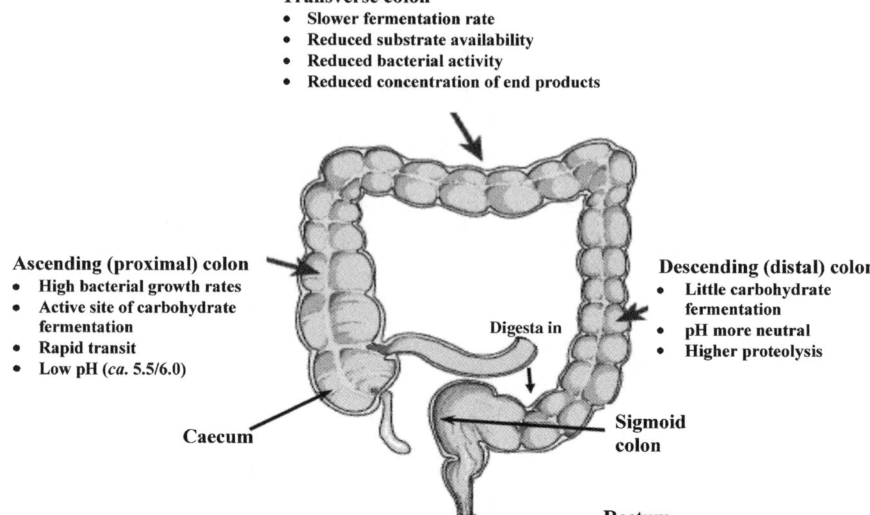

Figure 1.1 *Regions of the human large intestine with corresponding bacterial activities and physiological differences (Adapted from Cummings and Macfarlane, 1991). Modified with the permission of the authors from Journal of Applied Bacteriology, Vol 70, Cummings J.H. and Macfarlane, G.T., The control and consequences of bacterial fermentation in the human colon: a review, pp. 443–459, 1991, by permission of Blackwell Publishing*

and Roberfroid, 1995). The most obvious pathogens are strains of *E. coli* and clostridia. Pathogenic effects include diarrhoeal infections and putrefaction whereas beneficial aspects may be derived simply by improved the digestion/absorption of essential nutrients. This leads towards a consideration of factors that may influence the flora composition in a manner than can impact upon health.

The multiplicity of substrates is probably the single most important determinant for dynamics and stability of species existing in the large bowel (Gibson and Collins, 1999). Whilst these are mainly produced by dietary residues, there is appreciable contribution from host secretions like mucins. The colonic microflora derive substrates for growth from the diet (e.g. nondigestible oligosaccharides, dietary fibre, undigested protein reaching the colon) and from endogenous sources such as mucin, the main glycoprotein constituent of the mucus which lines the walls of the gastrointestinal tract (Rowland and Wise, 1985). The vast majority of bacteria in the colon are strict anaerobes and thus derive energy from fermentation (Macfarlane and McBain, 1999). The two main fermentative substrates of dietary origin are nondigestible carbohydrates (e.g. resistant starch, nonstarch polysaccharides and fibres of plant origin and nondigestible oligosaccharides) and protein which escapes digestion in the small intestine. Of these, carbohydrate fermentation is more energetically favourable, leading to a gradient of substrate utilisation spatially through the colon (Macfarlane *et al.*, 1992). The proximal colon is a saccharolytic environment with the majority of carbohydrate entering the colon being fermented in this region. As digesta moves through towards the distal colon, carbohydrate availability decreases and protein and amino acids become a more dominant metabolic energy source for bacteria in the distal colon (Macfarlane *et al.*, 1992). Overall

Table 1.2 *Description and nutrition of numerically predominant anaerobes in the human large intestine (Salminen et al., 1998a)*

Bacteria	Description	Numbers reported in faeces ($\log_{10}$ per g dry wt)		Nutrition	Fermentation products[a]
		Mean	Range		
Bacteroides	Gram −ve rods	11.3	9.2–13.5	Saccharolytic	A, P, S
Eubacteria	Gram +ve rods	10.7	5.0–13.3	Saccharolytic, some amino acid fermenters	A, B, L
Bifidobacteria	Gram +ve rods	10.2	4.9–13.4	Saccharolytic	A, L, f, e
Clostridia	Gram +ve rods	9.8	3.3–13.1	Saccharolytic, some amino acid fermenters	A, P, B, L, e
Lactobacilli	Gram +ve rods	9.6	3.6–12.5	Saccharolytic	L
Fusobacteria	Gram −ve rods	8.4	5.1–11.0	Amino acid fermenters, carbohydrate assimilated	B, A, L
Ruminococci	Gram +ve cocci	10.2	4.6–12.8	Saccharolytic	A
Peptostreptococci	Gram +ve cocci	10.1	3.8–12.6	Saccharolytic, some amino acid fermenters	A, L
Peptococci	Gram +ve cocci	10.0	5.1–12.9	Amino acid fermenters	A, B, L
Propionibacteria	Gram +ve rods	9.4	4.3–12.0	Saccharolytic, lactate fermenters	A, P
Actinomyces	Gram +ve rods	9.2	5.7–11.1	Saccharolytic	A, L, S
Streptococci	Gram +ve cocci	8.9	3.9–12.9	Carbohydrate and amino acid fermentation	L, A
Escherichia	Gram −ve rods	8.6	3.9–12.3	Carbohydrate and amino acid fermentation	Mixed acid
Desulfovibrios	Gram −ve rods	8.4	5.2–10.9	Various (e.g. SO_4, H_2, CO_2)	A
Methanobrevibacter	Gram +ve cocci bacilli	8.8	7.0–10.5	Chemolithotrophic	CH_4

[a]A, acetate; P, propionate; B, butyrate; L, lactate; S, succinate; f, formate; e, ethanol. +ve, positive; −ve, negative.

however, the principal substrates for bacterial growth are dietary carbohydrates. It has been estimated that about 10 to 60 g per day of dietary carbohydrate reaches the colon (Englyst and Cummings, 1986, 1987). A large proportion of this carbohydrate is made up of resistant starch (i.e. starch recalcitrant to the activities of human amylases). Resistant starch is readily fermented by a wide range of colonic bacterial species including members of the *Bacteroides* spp., *Eubacterium* spp. and the bifidobacteria (Englyst and Macfarlane, 1986). The remainder of the carbohydrate entering the colon is comprised of nonstarch polysaccharides (about 8–18 g per day), unabsorbed sugars, e.g. raffinose, stachyose and lactose (about 2–10 g per day) and oligosaccharides such as fructo-oligosaccharides, xylooligosaccharides, galactooligosaccharides (about 2–8 g per day) (Bingham *et al.*, 1990; Gibson *et al.*, 1990; Cummings and Macfarlane, 1991). The degree to which these carbohydrates are broken down by the gut microflora varies greatly. Unabsorbed sugars entering the colon are readily fermented and persist for only a short time in the proximal colon (Hudson and Marsh, 1995). Some sugars such as raffinose may have a more selective fermentation (being mainly assimilated by bifidobacteria and lactobacilli) while others support the growth of a range of colonic bacteria. Similarly, nondigestible oligosaccharides reaching the colon display different degrees of fermentation. Certain oligosaccharides such as fructooligosaccharides, galactooligosaccharides and lactulose may be fermented preferentially by bifidobacteria, which has given rise to the concept of prebiotics (discussed later) (Gibson and Roberfroid, 1995). Nonstarch polysaccharides include pectin, arabinogalactan, inulin, guar gum and hemicellulose, which are readily fermented by the colonic microflora, and lignin and cellulose, which are much less fermentable (Lewis *et al.*, 2001). Endogenous carbohydrates, chiefly from mucin and condroitin sulphate, contribute about 2–3 g per day of fermentable substrate (Quigley and Kelly, 1995). The main saccharolytic species in the colonic microflora belong to the genera *Bacteroides, Bifidobacterium, Ruminococcus, Eubacterium, Lactobacillus* and *Clostridium* (Gibson, 1998). Protein and amino acids are also available for bacterial fermentation in the colon. Approximately 25 g of protein enters the colon daily (Macfarlane and Macfarlane, 1997). Other sources of protein in the colon include bacterial secretions, sloughed epithelial cells, bacterial lysis products and mucins. The main proteolytic species belong to the bacteroides and clostridia groups.

Carbohydrates in the colon are fermented to short chain fatty acids (SCFA), principally, acetate, propionate and butyrate (Cummings, 1981, 1995) and a number of other metabolites such as the electron sink products lactate, pyruvate, ethanol, succinate as well as the gases H_2, CO_2, CH_4 and H_2S (Levitt *et al.*, 1995). SCFA are rapidly absorbed by the colonic mucosa and contribute towards energy requirements of the host (Cummings, 1981; Englehardt et al., 1991). Acetate is mainly metabolised in human muscle, kidney, heart and brain, while propionate is cleared by the liver, and is a possible gluceogenic precursor which suppresses cholesterol synthesis. Butyrate, on the other hand, is metabolised by the colonic epithelium where it serves as a regulator of cell growth and differentiation (Cummings, 1995). Protein reaching the colon is fermented to branched chain fatty acids such as isobutyrate, isovalerate and a range of nitrogenous compounds. Unlike carbohydrate fermentation, some of these end products may be toxic to the host, e.g. ammonia, amines and phenolic compounds (Macfarlane and Macfarlane, 1995). Excessive protein fermentation, especially in the distal colon, has therefore been linked with disease states such as colon cancer, which generally starts in this region of the colon before progressing proximally along the colon. Examples include bowel cancer and ulcerative colitis.

1.2.1 Influence of Diet on Microflora Activity and Health

Metchnikoff (1907) hypothesised that the onset of senility and shortening of life span resulted from putrefaction in the large bowel. In his opinion, the consumption of soured (fermented) milks was a progenitor for improved gastrointestinal health and the prolongation of life in Bulgarian populations. It is now known that bacterial activity in the human colon is involved in a number of disease states. The large intestine can harbour pathogens that are either part of the resident flora or exist as transient members (Gibson *et al.*, 1997). Attachment and overgrowth of the pathogens generally results in acute diarrhoeal infections, however more chronic forms of intestinal disease also occur (Gibson *et al.*, 1997; Gionchetti *et al.*, 2000). These include inflammatory bowel diseases (ulcerative colitis and Crohn's disease) (Chadwick and Anderson, 1995), colon cancer (Rowland, 1988; Rumney *et al.*, 1993) and pseudomembranous colitis (Duerden *et al.*, 1995). To varying extents, each has been linked into microflora composition and activities, and thereby diet as this provides the major source for their growth. The concept of probiotics was developed to influence the gut microbiota in a beneficial manner.

1.3 Probiotics

The word probiotic comes from the Greek 'for life' and is defined as 'a live microbial food supplement that is beneficial to host health' (Fuller, 1989). The definition of probiotics has evolved over the years, but the consensus designates probiotics as 'nonpathogenic, live microbial, mono- or mixed-culture preparations, which, when applied to humans or animals in high enough doses, beneficially affect the host by improving the intestinal microbial balance and its properties' (Fuller, 1989; Havenaar *et al.*, 1992; Havenaar and Huis in't Veld, 1992; Salminen *et al.*, 1998a). Accepted characteristics for probiotics are listed in Table 1.3. Hitherto, evaluating their ability to compete effectively with resident and established microorganisms for available nutrients in a multi-substrate gut environment, is one attribute not thoroughly investigated in the selection and implementation of probiotics. Their survivability may be enhanced greatly in the presence of prebiotic carbohydrates proven to select for useful species of *Bifidobacterium* and *Lactobacillus* (Kailasapathy and Chin, 2000). Such a mixture may improve therapeutic potential in the gastrointestinal tract, and are defined as synbiotics (see later).

The most widely used bacteria as probiotics are the lactobacilli and bifidobacteria but products incorporating other organisms such as Gram positive cocci, bacilli, yeasts and *E. coli* have also been applied (Holzapfel and Schillinger, 2002). Probiotic preparations are widely available to consumers as powders, tablets, drinks and fermented dairy products.

Safety is of utmost concern when selecting probiotics. Whilst many lactic acid bacteria, used in traditional fermented food products such as yoghurt, sauerkraut and kefir, have a long history of safe use, the recent explosion of probiotic-containing

Table 1.3 *Desirable properties of probiotic bacteria (Data sourced from Salminen et al., 1998a)*

Probiotic strain characteristics	Functional properties
Human origin, if intended for humans	Species-dependent health effects and maintained viability; applicability to functional and clinical foods.
Acid and bile stability	Survival in the intestine, maintaining adhesiveness and other colonization properties
Adherence to human intestinal cells and intestinal mucus glycoproteins (mucin)	Immune modulation, competitive exclusion of pathogens
Competitive exclusion and colonization of the human intestinal tract	Multiplication in the intestinal tract, competitive exclusion of pathogens, stimulation of beneficial microflora, immune modulation by contact with gut associated lymphoid tissue
Production of antimicrobial substances	Pathogen activation in the intestine, normalization of the gut flora
Antagonism against cariogenic and pathogenic bacteria	Pathogen exclusion, prevention of pathogen adhesion, normalization of gut flora, normalization of oral microflora
Safety in food and clinical use	Accurate strain identification (genus, species, strain) and chracterization, documented safety
Clinically validated and documented health effects	Dose–response data for minimum effective dosage in different products and population groups

foodstuffs incorporating a wide variety of different strains calls for investigations into safety and tolerance. The USA Food and Drug Administration (FDA) run a system for all food additives whereby a long history of safe use (prior to 1958) or substantial scientific evidence can lead to 'generally regarded as safe' (GRAS) status. Lactobacilli are commonly given this status due to the largely nonpathogenic nature of this genus (Salminen *et al.*, 1998b). This is also true for the bifidobacteria. To prove scientifically that a probiotic may be regarded as safe there are three levels of study: *in vitro* and animal studies can be of use in the first instance, but as these data will only pertain to the model system used, clinical trials are required. Controlled clinical trials have been used in safety assessments of probiotics and detailed measurements of many physiological parameters can be made. Thirdly, and possibly most reliably, history of safe use over a period of time can be studied retrospectively with the benefits of large data sets. One example of such a study investigated the consumption of *Lactobacillus rhamnosus* GG in Finland over a period of 11 years (Salminen *et al.*, 2002). Collection and comparison of clinical isolates of lactobacilli from bacteraemia cases showed no correlation between ingestion of *L. rhamnosus* GG and bacteraemia.

Another criterion for selecting probiotic strains is resistance of the strain to stresses it will encounter on its journey throughout the gastrointestinal tract. To reach the desired site of action (e.g. the colon) a probiotic microorganism must pass through the highly acidic stomach and survive bile secretions into the small intestine. Moreover, it should compete well with the resident flora. *In vitro* tests provide the easiest method of discerning tolerance in these situations of various candidate probiotics, as gaining samples from the human stomach and small intestine is ethically questionable and practically difficult.

As an alternative to these model studies it is possible to assess probiotic survival via identification of the strain in faeces, overcoming the shortfalls of the model systems described above. The difficulty with this sort of study is identification of the strain of interest in the complex microbial communities found in faeces. Traditional agar plate methods fail to identify only one species and if this approach is used, then further testing of colonies is required. Microscopic examination, testing for excreted peptides and carbohydrate fermentation patterns (for example as determined by the Analytical Profile Index system) have had to be employed to be confident that the target species has been selected (Wolf *et al.*, 1995). Yuki *et al.* (1999) developed monoclonal antibodies to detect their strain of interest and used these in an enzyme-linked imunosorbent assay (ELISA). Another approach is to add genetic markers to the strain, such that they can be selected in mixed culture. One study used a spontaneous mutant of the strain against rifampicin and streptomycin, with a transposon-encoded insertion for sucrose degradation (Klijn *et al.*, 1995). This combination of genetic elements made the strain easily selectable on agar plates, but tampering with the genetics of the bacteria can change its overall properties. In the same study, polymerase chain reaction (PCR) was used to check the identification of colonies, targeted to a nisin gene encoded on the transposon. Alternatives to PCR for confirmation of colony identity have also included randomly amplified polymorphic DNA-PCR (RAPD-PCR) (Alander *et al.*, 2001; Fujiwara *et al.*, 2001). In this approach, a single simple primer is used to generate many small amplicons from the same DNA molecule. When genetic differences are present (i.e. when the DNA is from a different bacterium) the primer will bind to different sites and different sised amplicons may occur. Running these amplicons on a gel and comparing the presence or absence of different sised bands will reveal whether the recovered species is the same as the fed species. To negate the need for culturing samples, Satokari *et al.* (2001), used PCR linked to denaturing gradient gel electrophoresis (PCR-DGGE) to detect the presence of their probiotic in faeces. The method amplifies a region of the 16S rDNA via PCR with a G-C rich ‘clamp’ at one end of the amplicon. The products are then ran on a polyacrylamide gel with a gradient of the denaturants formamide and urea (a temperature gradient can also be used in a similar manner and the technique is then known as TGGE). Amplicons are denatured as a function of their G-C content and sequence but the two strands are held together by the G-C clamp. The position on the gel can be used against markers, or the bands subsequently excised and sequenced to identify species. However, this approach can only be used qualitatively. A method by which probiotic bacteria can be quantified using molecular techniques is fluorescent *in situ* hybridisation (FISH). Oligonucleotide probes are hybridised to samples which are examined with the aid of fluorescence microscopy or flow cytometery (Langendijk *et al.*, 1995). This technique has been mainly used to quantify bacterial genera or groups of bacteria, but species-specific probes have been developed (McCartney, 2002). However, to use this technique successfully, the target organism must make up at least 1 % of the population due to the detection limits of this method. While important advances have been made in probiotic detection and discrimination, their recovery in faeces is not a definite indicator of gut colonisation or persistence.

Another factor that is important when selecting probiotics is the antibiotic resistance profile. Both benefits and risks are present when a strain is resistant to antimicrobial treatment. Resistance to drugs used to treat human infections could be advantageous, as the probiotic would not be affected whereas the resident microflora could suffer the loss of some of its members, allowing an overgrowth of potentially pathogenic organisms. However, the mobility of genetic elements on which antibiotic resistance genes can be carried could be problematic. Lactobacilli and enterococci can carry their antibiotic resistance genes on plasmids which are transferable to other bacteria (Salminen *et al.*, 1998b). Transfer of resistance to other organisms could be hazardous in that antimicrobials used to treat conditions caused by the transformed bacteria may become ineffective.

A further point for consideration is the ability of a strain to colonise the gut epithelium. In most trials with probiotics, cells are washed out and are no longer detectable in faeces after 1 week and following just a matter of days of cessation of treatment (Jacobsen *et al.*, 1999; Tannock *et al.*, 2000; Satokari *et al.*, 2001). This property can be investigated using cultured cells and applying probiotic bacteria. Using these methods it has been found that lactobacilli are generally more adherent than bifidobacteria (Dunne, 2001).

Delivery method should also be a consideration. Shelf life and survival in the given medium are critical for successful delivery of live cells to the colon. Milk products have been found not only to maintain viability but to support growth of bifidobacteria. A study by Shin *et al.* (2000), showed that some viability was maintained after 4 weeks at 4 °C in skimmed milk, and that the degree of viability could be greatly increased by adding various oligosaccharides.

Together with the barriers to product integrity (e.g. maintaining high counts of anaerobic bacteria and avoiding contamination) as well as within the intestinal tract, prebiotics have been developed. The attempt is to use nonviable food ingredients to fortify selected components of the indigenous microbiota, hence overcoming survivability issues. However, for prebiotics to be successful health advantages should be targeted. This is an advanced area of probiotic research and will be discussed below.

1.4 Health Benefits of Probiotics

Common definitions of probiotics often include references to health benefits, so success is ultimately likely to be dependent on such outcomes.

1.4.1 Inhibition of Pathogens

A concept known as colonisation resistance, in which the indigenous flora creates a barrier preventing new and possibly pathogenic organisms from invading, normally protects from disease (Macfarlane and McBain, 1999). Unfortunately, this fragile ecosystem can be disrupted, for example by treatment with antibiotics, allowing the growth of undesirable organisms. Payne *et al.* (2003) found that the addition of *Lactobacillus plantarum* to an *in vitro* model gut challenged with tetracycline, caused resistance to *Candida albicans* to increase. This resistance may have been due to restoration of normal colonisation resistance by the addition of lactobacilli to the depleted flora, or via a direct antimicrobial action. Protection from various gastrointestinal pathogens has been widely reported. Co-culture experiments have shown bifidobacteria to inhibit various gut pathogens (Bruno and Shah, 2002). In this research, the presence of supernatant from culture medium of the bifidobacteria was sufficient to cause inhibition but when the pH was adjusted to 7, antimicrobial effects diminished. Lactic acid bacteria produce organic acids (Alvarez-Olmos and Oberhelman, 2001) such as lactate and acetate which acidify the surroundings to a pH at which these pathogenic organisms are unable to effectively compete. However, there are other mechanisms by which probiotics can be antipathogenic. Various lactic acid bacteria produce antimicrobial peptides, which are secreted into the growth medium (Anderssen *et al.*, 1998). Competition for nutrients can also decrease pathogenic bacteria in the gastrointestinal tract. This was shown by Yamamoto-Osaki *et al.* (1994), where amino acids were depleted in cultures where inhibition of *Clostridium difficile* was noted and not where *C. difficile* was not inhibited by the addition of protective faecal flora from infants. *Clostridium botulinum* is another member of the clostridia that can cause disease that can be treated with probiotics (Sullivan *et al.*, 1988). Infant botulism is a disorder whereby botulinal neurotoxin is produced by *C. botulinum* spores that have been ingested and colonised the infant gut, before a more protective adult flora becomes established (Salyers and Whitt, 1994). This is different to food-borne botulism where only the toxin is ingested. *In vitro* data have shown that bifidobacteria and *Enterococcus faecalis* isolated from human infant faeces can be inhibitory to *C. botulinum* (Sullivan *et al.*, 1988).

Other conditions have also been successfully treated through probiotics. Rotavirus is a common cause of acute diarrhoea in children and is a serious problem worldwide. Probiotic therapy using species such as *Lactobacillus rhamnosus* GG, *L. casei* subsp. Shirota and *Bifidobacterium lactis* Bb-12 have shortened symptoms (Gorbach, 2002; Ouwehand *et al.*, 2002; Saarela *et.al.*, 2002). It has also been shown that application of *Bifidobacterium bifidum* can also act prophylactically against rotavirus in hospitalised children.

E.coli is a normal resident of the human gastrointestinal tract. Some strains can cause disease in the gut or be transferred to the urinary tract where infection may develop. In a clinically reported case, Gerasimov (2004) successfully used *Lactobacillus acidophilus* to prevent further recurrence of *E. coli* urinary tract infection in a frequently affected child. *E. coli* can also be pathogenic in the gut environment when exogenous species appear. Enteropathogenic strains of *E.coli* are frequently responsible for travellers' diarrhoea. Two strains of *Lactobacillus rhamnosus* reduced the ability of these *E. coli* to adhere to cultured cells. Other enteropathogenic microorgamisms were also inhibited. *Salmonella typhimurium* adherence was inhibited by *Lactobacillus johnsonii* and *Lactobacillus casei* subsp. Shirota (Sullivan and Nord, 2002).

1.4.2 Immune Stimulation

Probiotic use can also stimulate the immune system. Cells in the colon regularly sample the microflora and there is an inherent tolerance to the commensal flora (Schiffrin and Brassart, 1999). However, the presence of exogenous Gram positive bacteria can induce the secretion of cytokines of a pro- or anti-inflammatory nature dependent upon the species (Cross, 2002). Pro-inflammatory cytokines include TNFα, IL-12 and IFN-γ which encourage migration of immune cells (Roitt *et al.*, 1998). This type of reaction may be useful in the case of cancer where the immune response helps eliminate cancerous cells. Secretion of anti-inflammatory cytokines, such as IL-10, which inhibit the production of other cytokines may promote inflammation (Roitt *et al.*, 1998) would be of help in the case of hypersensitivity, allergy and inflammatory bowel disease, whereby the inflammatory response is overactive (Saarela *et al.*, 2002). IL-10 deficient mice develop colitis as a consequence of their inability to down-regulate the inflammatory response and it has been found that the feeding of lactobacilli can restore IL-10 levels and prevent onset of colitis (Sullivan and Nord, 2002). Severity of atopic dermatitis has been shown to decrease when lactobacilli are given. Other components of the immune system may be responsible for this, as in a study by Rosenfeldt *et al.* (2003) interleukins and IFN-γ were unaffected but serum eosinophil cationic protein (released by granules in eosinophils during inflammation) levels decreased. Pathogens in the gut can be cleared with the help of immune stimulation as caused by probiotic organisms. In murine models, *E. coli* and *Salmonella typhimurium* clearance occurred more quickly upon treatment with *L. casei*, with a concomitant increase in intestinal IgA specific to the pathogen (Cross, 2002). Cross (2002) also noted that effects of probiotic administration on the immune system can reach further than the gastrointestinal tract and into other systems of the body, as illustrated by a faster clearance of bacterial and viral pathogens from the respiratory tract following probiotic feeding, accompanied by an increase in nonspecific phagocytic activity of alveolar macrophages and pathogen-specific serum IgG.

1.4.3 Cholesterol Reduction and Cardiovascular Disease Risk

In vitro and *in vivo* research has shown that probiotics may be able to lower serum cholesterol levels, although this is still a debatable area of research (Naruszewicz *et al.*, 2002; Pereira *et al.*, 2003). The mechanisms for this are currently unclear with many different hypotheses being proposed. Pereira and Gibson (2002) suggested four possible mechanisms: the production of propionate, assimilation of cholesterol by bacteria, binding of cholesterol to bacterial cell walls and enzymatic degradation. In experiments by Pereira *et al.* (2003) *in vitro* cholesterol levels were lowered by *Lactobacillus fermentum*. The authors suggested that in this case a high level of propionate and/or bile acid deconjugation were probable mechanisms. In a human study of cholesterol and cardiovascular disease risk factors, Naruszewicz *et al.* (2002) found that *Lactobacillus plantarum* was able to lower blood pressure, fibrinogen and LDL cholesterol and raise HDL cholesterol.

1.4.4 Cancer

Large bowel cancer is a leading cause of death in the western world, and whilst genetic predisposition can be a factor, diet can also play an important role in this disease (Rao, 1999). Meat can be converted to heterocyclic amines during the cooking process and the bacterial fermentation of protein produces amines and ammonia, which are toxigenic. The ingestion of vegetables can offer protection against colorectal cancer from compounds such as flavols, lycopenes, sulphur compounds, isoflavones, lignans, and saponins that can stimulate immunity, are antioxidants, or detoxify genotoxic compounds (Rao, 1999). Two enzymes that can produce carcinogens are β-glucuronidase in the host and flora and β-glucosidase in the microflora, although the latter is also responsible for catalysing the production of antimutagenic substances (Burns and Rowland, 2000). Lactic acid bacteria have been shown to reduce levels of both of these enzymes in faeces (Burns and Rowland, 2000). *B. longum* has also been shown to decrease the incidence of some tumours and completely inhibit other types of tumours in rat models (Reddy, 1999). This may be due to inhibition of an enzyme, orthinine decarboxylase, present in high levels in cancer or the blocking of expression of a tumour promoting gene, *ras*-21. Burns and Rowland (2000) also noted that, *in vitro*, some probiotic bacteria have the ability to bind carcinogens such as heterocyclic amines, but this seems to have no effect *in vivo*. In addition to these mechanisms immune stimulation, as discussed in section above, could be another mechanism of anticancer activity of probiotics.

1.4.5 Other Health Benefits and Future Directions for Research

In the case of lactose maldigestion (deficiency in β-galactosidase), fermented milk products are better tolerated than milk. This is due to the presence of lactic acid bacteria, which may posses this enzyme, in the fermented milk. The bacteria are lysed in the gastrointestinal tract and the enzyme may be released, allowing better digestion of lactose in the gut (Ouwehand *et al.*, 2002). This mechanism could also be of use in the case of sucrose digestion in sucrase-deficient infants where *Saccharomyces cerevisiae* has been shown to act similarly (Marteau *et al.*, 2001).

Irritable bowel syndrome (IBS) is a poorly defined disorder characterised by abdominal pain and a change in bowel habit with disordered defaecation and distension. Many causes have been suggested including an imbalance in the intestinal microflora. Studies in IBS patients have shown that a variety of probiotic organisms, including *L. acidophilus* (viable and heat-killed), *Enterococcus faecium* and *L. plantarum* to improve symptoms (Marteau *et al.*, 2001; Saarela *et al.*, 2002).

Due to their inherent tolerability it is thought that in the future probiotic bacteria could be used as vectors, expressing vitamins or insulin to those with deficiencies or genetically modified to act as oral vaccines, expressing viral antigens (Gorbach, 2002). Mucosal vaccination with lactobacilli may also be possible to stop the spread of HIV and other sexually transmitted diseases as research has shown a lack of lactobacilli populations in patients suffering from such diseases (Alvarez-Olmos and Oberhelman, 2001).

It is also hoped that the treatment of bacterial infections with probiotics instead of traditional antibiotics may reduce the increasing problem of multidrug resistance (Bengmark, 1998; Alvarez-Olmos and Oberhelman, 2001).

1.5 Prebiotics

A prebiotic is defined as 'a nondigestible food ingredient that beneficially affects the host by selectively stimulating the growth and/or activity of one or a limited number of bacteria in the colon and thus improves host health' (Gibson and Roberfroid, 1995). The stimulated bacteria should be of a beneficial nature, namely bifidobacteria and lactobacilli (Gibson *et al.*, 1999). To have these effects, prebiotics must be able to withstand digestive processes before they reach the colon and preferably persist throughout the large intestine such that benefits are apparent distally (Gibson *et al.*, 2004).

Lower molecular weight oligosaccharides have been the subject of recent interest because, apart from the nonstarch polysaccharides, they present themselves as the most portable source of carbon for colonic bacteria. When ingested, these carbohydrates are not digested in the small intestine and reach the ileocaecal region in a relatively unmodified form (Oku *et al.*, 1984; Nilsson *et al.*, 1988; Ellegard *et al.*, 1997). In the colon they can act as dietary bulking agents or make up a percentage of available substrate for resident colonic bacteria, consequently, contributing towards a decrease in pH and the production of SCFA, effects that may result in reduced numbers of pathogenic microorganisms (Morisse *et al.*, 1993).

Oligosaccharides are relatively short chain carbohydrates that occur widely in nature. They are typically found in the plant kingdom but have also been detected in relatively smaller quantities, as free sugars or glycoconjugates, in human milk and the colostrum of various animals (Bucke and Rastall, 1990). Primarily, they are retained as reserve sugars in seeds and tubers and used when growth begins, however their effects on gastrointestinal physiology in recent years has highlighted renewed significance to human health (Van Loo *et al.*, 1999).

Certain oligosaccharides (Table 1.4) have all been reported, at varying concentrations, to have an ability to promote the growth of organisms whose metabolism have positive physiological consequences (Fuller and Gibson, 1998; Gibson *et al.*, 2000). It may be possible to develop a range of such carbohydrates for incorporation into foods in a bid to improve their 'prebiotic effect'. This is especially true in the West where conventional diets only have relatively small quantities of oligosaccharides (ca. 2–5 g per day) consumed daily (Macfarlane and Cummings, 1991; Roberfroid *et al.*, 1993). Fructooligosaccharides are well researched prebiotics which occur naturally in the diet (Table 1.5). A fuller description of prebiotic oligosaccharides is given elsewhere in this book.

Candidate prebiotics are reported to be particularly suited to the growth and activities of the bifidobacteria and lactobacilli (Rowland and Tanaka, 1993; Fuller and Gibson,

Table 1.4 *Some candidate prebiotic compounds*

Prebiotic	Production method	Reference
Inulin [Fructooligosaccharide (FOS)]	Hot water extraction from chicory root (followed by enzymatic hydrolysis) or polymerization of fructose monomers	Bornet *et al.*, 2002
Galactooligosaccharide (GOS)	Enzymatic lactose transgalactosylation	Teuri and Korpel, 1998
Xylooligosaccharide (XOS)	Enzymatic hydrolysis of plant xylans	Imaizumi *et al.*, 1991
Isomaltooligosacchairde (IMO)	Transglucosylation of liquefied starch	Morgan *et al.*, 1992
Lactulose	Isomerization of lactose	Salminen and Salminen, 1997

Table 1.5 Natural occurrence of fructooligosaccharides (Data sourced from Mitsuoka et al., 1987; Roberfroid et al., 1993; Modler, 1994)

Source	Scientific name	Fructose units	Fructooligosaccharides (%) in fresh material
Banana	*Musa* spp.	2	0.3–0.7
Rye	*Secale cereale*	2	0.5–1.0
Leek	*Allium ampeloprasmus*	*n*[a]	2.0–10.0
Wheat	*Triticum asetivum*	*n*	0.8–4.0
Garlic	*Allium sativum*	*n*	1.0–16.0
Chicory roots	*Cichorium intybus*	*n*	15.0–24.0
Asparagus shoot	*Asparagus officinalis*	2–4	2.0–3.0
Jerusalem artichoke	*Heliantus tuberosus*	2	16.0–22.0
Globe artichoke	*Cynara scolymus*	2	3.0–10.0
Onions	*Allium cepa*	2–4	1.1–7.5
Salisfy	*Scorzonera hispanica*	*n*	4.0–11.0
Dandelion	*Taraxacum officinale*	*n*	12.0–15.0
Dahlia	—	*n*	13.0
Burdock	—	2–4	3.6

[a]*n* is either >4 or number of individual fructose units not described as yet.

1998; Bouhnik *et al.*, 1999). These are classed as beneficial microorganisms because species within these groups have been reported to exert therapeutic and prophylactic influences on the health of infants and adults (Goldin and Gorbach, 1992; Salminen *et al.*, 1998a). The high incidence of *Bifidobacterium* spp. in breast-fed babies for example may be instrumental in protection against childhood diseases and prevent the colonisation of transient pathogens (Fuller, 1991). Likewise, reduced levels of bifidobacteria may, at least partly, explain increased susceptibility to disorders in the elderly (Mitsuoka, 1990; Rowland and Tanaka, 1993; Gibson *et al.*, 1995; Buddington *et al.*, 1996). A potential correlation therefore exists with reduced pathogen resistance, decreased numbers of bifidobacteria in the elderly and the production of natural resistance factors. In essence, the natural gut flora may be compromised through reduced bifidobacterial numbers and have a reduced ability to deal with pathogens. If prebiotics are used to increase bifidobacteria or lactobacilli towards being the numerically predominant genus in the colon, an improved colonisation resistance ought to result, but has not yet been proven.

By definition, a prebiotic must stimulate the growth of a limited number of bacteria and thus will lead to a change in the overall microbial balance in the colon. Static batch culture fermentations with human faecal bacteria have shown that fructooligosaccharides (FOS), galactooligosaccharides (GOS), xylooligosaccharides (XOS), isomaltooligosaccharides (IMO) and lactulose alter the microflora, increasing the level of bifidobacteria and/or lactobacilli and in some cases causing clostridia and bacteroides to decline (Rycroft *et al.*, 2001). Another *in vitro* study, using a more complex model for fermentation in the colon, found that inulin increased lactobacilli and to a lesser extent bifidobacteria, in simulations of proximal regions of the model (McBain and Macfarlane, 2001). This study also investigated GOS in the same way and found that adding it to the model increased bifidobacteria and lactobacilli in regions of the model representing the proximal and transverse colon. Similar results have also been recorded in feeding studies on rodents. When inulin was given in the diet of mice caecal bacteria increased as did the production of SCFA, causing a reduction in pH (Apajalahti *et al.*, 2002). Molecular analysis of caecal contents by GC profiling showed large changes in populations. Sequencing showed that inulin increased bifidobacteria and decreased the less desirable clostridia and desulfovibrios as well as causing changes in the levels of unidentified species. In another study, resistant starch was administered to human flora associated rats to assess its prebiotic effects (Silvi *et al.*, 1999). In this case, lactic acid bacteria increased and enterobacteria decreased. Modulation of the flora can also be seen in human feeding studies. GOS and FOS have been shown to increase faecal bifidobacteria (Ito *et al.*, 1990; Bouhnik *et al.*, 1999) This effect has been shown to be both dose-dependent and related to the initial level of bifidobacteria, with individuals with the lowest starting populations showing the greatest increase, also reflecting *in vitro* observations by Rycroft *et al.* (2001).

Three oligosaccharides are available in usable quantities in Europe and seem to have proven efficacy. These are fructooligosaccharides (including inulin), *trans*-galactooligosaccharides and lactulose. Molecular based methodologies in human trials to confirm the prebiotic effect of FOS and lactulose (Tuohy *et al.*, 2001a, 2002). The former has also been seen to be a highly effective prebiotic when incorporated into a biscuit product at 8 g per day (Tuohy *et al.*, 2001b). Other prebiotics, as mentioned above, are widely used in Japan. Some may have more desirable attributes than the currently recognised European forms and this is currently under evaluation. Moreover, the use of 'glycobiology' offers the deliberate manufacture of multi-functional prebiotics. This could include forms that have anti-adhesive capacities against common food borne pathogens, types that persist to the distal bowel (the main site of colonic disorder), carbohydrates which attenuate the virulent properties of certain microorganisms and prebiotics that target individual species, not genera, of gut bacteria (Gibson *et al.*, 2000).

1.5.1 Evaluating Prebiotic Functionality

A variety of model systems have been developed to investigate colonic fermentation of prebiotics (Rumney and Rowland, 1992; Conway, 1995; Rycroft *et al.*, 1999). Screening usually starts by looking at the relative fermentability of prebiotics with *in vitro* static batch culture fermenters. In its simplest form, the effect is investigated with selected pure cultures of gut bacteria in anaerobic Hungate tubes sealed under nitrogen. This is advantageous because results can be generated quickly and become especially useful, e.g. in some cases when only low amounts of oligosaccharides are available. However, this method does not identify true selectivity of a substrate. A modification is the use of defined mixed- or co-cultures as inocula (Rycroft *et al.*, 1999). This may introduce competition between microbes but still does not fully represent the complex interactions present in the human large intestine. Further modifications have thus used faecal suspensions to increase diversity, larger vessels and volumes, stirred conditions with pH and temperature control allowing more comparative determinations to be made on fermentability. The prebiotic effect is determined at intervals by removing samples and assessing growth through cultural microbiological techniques or molecular based methods. However, these are all closed systems in which substrate availability soon becomes the limiting factor, and there is an inevitable build up of acid and metabolites that can affect carbohydrate utilisation.

Conversely, parameters in continuous cultures (chemostats) such as varying dilution rate and pH can be optimised to conditions, physiologically more similar to events that occur in the large gut (Gibson and Wang, 1994a). There is a continuous input of growth media at one end whilst spent culture plus bacteria is steadily removed at another end, thus enabling continual biomass production and eventually reaching steady state. At steady state, equilibrium is reached, growth rate is equal to dilution rate and this enables the composition of a diverse microflora community to be maintained whilst reproducible measurements can be made. In their simplest form, continuous culture systems usually consist of one reaction vessel, but variations occur. One study to examine the prebiotic effect of oligofructose (FOS) used both single- and a three-stage continuous culture models of the human large intestine (Gibson and Wang, 1994a,b). The multiple-stage system was set up with vessels in succession; the first, relatively nutrient-rich, acidic pH and faster transit (due to a small operating volume) than the third with a more neutral pH and comparatively less substrate, thus mimicking the proximal and the distal colonic environments, respectively. Conditions for the transverse colon were represented in vessel two. Parameters and events in the three-stage gut model have been validated against the colonic contents of sudden death victims (Macfarlane *et al.*, 1998). Species composition in the three compartments representing different regions of the human large intestine, population levels and chemical measurements correlated well with the *in vivo* samples, proving a useful model for studying nutritional effects on intestinal ecology and physiology. Another model attempted to study digestive events from the jejunum to the distal regions of the colon has been developed, consisting of five vessels sequentially fed with growth media (Molly *et al.*, 1993). Semi-continuous variations have also been used when medium is added and spent culture removed at timed intervals (Rumney and Rowland, 1992; Miller and Wolin, 1981). Although these models are designed and controlled to represent metabolic events that take place in the lower gut environments, limitations arise, especially from the inability to study contributions from biofilm communities and the absence of a physiological absorptive mechanism.

These can be overcome by using animal models, like rats and mice, but large differences do occur in microflora between the rodent's intestinal system and humans. The use of germ-free (gnotobiotic) rats associated with human faecal flora, known as human flora associated rats may further simulate, but still not fully represent, the human intestinal physiology. The use of human volunteers in properly controlled (dose, duration, diet, blinded) studies is the ultimate test for prebiotic functionality, however, analysis is carried out on faecal material and this may poorly represent events in proximal regions. It has been suggested that a consolidation of all these approaches, systematically performed, should clearly indicate the potential of candidate prebiotics to be of beneficial value (Rycroft *et al.*, 1999).

1.6 Health Outcomes Associated with Prebiotic Intake

The health outcomes of prebiotic intake can be similar to those described for probiotics if the selectively stimulated bacteria possess probiotic properties. In addition, there are benefits of dietary prebiotic supplementation that stand alone from effects attributable to an increase in the numbers of lactic acid bacteria, and are instead due to a switch in the metabolism of these organisms.

1.6.1 Acute Gastroenteritis

Acute gastroenteritis is something that probably affects everyone at one time or another. Usually it involves the ingestion of food or water contaminated with pathogenic microorganisms and/or their toxins (Hui *et al.*, 1994). The economic costs and medical aspects are therefore huge, with food safety incidence still increasing in most civilisations. Typical causative agents include shigellae, salmonellae, *Yersinia enterocolitica*, *Campylobacter jejuni*, *E. coli*, *Vibrio cholera* and *Clostridium perfringens*. Pathogens may either, colonise and grow within the gastrointestinal tract and then invade host tissue, or they may secrete toxins contaminating food prior to its ingestion. Such toxins disrupt function of the intestinal mucosa, causing nausea, vomiting and diarrhoea (Hui *et al.*, 1994). The principal human intestinal bacterial pathogens can be characterised according to the virulence factors that enable them to overcome host defences. These include invasion which enables bacterial multiplication within enterocytes or colonocytes, for example *E. coli*, *Shigella* spp., salmonellae and yersinae. Cytotoxic bacteria which include enteropathogenic and enterohaemorrhagic strains of *E. coli* as well as some shigellae are able to produce substances which can directly cause cell injury. Toxigenic bacteria such as *V. cholerae* and some shigellae are capable of producing enterotoxins, which affect salt and water secretion in the host. Lastly, enteroaggregative *E. coli* have the ability to tightly adhere to the colonic mucosa. Such mechanisms enable potentially pathogenic bacteria to establish infections in the gastrointestinal tract, evade the immune system and surmount colonisation resistance afforded by the indigenous gut microflora.

The gut microflora and the mucosa itself act as a barrier against invasion by potentially pathogens. Bifidobacteria and lactobacilli may inhibit pathogens like *E. coli*, *Campylobacter* and *Salmonella* spp. (Gibson and Wang, 1994b). The lactic microflora of the human gastrointestinal tract is thought to play a significant role in improved colonisation resistance (Gibson *et al.*, 1997). There are a number of possible mechanisms in operation:

- metabolic end products such as acids excreted by these microorganisms may lower the gut pH, in a microniche, to levels below those at which pathogens are able effectively compete;
- competitive effects from occupation of normal colonisation sites;
- direct antagonism through natural antimicrobial excretion;
- competition for nutrients.

The possibility exists therefore, that increased levels of beneficial bacteria in the large gut may, along with other factors such as immune status, offer improved protection. This is in a similar manner to probiotics, but may well be more efficacious given the comparative survivability issues.

The idea of combining prebiotic properties with anti-adhesive activities is currently under investigation. This would add major functionality to the approach of altering gut pathogenesis. Many intestinal pathogens utilise monosaccharides or short oligosaccharide sequences as receptors and knowledge of these receptor sites has relevance for biologically enhanced prebiotics. Binding of pathogens to these receptors is the first step in the colonisation process (Finlay and Falkow, 1989; Karlsson, 1989). There are currently several pharmaceutical preparations based upon such oligosaccharides in clinical trials. These agents are multivalent derivatives of the sugars and act as 'blocking factors', dislodging the adherent pathogen (Heerze *et al.*, 1994; Jayaraman *et al.*, 1997). There is much potential for developing prebiotics, which incorporate such a receptor monosaccharide or oligosaccharide sequence. These molecules should have enough anti-adhesive activity to inhibit binding of low levels of pathogens.

The prebiotic concept may be extrapolated further by considering an attenuation of virulence in certain food-borne pathogens. For example, the plant derived carbohydrate cellobiose is able to repress pathogenicity in *Listeria monocytogenes* through down regulation of its virulence factors (Park and Kroll, 1993). As such, this organism is avirulent in its natural habitat of soil, where it is exposed to rotting vegetation and therefore cellobiose. In the human body, an absence of cellobiose may allow the virulence factors to be expressed, and it is possible that further incorporation of this disaccharide to foods susceptible to *Listeria* contamination could reduce this virulence.

1.6.2 Reduction of Cancer Risk

Genotoxic enzyme activity has been seen to reduce on the administration of prebiotics. An early study on feeding GOS to humans resulted in a decrease in nitroreductase (a metabolic activator or mutagenic/carcinogenic substances) and also decreased levels of indole and isovaleric acid (produced as products of proteolysis and deamination and markers of putrefaction) (Ito *et al.*, 1990). When a model system of the human gut was used to investigate the effect of GOS on genotoxic enzymes it was found that β-glucosidase, β-glucuronidase and arylsulphatase were strongly inhibited but azo- and nitroreductase were stimulated (McBain and Macfarlane, 2001). As these effects occurred rapidly on the addition of GOS to the system, changes attributable to population levels can be ruled out and it is more feasible that direct inhibition by GOS or the production of repressors or deactivators by bacteria was responsible. However, increasing the proportion of bifidobacteria and lactobacilli at the expense of bacteroides and clostridia may also decrease genotoxic enzyme production, as the former produce lower levels of such enzymes than the latter (Burns and Rowland, 2000).

Another study looked at the effects of resistant starch administration to human-flora associated rats (Silvi *et al.*, 1999). Although β-glucosidase increased, β-glucuronidase and ammonia levels decreased. A further observation important to the reduction of cancer was a high level of caecal butyrate. Not only is butyrate the major source of energy for colonocytes and helps maintain a healthy epithelium (Topping and Clifton, 2001), it can also play an important role in preventing cancer. Several cellular processes are affected by butyrate, largely by interaction with DNA and its surrounding proteins (Kruh, 1982). These processes include induction of apoptosis, a process which is deactivated in cancer cells which would normally lead to their elimination and an increase in immunogenicity of cancer cells due to an increase in expression of cell surface proteins (Bornet *et al.*, 2002). However, it should be pointed out that the usual target bacteria for prebiotic use (bifidobacteria, lactobacilli) are not butyrate producers. Hence, there could be rationale for fortifying other gut flora components (e.g. eubacteria).

1.6.3 Mineral Absorption

Uptake of calcium and magnesium is crucial for bone structure and increasing absorption can prevent conditions such as osteoporosis. Chonan *et al.* (2001) have shown that adding GOS to the diet of rats can increase calcium and magnesium absorption. The mechanism for this is unclear but in this case the presence of a colonic flora is required for GOS to have this effect, though the authors acknowledged that microbial mediated and nonmicrobial mediated mechanisms probably exist. FOS can also affect mineral absorption and in human studies 15 g per day oligofructose or 40 g per day inulin increased the apparent calcium absorption (Roberfroid, 2002). Magnesium absorption has also been shown to increase when ingesting FOS (Bornet *et al.*, 2002).

1.6.4 Lipid Regulation

Prebiotics may also have an effect on lipid regulation. Although the mechanism is currently unknown, studies have shown positive results and mechanistic hypotheses have been developed. A study on diabetic rats found that when XOS replaced simple carbohydrates in the diet, the serum cholesterol and triglyceride increases observed in diabetes were reduced and liver triglycerides increased to a comparable level seen in healthy rats (Imaizumi *et al.*, 1991). Other studies have examined FOS, which was also found to reduce blood lipids (Bornet *et al.*, 2002; Roberfroid, 2002). This was thought to be due to an inhibition of lipogenic enzymes in the liver, which may be a result of the action of propionate produced from the fermentation of prebiotics by gut bacteria (Wolever *et al.*, 1991). Whilst prebiotics can be of use in correcting hyperlipidaemia brought about by diabetes and other conditions, decreases in lipids have not been observed in healthy subjects (Bornet *et al.*, 2002), which is a useful safety feature as misuse or overdose does not seem to have negative effects.

1.6.5 Development of New Prebiotics

As our understanding of the interactions of the gut flora with its host improves, more prebiotics are being designed with specific health outcomes in mind, and from different materials. Native polymers or those that have been made into oligosaccharides that may have prebiotic effects include dextran (Olano-Martin *et al.*, 2000), bacterial exopolysaccharide (Korakli *et al.*, 2002), chitin and chitosan (Lee *et al.*, 2002) and components of the dietary fibre found in cereal grains (Charalampopoulos *et al.*, 2002). These materials are all abundant and inexpensive; another important quality when developing potential prebiotics.

1.7 Synbiotics

Synbiotics can be described as 'a mixture of probiotics and prebiotics that beneficially affects the host by improving the survival and implantation of live microbial dietary supplements in the gastrointestinal tract, by selectively stimulating the growth and/or activating the metabolism of one or a limited number of health promoting bacteria, and thus improving host welfare' (Gibson and Roberfroid, 1995). By combining the probiotic and prebiotic strategies previously described, additive or synergistic effects may be observed. There are numerous mechanisms by which this may occur. Increased survival of probiotic bacteria and hence shelf life in consumer products as a result of prebiotic addition would lead to an increased ingestion of viable cells (Shin *et al.*, 2000). Higher probiotic numbers may also be achieved by simultaneous feeding with a prebiotic which can be competitively utilised by the probiotic (Holzapfel and Schillinger, 2002). In addition, the presence of a prebiotic may not only stimulate the growth or activity of the fed probiotic strain but also selected indigenous bacteria in the colon that are considered beneficial (Roberfroid, 1998). Furthermore, it may be possible to target a synbiotic product to two different regions of the gastrointestinal tract, for example the small and large intestine (Holzapfel and Schillinger, 2002).

In a study on weanling pigs, Bomba *et al.* (2002) found that when FOS was given at the same time as the probiotic *Lactobacillus paracasei* there was a significantly greater increase in lactobacilli and bifidobacteria than was observed with the probiotic alone.

Other studies have shown an increased persistence of probiotics in synbiotic preparations both in terms of the location in the colon (Rastall and Maitin, 2002) and how long the effects can be seen following cessation of taking the product (Roberfroid, 1998), the latter being a possible indicator of better implantation of the probiotic strain into the indigenous flora or a more general increase in indigenous bifidobacteria caused by the prebiotic.

Colon cancer has been studied as a situation where synbiotics could be of benefit. Most studies have been carried out in rats and looked at the reduction in number of aberrant crypt foci (cancer precursors) in rats treated with azoxymethane (a tumour promoter). Gallaher and Khil (1999) found that FOS and bifidobacteria had no effect when administered alone but as a synbiotic five out of six subjects had decreased aberrant crypt foci. Another study has shown that whilst a mixture of short and long chain FOS can decrease the number of adenomas and malignant cancers, a mixture of *Lactobacillus reuteri* GG and *Bifidobacterium lactis* Bb12 only had an effect in reducing numbers of malignant tumours. The results were improved with the pro and prebiotics combined but this time the effects were seen to be additive and not synergistic (Fermia *et al.*, 2002).

Immune modulation may also be more effective with synbiotics, and may be an example whereby different components of the synbiotic act at different sites. The same combination of pro and prebiotics were as used by Fermia *et al.* (2002) and immune parameters measured. The results showed that peripheral blood mononuclear cells were specifically affected by probiotics and prebiotics but in some immune compartments a greater effect was shown by the synbiotic (Roller *et al.*, 2004).

References

Alander, M., Matto, J., Kniefel, W., Johansson, M., Kogler, B., Crittenden, R., Mattila-Sandholm, T. and Saarela, M. (2001) Effect of galacto-oligosaccharide suplementation on human faecal microflora and on survival and persistance of *Bifidobacterium lactis* Bb-12 in the gastrointestinal tract. *International Dairy Journal*, **11**, 817–825.

Alvarez-Olmos, M. I. and Oberhelman, R. A. (2001) Probiotic agents and infectious diseases: a modern perspective on a traditional therapy. *Clinical Infectious Diseases*, **32**, 1567–1576.

Anderssen, E. L., Diep, D. B., Nes, I. F., Eijsink, V. G. H. and Nissen-Meyer, J. (1998) Antagonistic activity of *Lactobacillus plantarum* C11: two new two-peptide bacteriocins, plantaricins EFand JK, and the induction factor plantaricin A. *Applied and Environmental Microbiology*, **64**, 2269–2272.

Apajalahti, J. H., Kettunen, H., Kettunen, A., Holben, W. E., Nurminen, P. H., Rautonen, N. and Mutanen, M. (2002) Culture-independent microbial community analysis reveals that inulin in the diet primarily affects previously unknown bacteria in the mouse cecum. *Applied and Environmental Microbiology*, **68**, 4986–4995.

Bengmark, S. (1998) Ecological control of the gastrointestinal tract. The role of probiotic flora. *Gut*, **42**, 2–7.

Beritzoglou, E. (1997) The intestinal flora during the first weeks of life. *Anaerobe*, **3**, 173–177.

Bingham, S. A., Pett, S. and Day, K. C. (1990) NSP intake of a representative sample of British adults. *Journal of Human Nutrition and Diet*, **3**, 339–344.

Bomba, A., Nemcova, R., Gancarcikova, S., Herich, R., Guba, P. and Mudronova, D. (2002) Improvement of the probiotic effect of microorganisms by their combination with maltodextrins, fructo-oligosaccharides and polyunsaturated fatty acids. *British Journal of Nutrition*, **88**, 95S–99S.

Bornet, F. R. J., Brouns, F., Tashiro, Y. and Duviller, V. (2002) Nutritional aspect of short-chain fructooligosaccharides: natural occurrence, chemistry, physiology and health implications. *Digestive and Liver Disease*, **34**, S111–S120.

Bouhnik, Y., Vahedi, K., Achour, L., Attar, A., Salfati, J., Pochart, P., Marteau, P., Flourie, B., Bornet, F. and Rambaud, J.-C. (1999) Short-chain fructo-oligosaccharide administration dose-dependently increases faecal bifidobactera in humans. *Journal of Nutrition*, **129**, 113–116.

Bruno, F. A. and Shah, N. P. (2002) Inhibition of pathogenic and putrefactive microorganisms by *Bifidobacterium* sp. *Milchwissenschaft*, **57**, 617–621.

Burns, A. J. and Rowland, I. R. (2000) Anti-carcinogenicity of probiotics and prebiotics. *Current Issues in Intestinal Microbiology*, **1**, 13–24.

Bucke, C. and Rastall, R. A. (1990) Synthesising sugars by enzymes in reverse. *Chemistry Britain*, **26**, 675–678.

Buddington, R. K., Williams, C. H., Chen, S. and Witherly, S. A. (1996) Dietary supplement of neosugar alters the faecal flora and decreases activities of some reductive enzymes in human subjects. *American Journal of Clinical Nutration*, **63**, 709–716.

Burns, A. J. and Rowland, I. R. (2000) Anti-carcinogenicity of probiotics and prebiotics. *Current Issues in Intestinal Microbiology*, **1**, 13–24.

Chadwick, V. S. and Anderson, R. P. (1995) The role of intestinal bacteria in etiology and maintenance of inflammatory bowel diseases. In: *Human Colonic Bacteria: Role in Nutrition, Physiology and Pathology*. G. R. Gibson and G. T. Macfarlane (Eds). CRC Press, Boca Raton, FL, pp. 227–256.

Charalampopoulos, D., Wang, R., Pandiella, S. S. and Webb, C. (2002) Application of cereals and cereal components in functional foods: a review. *International Journal of Food Microbiology*, **79**, 131–141.

Chonan, O., Takahashi, R. and Watanuki, M. (2001) Role and activity of gastrointestinal microflora in absorption of calcium and magnesuim in rats fed b1-4 linked galactooligosaccarides. *Bioscience, Biotechnology and Biochemistry*, **65**, 1872–1875.

Conway, P. L. (1995) Microbial ecology of the human large intestine. In: *Human Colonic Bacteria: Role in Nutrition, Physiology and Pathology*. G. R. Gibson and G. T. Macfarlane (Eds). CRC Press, Boca Raton, FL, pp. 1–24.

Cross, M. (2002) Microbes versus microbes: immune signals generated by probiotic lactobacilli and their role in protection against microbial pathogens. *FEMS Immunology and Medical Microbiology*, **34**, 245–253.

Cummings, J. H. (1981) Cellulose and the human gut, *Gut*, **25**, 805–810.

Cummings, J. H. (1995) Short chain fatty acids. In: *Human Colonic Bacteria: Role in Nutrition, Physiology and Pathology*. G. R. Gibson and G. T. Macfarlane (Eds). CRC Press, Boca Raton, FL, pp. 101–130.

Cummings, J. H. and Macfarlane, G. T. (1991) The control and consequences of bacterial fermentation in the human colon: a review. *Journal of Applied Bacteriology*, **70**, 443–459.

Duerden, B. I., Wade, W. G., Brazier, J. S., Eley, A., Wren, B. and Hudson, M. J. (Eds) (1995) Ecology and epidemiology of *Clostridium difficile*. In: *Medical and Dental Aspects of Anaerobes*. Science Reviews, Middlesex, pp. 153–248.

Dunne, C. (2001) Adaptation of bacteria to the intestinal niche: probiotics and gut disorder. *Inflammatory Bowel Disease*, **7**, 136–145.

Ellegård, L., Andersson, H. and Bosaeus, I. (1997) Inulin and oligofructose do not influence the absorption of cholesterol, or the excretion of cholesterol, Ca, Mg, Zn, Fe or bile acids but increases energy excretion in ileostomy subjects. *European Journal of Clinical Nutrition*, **51**, 1–5.

Engelhardt, W., Busche, R., Gros, G. and Rechkemmer, G. (1991) Absorption of short-chain fatty acids: mechanisms and regional differences in the large intestine. In: *Short-chain Fatty Acids: Metabolism and Clinical Importance*. J. H. Cummings, J. L. Rombeau and T. Sakata (Eds). Ross Laboratories Press, Columbus, pp. 60–62.

Englyst, H. N. and Cummings, J. H. (1986) Digestion of the carbohydrates of banana (*Musa paradisiaca sapientum*) in the human small intestine. *American Journal of Clinical Nutrition*, **44**, 42–50.

Englyst, H. N. and Cummings, J. H. (1987) Digestion of polysaccharides of potato in the small intestine of man. *American Journal of Clinical Nutrition*, **45**, 423–431.

Englyst, H. N. and Macfarlane, G. T. (1986) Breakdown of resistant and readily digestible starch by human gut bacteria. *Journal of Science and Food Agriculture*, **37**, 699–706.

Fermia, A. P., Luceri, C., Dorola, P., Giannini, A., Biggeri, A., Salvadori, M., Clune, Y., Collins, K. J., Paglierani, M. and Caderni, G. (2002) Antitumorigenic activity of the prebiotic inulin enriched with oligofructose in combination with the prebiotics *Lactobacillus rhamnosus* and *Bifidobacterium lactis* on azoxymethane-induced colon carcinogenesis in rats. *Carcinogenesis*, **23**, 1953–1960.

Finlay, B. B. and Falkow, S. (1989) Common themes in microbial pathogenicity. *Microbiological Reviews*, **53**, 210–230.

Fujiwara, S., Seto, Y., Kimura, A. and Hashiba, H. (2001) Intestinal transit of an orally administered strptomycin-rifampcin-resistant variant of *Bifidobacterium longum* 2928: its long-term survival and effect on the intestinal microflora and metabolism. *Journal of Applied Microbiology*, **29**, 43–52.

Fuller, R. (1989) Probiotics in man and animals. *Journal of Applied Bacteriology*, **66**, 365–378.

Fuller, R. (1991) Factors affecting the composition of the intestinal microflora of the human infant. In: *Nutritional Needs of the Six to Twelve Month Old Infant*. W. C. Heird (Ed.). Carnation Nutrition Education Glendale / Raven Press Ltd, New York, Series **2**, pp. 121–130.

Fuller, R. and Gibson, G. R. (1998) Probiotics and prebiotics: microflora management for improved gut health. *Clincinal Microbiology and Infection*, **4**, 477–480.

Gallaher, D. D. and Khil, J. (1999) The effect of synbiotics on colon carcinogenesis in rats. *Journal of Nutrition*, **129** (Suppl.), 1483S–1487S.

Gerasimov, S. V. (2004) Probiotic prophylaxis in paediatric recurrent urinary tract infections. *Clinical Paediatrics*, **43**, 95–98.

Gibson, G. R. (1998) Dietary modulation of the human gut microflora using prebiotics. *British Journal of Nutrition*, **80**, S209–S212.

Gibson, G. R. and Collins, M. D. (1999) Concept of balanced colonic microbiota, prebiotics and synbiotics. In: *Probiotics, Other Nutritional Factors and Intestinal Microflora*. L. A. Hanson and R. H. Yolken (Eds). Nestle Nutrition Workshop Series, vol. 42. Raven Publishers, Philadelphia, pp. 139–153.

Gibson, G. R. and Roberfroid, M. B. (1995) Dietary modulation of the human colonic microbiota: introducing the concept of prebiotics. *Journal of Nutrition*, **125**, 1401–1412.

Gibson, G. R. and Wang, X. (1994a) Enrichment of bifidobacteria from human gut contents by oligofructose using continuous culture. *FEMS Microbiological Letters*, **118**, 121–128.

Gibson, G. R. and Wang, X. (1994b) Regulatory effects of bifidobacteria on the growth of other colonic bacteria. *Journal of Applied Bacteriology,* **77**, 412–420.

Gibson, G. R., Macfarlane, S. and Cummings, J. H. (1990) The fermentability of polysaccharides by mixed human faecal bacteria in relation to their suitability as bulk-forming laxatives. *Letters in Applied Microbiology,* **11**, 251–254.

Gibson, G. R., Beatty, E. R., Wang, X. and Cummings, J. H. (1995) Selective stimulation of bifidobacteria in the human colon by oligofructose and inulin. *Gastroenterology,* **108**, 975–982.

Gibson, G. R., Saavedra, J. M., Macfarlane, S. and Macfarlane, G. T. (1997) Probiotics and intestinal infections. In: *Probiotics 2: Applications and Practical Aspects.* R. Fuller (Ed.). Chapman and Hall, London, pp. 10–39.

Gibson, G. R., Rastall, R. A. and Roberfroid, M. B. (1999) Prebiotics. In: *Colonic Microbiota: Nutrition and Health.* G. R. Gibson and M. B. Roberfroid (Eds). Kluwer Academic Press, Dordrecht, pp. 101–124.

Gibson, G. R., Berry Ottaway, P. and Rastall, R. A. (2000) *Prebiotics: New Developments in Functional Foods.* Chandos Publishing Limited, Oxford.

Gibson, G. R., Probert, H. M., van Loo, J. A. E., Rastall, R. A. and Roberfroid, M. B. (2004) Dietary modulation of the human colonic microbiota: Updating the concept of prebiotics. *Nutrition Research Reviews,* **17**, 259–275.

Gionchetti, P., Rizzello, F., Venturi, A. and Campieri, M. (2000) Probiotics in infective diarrhoea and inflammatory bowel diseases. *Journal of Gastroenterology and Hepatology,* **15**, 489–493.

Goldin, B. R. and Gorbach, S. L. (1992) Probiotics for humans. In: *Probiotics. The Scientific Basis.* Fuller R. (Ed.). Chapman and Hall, London, pp. 355–376.

Gorbach, S. L. (2002) Probiotics in the third millennium. *Digestive and Liver Disease,* **34**, S2–S7.

Gronlund, M. M., Lehtonen, O. P., Eerola, E. and Kero, P. (1999) Faecal microflora in healthy infants born by different methods of delivery: permanent changes in intestinal flora after Cesarean delivery. *Journal of Paediatric Gastroenterology and Nutrition,* **28**, 19–25.

Havenaar, R. and Huis in't Veld, J. H. J. (1992) Probiotics: a general view. In: *The Lactic Acid Bacteria. Vol 1. The Lactic Acid Bacteria in Health and Disease.* B. J. B Wood (Ed.). Elsevier Applied Science, Barking, pp. 151–170.

Havenaar R., Ten Brink, B. and Huis in't Veld, J. H. J. (1992) In: *Probiotics. The Scientific Basis.* R. Fuller (Ed.). Chapman and Hall, London, pp. 209–224.

Heerze, L. D., Kelm, M. A. and Talbot, J. A. (1994) Oligosaccharide sequences attached to an inert support (SYNSORB) as potential therapy for antibiotic-associated diarrhoea and pseudomembranous colitis. *Journal of Infectious Diseases,* **169**, 1291–1296.

Holzapfel, W. H. and Schillinger, U. (2002) Introduction to pre- and probiotics. *Food Research International,* **35**, 109–116.

Hudson, M. J. and Marsh, P. D. (1995) Carbohydrate metabolism in the colon. In*: Human Colonic Bacteria: Role in Nutrition, Physiology and Pathology.* G. R. Gibson and G. T. Macfarlane (Eds). CRC Press, Boca Raton, FL, pp. 61–73.

Hui, Y. H., Gorham, J. R., Murrell, K. D. and Cliver, D. O. (Eds) (1994) *Foodborne diseases Handbook–Diseases Caused by Bacteria.* Marcel Dekker Inc., New York.

Imaizumi, K., Nakatsu, Y., Sato, M., Sedarnawati, Y. and Sugano, M. (1991) Effects of xylooligosaccharides on blood glucose, serum and liver lipids and caecum short-chain fatty acids in diabetic rats. *Agriculture Biology and Biochemistry,* **55**, 199–205.

Ito, M., Deguchi, Y., Miyamori, A., Matsumoto, K., Kikuchi, H., Matsumoto, K., Kobayashi, Y., Yajima, T. and Kan, T. (1990) Effects of administration of galactooligosaccharide on the human feacal microflora, stool weight and abdominal sensation. *Microbial Ecology in Health and Disease,* **3**, 285–292.

Jacobsen, C. N., Nielsen, V. R., Hayford, A. E., Moller, P. L., Michaelson, K. F., Paeregaard, A., Sandstrom, B., Tvede, M. and Jakobsen, M. (1999) Screening of the probiotic activities of forty-seven strains of *Lactobacillus* spp. by in vitro techniques and evaluation of the colonization ability of five selected strains in humans. *Applied and Environmental Microbiology,* **65**, 4949–4956.

Jayaraman, N., Nepogodiev, S. A. and Stoddart, J. F. (1997) Synthetic carbohydrate-containing dendimers. *Chemistry in Europe Journal,* **3**, 1193–1199.

Kailasapathy, K. and Chin, J. (2000) Survival and therapeutic potential of probiotic organisms with reference to *Lactobacillus acidophilus* and *Bifidobacterium* spp. *Immunology and Cell Biology*, **78**, 80–88.

Karlsson, K.-A. (1989) Animal glycosphingolipids as membrane attachment sites for bacteria. *Annual Review of Biochemistry*, **58**, 309–50.

Klijn, N., Weerkamp, A. H. and de Vos, W. M. (1995) Genetic marking of *Lactococcus lactis* shows its survival in the human gastrointestinal tract. *Applied and Environmental Microbiology*, **61**, 2771–2774.

Korakli, M., Ganzle, M. G. and Vogel, R. F. (2002) Metabolism by Bifidobacteria and lactic acid bacteria of polysaccharides from wheat, rye and exopolysaccharides produced by *Lactobacillus sanfranciencis*. *Journal of Applied Microbiology*, **92**, 958–965.

Kruh, J. (1982) Effects of sodium butyrate, a new pharmacological agent on cells in culture. *Molecular and Cellular Biochemistry*, **42**, 65–82.

Kunz, C. and Rudloff, S. (1993) Biological functions of oligosaccharides in human milk. *Acta Paediatric*, **82**, 903–912.

Langendijk, P. S., Schut, F., Jansen, G. J., Raangs, G. C., Kamphuis, G. R., Wilkinson, M. H. F. and Welling, G. W. (1995) Quantitative fluorescence in situ hybridisation of *Bifidobacterium* spp. with genus-specific 16S rRNA-targeted probes and its application in faecal samples. *Applied and Environmental Microbiology*, **61**, 3069–3075.

Levitt, M. D., Gibson, G. R. and Christl, S. U. (1995) Gas metabolism in the large intestine. In: *Human Colonic Bacteria: Role in Nutrition, Physiology and Pathology*. G. R. Gibson and G. T. Macfarlane (Eds). CRC Press, Boca Raton, FL, pp. 131–153.

Lewis, B. A., Hall, M. B. and Van Soest, P. J. (2001) Interaction between human gut bacteria and dietary fiber substrates. In: *CRC Handbook of Dietary Fiber in Human Nutrition, 3rd Edition*. G. A. Spiller (Ed.). CRC Press, Boca Raton, FL, pp. 271–276.

Macfarlane, G. T. and Cummings, J. H. (1991) The colonic flora, fermentation and large bowel digestive function. In: *The Large Intestine: Physiology, Pathophysiology and Disease*. S. F. Phillips, J. H. Pemberton and R. G. Shorter (Eds). Raven Press, New York, pp. 51–88.

Macfarlane, G. T. and Gibson, G. R. (1994) Metabolic activities of the normal colonic flora. In: *Human Health: The Contribution of Microorganisms*. S. A. W. Gibson (Ed.). Springer, London, pp. 17–52.

Macfarlane, G. T. and Macfarlane, S. (1995) Proteolysis and amino acid fermentation. In: *Human Colonic Bacteria: Role in Nutrition, Physiology and Pathology*. G. R. Gibson and G. T. Macfarlane (Eds). CRC Press, Boca Raton, FL, pp. 75–100.

Macfarlane, G. T. and Macfarlane, S. (1997) Human colonic microbiota: ecology, physiology and metabolic potential of intestinal bacteria. *Scandinowian Journal of Gastroenterology*, **32** (Suppl. 222), 3–9.

Macfarlane, G. T. and McBain, A. J. (1999) The human colonic microbiota. In: *Colonic Microbiota, Nutrition and Health*. G. R. Gibson and M. Roberfroid (Eds). Kluwer Academic Publishers, London, pp. 1–26.

Macfarlane, G. T., Gibson, G. R. and Cummings, J. H. (1992) Comparison of fermentation reactions in different regions of the human colon. *Journal of Applied Bacteriology*, **72**, 57–64.

Macfarlane, S., McBain, A. J. and Macfarlane, G. T. (1997) Consequences of biofilm and sessile growth in the large intestine. *Advances in Dental Research*, **11**, 59–68.

Macfarlane, G. T., Macfarlane, S. and Gibson, G. R. (1998) Validation of a three-stage compound continuous culture system for investigating the effect of retention time on the ecology and metabolism of bacteria in the human colon. *Microbial Ecology*, **35**, 180–187.

Marteau, P., de Vrese, M., Cellier, C. J. and Schrezenmeir, J. (2001) Protection from gastrointestinal diseases with the use of probiotics. *American Journal of Clinical Nutrition*, **73** (Suppl.), 430S–436S.

McBain, A. J. and Macfarlane, G. T. (2001) Modulation of genotoxic enzyme activities by non-digestible oligosaccharide metabolism in *in-vitro* human gut bacterial systems. *Journal of Medical Microbiology*, **50**, 833–842.

McCartney, A. L. (2002) Application of molecular biological methods for studying probiotcs and the gut flora. *British Journal of Nutrition*, **88**, S29–S37.

Metchnikoff, E. (1907) *The Prolongation of life*. Heinemann, London.

Miller, T. L. and Wolin, M. J. (1981) Fermentation by the human large intestine microbial community in an *in vitro* semicontinuous culture system. *Applied Environmental Microbiology*, **42**, 400–407.

Mitsuoka, T. (1990) Bifidobacteria and their role in human health. *Journal of Industrial Microbiology*, **6**, 263–268.

Mitsuoka, T., Hidaka, H. and Eida, T. (1987) Effect of fructooligosaccharides on intestinal microflora. *Die Nahrung*, **31**, 427–436.

Modler, H. W. (1994) Bifidogenic factors – sources, metabolism and applications. *International Dairy Journal*, pp. 383–407.

Molly, K., Vande Woestyne, M. and Verstraete, W. (1993) Development of a 5-step multi-chamber reactor as a simulation of the human intestinal microbial ecosystem. *Applied Microbiology and Biotechnology*, **39**, 254–258.

Moore, W. E. C., Cato, E. P. and Holdeman, L. V. (1978) Some current concepts in intestinal bacteriology. *American Journal of Clinical Nutrition*, **31**, S33–S42.

Morgan, A. J., Mul, A. J., Beldman, G. and Voragen, A. G. J. (1992) Dietary oligosaccharides – new insights. *AGRO Food Industry High Technology*, **Nov/Dec**, 35–38.

Morisse, J. P., Maurice, R., Boilletot, E. and Cotte, J. P. (1993) Assessment of the activity of a fructo-oligosaccharide on different caecal parameters in rabbits experimentally infected with *E. coli* 0.103. *Annals of Zootechnology*, **42**, 81–87.

Naruszewicz, M., Johansson, M.-L., Zapolska-Downar, D. and Bukowska, H. (2002) Effect of *Lactobaillus plantarum* 299v on cardivascular disease risk factors in smokers. *American Journal of Clinical Nutrition*, **76**, 1249–1255.

Nilsson, U., Oste, R., Jagerstad, M. and Birkhed, D. (1988) Cereal fructans: *in vitro* and *in vivo* studies on availability in rats and humans. *Journal of Nutrition*, **118**, 1325–1330.

Oku, T., Tokunaga, T. and Hosoya, N. (1984) Nondigestibility of a new sweetener, 'Neosugar' in the rat. *Journal of Nutrition*, **114**, 1574–1581.

Olano-Martin, E., Mountzouris, K. C., Gibson, G. R. and Rastall, R. A. (2000) *In vitro* fermentability of dextran, oligodextran and maltodextrin by human gut bacteria. *British Journal of Nutrition*, **83**, 247–255.

O'Sullivan, M. G. (1996) Metabolism of bifidogenic factors by gut flora – an overview. *Bulletin of International Dairy Foundation*, **313**, 23–25.

Ouwehand, A. C., Salmien, S. and Isolauri, E. (2002) Probiotics: an overview of the beneficial effects. *Antonie van Leeuwenhoek*, **82**, 279–289.

Park, S. F. and Kroll, R. G. (1993) Expression of listeriolysin and phosphatidylinositol-specific phospholipase C is repressed by the plant-derived molecule cellobiose in *Listeria monocytogenes*. *Molecular Microbiology*, **8**, 653–661.

Payne, S., Gibson, G., Wynne, A., Hudspith, B., Brostoff, J. and Tuohy, K. (2003) *In vitro* studies on colonization resistance of the human but microbiota to *Candida albicans* and the effects of tetracycline and *Lactobacillus plantarum* LPK. *Current Issues in Intestinal Microbiology*, **4**, 1–8.

Pereira, D. I. A. and Gibson, G. (2002) Cholesterol assimilation by lactic acid bacteria and bifidobacteria isolated from the human gut. *Applied and Environmental Microbiology*, **68**, 4689–4693.

Pereira, D. I. A., McCartney, A. L. and Gibson, G. R. (2003) An *in vitro* study of the probiotic potential of a bile-salt-hydolysing *Lactobacillus* fermentum strain, and determination of its cholesterol-lowering properties. *Applied and Environmental Microbiology*, **69**, 4743–4752.

Quigley, M. E. and Kelly, S. M. (1995) Structure, function and metabolism of host mucus glycoproteins. In: *Human Colonic Bacteria – Role in Nutrition, Physiology and Pathology*. G. R. Gibson and G. T. Macfarlane (Eds). CRC Press, London, pp. 175–199.

Rao, A. V. (1999) Large bowel cancer and colonic foods. In: *Colonic Microbiotia, Nutrition and Health*. G. R. Gibson and M. Roberfroid (Eds). Kluwer Academic Publishers, London, pp. 257–268.

Rastall, R. A. and Maitin, V. (2002) Prebiotics and synbiotics: towards the next generation. *Current Opinion in Biotechnology*, **13**, 490–496.

Reddy, B. S. (1999) Possible mechanisms by which pro- and prebiotics influence colon carcinogenesis and tumour growth. *Journal of Nutrition*, **129**, 1478S–1482S.

Roberfroid, M. (1998) Prebiotics and synbiotics: concepts and nutritional properties. *British Journal of Nutrition*, **80** (Suppl. 2), 197S–202S.

Roberfroid, M. (2002) Functional food concept and its application to prebiotics. *Digestive and Liver Disease*, **34**, S105–S110.

Roberfroid, M., Gibson, G. R. and Delzenne, N. (1993) The biochemistry of oligofructose, a non-digestible fibre: an approach to calculate its calorific value. *Nutrition Reviews*, **51**, 137–146.

Roitt, I., Brostoff, J. and Male, D. (1998) *Immunology.* Mosby, London.

Roller, M., Rechkemmer, G. and Watzl, B. (2004) Prebiotic inulin enriched with oligofructose in combination with the probiotics *Lactobacillus rhamnosus* and *Bifidobacterium lactis* modulates immune function in rats. *Journal of Nutrition*, **134**, 153–156.

Rosenfeldt, V., Benfeldt, E., Nielsen, S. D., F., M. K., Jeppesen, D. L., Valerius, N. H. and Paeregaard, A. (2003) Effect of probiotic *Lactobacillus* strains in children with atopic dermatitis. *Journal of Allergy and Clinical Immunology*, **111**, 389–395.

Rowland, I. R. (Ed.) (1988) *Role of the Gut Flora in Toxicity and Cancer.* Academic Press, London.

Rowland, I. R. and Mallett, A. K. (1990) The effect of diet on mammalian gut flora and its metabolic activities. *CRC Critical Reviews in Toxicology*, **16**, 31–103.

Rowland, I. R. and Tanaka, R. (1993) The effects of transgalactosylated oligosaccharides on gut flora metabolism in rats associated with human faecal microflora. *Journal of Applied Bacteriology*, **74**, 667–674.

Rowland, I. R. and Wise, A. (1985) The effect of diet on the mammalian gut flora and its metabolic activities. *CRC Critical Reviews in Toxicology*, **16**, 31–103.

Rumney, C. J. and Rowland, I. R. (1992) *In vivo* and *in vitro* models of the human colonic flora. *Critical Reviews in Food Science and Nutrition*, **34**, 229–311.

Rumney, C. J., Rowland, I. R., Coutts, C. M., Randerath, K., Reddy, R., Shah, A. B., Ellul, A. and O'Neil, I. K. (1993) Effects of risk-associated human dietary macrocomponents on processes related to carcinogenesis in human-flora-associated (HFA) rats. *Carcinogenesis*, **14**, 79–84.

Rycroft, C. E., Fooks, L. J. and Gibson, G. R. (1999) Methods for assessing the potential of prebiotics and probiotics. *Current Opinion in Clinical Nutrition and Metabolic Care*, **2**, 481–484.

Rycroft, C. E., Jones, M. R., Gibson, G. R. and Rastall, R. A. (2001) A comparative *in vitro* evaluation of the fermantation properties of prebiotic oligosaccharides. *Journal of Applied Microbiology*, **91**, 878–887.

Saarela, M., Lahteenmaki, L., Crittenden, R., Salmien, S. and Mattila-Sandholm, T. (2002) Gut bacteria and health foods – the European perspective. *International Journal of Food Microbiology*, **78**, 99–117.

Salminen, S. and Salminen, E. (1997) Lactulose, lactic acid bacteria, intestinal microecology and mucosal protection. *Scandinavian Journal of Gastroenterology*, **32** (Suppl.), 45–48.

Salminen, S., Bouley, C., Boutron-Ruault, M-C., Cummings, J. H., Franck, A., Gibson, G. R., Isolauri, E., Moreau, M. C., Roberfroid, M. B. and Rowland, I. R. (1998a) Functional food science and gastrointestinal physiology and function. *British Journal of Nutrition*, **80**, S147–S171.

Salminen, S., von Wright, A., Morelli, L., Marteau, P., Brassart, D., de Vos, W. M., Fonden, R., Saxelin, M., Collins, K., Mogensen, G., Birkeland, S.-E. and Mattila-Sandholm, T. (1998b) Demonstation of the safety of probiotics – a review. *International Journal of Food Microbiology*, **44**, 93–106.

Salminen, M. K., Tynkkynen, S., Rautelin, H., Saxelin, M., Vaara, M., Ruutu, P., Sarna, S., Valtonen, V. and Jarvinen, A. (2002) *Lactobacillus* bacteremia during a rapid increase in probiotic use if *Lactobacillus rhamnosus* GG in Finland. *Clinical Infectious Diseases*, **35**, 1155–1160.

Salyers, A. A. and Whitt, D. D. (1994) In: *Bacterial Pathogenesis: a Molecular Approach.* ASM Press, Washington, DC, pp. 130–140.

Satokari, R. M., Vaughn, E. E., Akkermans, A. D. L., Saarela, M. and de Vos, W. M. (2001) Polymerase chain reaction and denaturing gradient gel electrophoesis monitoring of faecal *Bifidobacterium* populations in a prebiotic and probiotic feeding trial. *Systemic and Applied Microbiology*, **24**, 227–231.

Schriffin, E. J. and Brassart, D. (1999) Intestinal microflora and the mucosal mechanisms of protection. In: *Colonic Microbiota, Health and Disease*. G. R. Gibson and M. Roberfroid (EDs). Kluwer Academic Publishers, London, pp. 201–212.

Shin, H.-S., Lee, J.-H., Pestka, J. J. and Ustunol, Z. (2000) Growth and viability of commercail *Bifidobacterium* spp. in skim milk containing oligosaccharides and inulin. *Journal of Food Science*, **65**, 884–887.

Silvi, S., Rumney, C. J., Cresci, A. and Rowland, I. R. (1999) Resistant starch modifies gut flora and microbial metabolismin human flora-associated rats inoculated with feaces from Italian and UK donors. *Journal of Applied Microbiology*, **86**, 521–530.

Simon, G. L. and Gorbach, S. L. (1984) Intestinal flora in health and disease. *Gastroenterology*, **86**, 174–193.

Sullivan, A. and Nord, C. E. (2002) The place of probiotics in human intestinal infections. *International Journal of Antimicrobial Agents*, **20**, 313–319.

Sullivan, N. M., Mills, D. C., Riemann, H. P. and Arnon, S. S. (1988) Inhibition of growth of *Clostriduim botulinum* by intestinal microflora isolated from healthy infants. *Microbial Ecology in Health and Disease*, **1**, 179–192.

Tannock, G. W., Munro, K., Harmslen, H. J. M., Welling, G. W., Smart, J. and Gopal, P. K. (2000) Analysis of the faecal microflora of human subjects consuming a probiotic product containing *Lactobacillus rhamnosus* DR20. *Applied and Environmental Microbiology*, **66**, 2578–2588.

Teuri, U. and Korpel, R. (1998) Galacto-oligosaccharides relieve constipation in elderly people. *Annals of Nutrition and Metabolism*, **42**, 319–327.

Topping, D. L. and Clifton, P. M. (2001) Short-chain fatty acids and human colonic function: roles of resistant starch and nonstarch polysaccharides. *Physiological Reviews*, **81**, 1031–1064.

Tuohy, K. M., Finlay, R. K., Wynne, A. G. and Gibson, G. R. (2001a) A human volunteer study on the prebiotic effects of HP-inulin – gut bacteria enumerated using fluorescent *in situ* hybridisation (FISH). *Anaerobe*, **7**, 113–118.

Tuohy, K. M., Kolida, S., Lustenberger, A. and Gibson, G. R. (2001b) The prebiotic effects of biscuits containing partially hydrolyzed guar gum and fructooligosaccharides – a human volunteer study. *British Journal of Nutrition*, **86**, 341–348.

Tuohy, K. M., Ziemer, C. J., Klinder, A., Knobel, Y., Pool-Zobel, B. L. and Gibson, G. R. (2002) A human volunteer study to determine the prebiotic effects of lactulose powder on human colonic bacteria. *Microbial Ecology in Health and Disease*, **14**, 165–173.

Van Loo, J., Cummings, J., Delzenne, N., Englyst, H., Franck, A., Hopkins, M., Kok, N., Macfarlane, G., Newton, D., Quigley, M., Roberfroid, M., Van Vliet, T. and Van den Heuvel, E. (1999) Functional food properties of non-digestible oligosaccharides: a consensus report from the ENDO project (DGXII AIRII-CT94-1095. *British Journal of Nutrition*, **81**, 121–132.

Wolever, T. M. S., Spadafora, P. and Eshuis, H. (1991) Interaction between colonic acetate and propionate in humans. *American Journal of Clinical Nutrition*, **53**, 681–687.

Wolf, B. W., Garleb, K. A., Ataya, D. G. and Casas, I. A. (1995) Safety and tolerance of *Lactobacillus reuteri* in healthy adult male subjests. *Microbial Ecology in Health and Disease*, **8**, 41–50.

Yamamoto-Osaki, T., Kamiya, S., Sawamura, S., Kai, M. and Ozawa, A. (1994) Growth inhibition of *Clostridium difficile* by intestinal flora of infant faeces in continuous flow culture. *Journal of Medical Microbiology*, **40**, 179–187.

Yuki, N., Watanabe, K., Mike, A., Tagami, Y., Tanaka, R., Ohwaki, M. and Morotomi, M. (1999) Survival of a probiotic, *Lactobacillus casei* strain Shirota, in the gastrointestinal tract: selective isolation from faeces and identification using monoclonal antibodies. *International Journal of Food Microbiology*, **48**, 51–57.

Zetterström, R., Bennett, R. and Nord, K.-E. (1994) Early infant feeding and micro-ecology of the gut. *Acta Paediatric Japonica*, **36**, 562–571.

2

Manufacture of Prebiotic Oligosaccharides

Tamara Casci and Robert A. Rastall

2.1 Introduction

The production of prebiotic oligosaccharides is usually achieved through one of three general processes (Grizard and Barthomeuf, 1999a): (1) direct extraction of natural oligosaccharides from plants; (2) controlled hydrolysis of natural polysaccharides; and (3) enzymatic synthesis using hydrolases and/or glycosyl transferases from plant or of microbial origin (L'Hocine *et al.*, 2000).

Most prebiotics (Table 2.1) are enzymatically produced, e.g. xylooligosaccharides (XOS) and isomaltooligosaccharides (IMO) are produced by enzymatic hydrolysis of polysaccharides, fructooligosaccharides (FOS), lactosucrose and galactooligosaccharides (GOS) are produced by transgalactosylation (Nakakuki, 1993; Sako *et al.*, 1999); gentiooligosaccharides are produced by condensation reactions, inulin and soybean oligosaccharide, instead, are extracted from source and lactulose is produced chemically through isomerisation.

The main types of enzymes used to catalyse oligosaccharide synthesis are the hydrolases (EC 3.2.) and the transferases (glycosyl-transferases, EC 2.4.) (Monsan and Paul, 1995). Glycosidases are able to catalyse either the direct coupling of glycosyl moieties by reversion of the hydrolysis reaction (also known as reverse hydrolysis), or the transfer of a glycosyl residue from an activated donor onto an acceptor (transglycosylation) (Monsan and Paul, 1995).

Prebiotics: Development and Application Edited by G.R. Gibson and R.A. Rastall

Table 2.1 *Prebiotic oligosaccharides. (Adapted from Crittenden and Playne, 1996)*

	Class of oligosaccharide	Molecular structure[a] (Sako *et al.*, 1999)	Physiological features (Sako *et al.*, 1999)	Major manufacturers	Trade names
1	Fructooligosaccharides/ Inulin	$(Fr)_n$- Gu	β-1,2	Meiji Seika Kaisha (Japan)	Meioligo
				Beghin-Meiji Industries (France)	Actilight
				Orafti (Belgium)	Raftilose
2	Galactooligosaccharides	$(Ga)_n$-Gu	β-1,4, β-1,6	Yakult Honsha (Japan)	Oligomate Cup-Oligo
				Nissin Sugar Manufacturing Company (Japan)	
3	Isomaltooligosaccharides	$(Gu)_n$	α-1,6	Showa Sangyo (Japan)	Isomalto-900
4	Lactulose	Ga-Fr	β-1,4	Morinaga Milk Industry Co. (Japan)	MLS/P/C
				Solvay (Germany)	
5	Soybean oligosaccharides	$(Ga)_n$–Gu-Fr	α-1,6	The Calpis Food Industry Co. (Japan)	Soya-oligo
6	Lactosucrose	Ga-Gu-Fr	β-1,4	Ensuiko Sugar Refining Co. (Japan)	Nyuka-Origo
7	Xylooligosaccharides	$(Xy)_n$	β-1,4	Suntory Ltd (Japan)	Xylo-oligo
8	Gentiooligosaccharides	$(Gu)_n$	β-1,6	Nihon Shokuhin Kako (Japan)	Gentose

[a]Ga, galactose; Gu, glucose; Fr, fructose; Xy, xylose.

2.2 Fructooligosaccharides and Inulin-type Fructans

2.2.1 Introduction

Fructan is a general name used for any carbohydrate in which one or more fructosyl-fructose links constitutes the majority of glycosidic bonds. They are synthesised in plants from sucrose by repeated fructosyl transfer and therefore usually exhibit a terminal glucose unit (Grizard and Barthomeuf, 1999b; Hidaka *et al.*, 1988; L'Hocine *et al.*, 2000). They are mainly composed of 1-kestose (GF_2), nystose (GF_3) and 1^F- fructofuranosyl nystose (GF_4) in which fructosyl units (F) are bound at the β-2,1 position of sucrose (GF) (Yun, 1996). FOS is bifidogenic and is a well established prebiotic (Fuller and Gibson, 1997; Gibson *et al.*, 1995, 2000; O'Sullivan, 1996).

Fructans can be of the inulin and laevan types. The inulin type is that with β-2,1-D fructofuranosyl units found in plant and synthesised by fungi. The laevan type is that with β-6,2-D fructofuranosyl units found in plants and synthesised by bacteria such as *Streptococcus mutans* (Patel *et al.*, 1994).

Inulin-type fructans have a degree of polymerisation (DP) varying between 2 and 70 (Gibson *et al.*, 2000). FOS have a lower molecular weight than inulin ($DP > 30$) (Kosaric *et al.*, 1984).

There are many sources of fructans, principally from plants (chicory, Jerusalem artichoke, Dahlia, onion, garlic, leek and grains including wheat). The amount of fructan and the DP vary greatly with source and storage conditions (Modler, 1994).

FOS may be produced from sucrose (Figure 2.1) (Hidaka *et al.*, 1988) through the transfructosylating action of either β-fructofuranosidases (β-FFase, EC 3.2.1.26) or β-D-fructosyltransferases (β-FTase, EC 2.4.1.9). The Japanese FOS product *Neosugar* is a commercial FOS mixture of GF_2, GF_3, GF_4 produced using the enzyme from *Aspergillus niger* ATCC 20611. The same product is also manufactured in the EU under the trade name Actilight.

Industrial enzymatic FOS production processes from sucrose are via one of two routes (Yun, 1996): (1) batch systems using soluble enzymes; and (2) continuous systems using immobilised enzymes or whole cells (Figure 2.2). The enzymes involved may be intra- or extracellular.

Commercial FOS syrups contain 25–30 % (w/w) 1-kestose (GF_2), 10–15 % nystose (GF_3) and 5–10 % 1^F-fructofuranosylnystose (GF_4) and 25–30 % glucose as a by-product (Kim *et al.*, 2000).

The nomenclature of FOS producing enzymes remains in dispute: some workers use the term β-fructofuranosidase, a hydrolase, whereas others use the term fructosyltransferase, emphasising the nature of transfructosylation of the enzyme and to distinguish it from hydrolytic enzyme nomenclature (L'Hocine *et al.*, 2000; Yun, 1996). Throughout this chapter, the term β-fructosyltransferase (FTS) will be used. The enzyme acts on sucrose in a disproportionation type reaction where one molecule of sucrose serves as a donor and the other as an acceptor.

2.2.2 Enzyme Characteristics and Sources

There are two main sources of fructosyltransferases: of plant origin, and from microorganisms, especially fungi. The species of microorganism most frequently used as a source of enzyme to produce FOS are *A. niger, Aureobasidium pullulans* and *Aspergillus japonicus*, which preferably produce 1^F-type FOS (L'Hocine *et al.*, 2000; Madlova *et al.*, 2000; Yun, 1996).

β-Fructosyltransferase has both a hydrolysis (Uh) and a fructosyl transfer (Ut) activity but the ratio (Ut/Uh) varies with each enzyme (Yanai *et al.*, 2001). The enzyme has either a preference for water as an acceptor (in a hydrolysis reaction) or for another sugar or alcohol (in a transfer reaction). It is the transfer reaction that produces FOS. Many fungal strains have been screened in an attempt to maximise the production of FOS. *Aureobasidium pullulans* and *A. niger* have the highest transferase activity (Madlova *et al.*, 1999). Many researchers have characterised the FOS-producing fructosyltransferase in an attempt to find enzymes capable of producing high yields of a desired product or acceptable yields of a desired product profile.

In general, enzymes from different sources have sizes ranging from 232 kDa (Imamura *et al.*, 1994) to 346 kDa (Hayashi *et al.*, 1991) and can consist of more than one subunit or have carbohydrates attached to it. The optimum pH and temperature for FOS production are generally between pH 5.0 and 6.0, and between 50 and 60 °C, respectively. These values can vary depending upon the enzyme source. The reported maximum yields of FOS vary, according to source and reaction setup, from 42 % (Kim *et al.*, 1998) from the *Bacillus macerans* enzyme to 80 % (Madlova *et al.*, 1999) from *Auerobasidium pullulans* and *A. niger* enzymes.

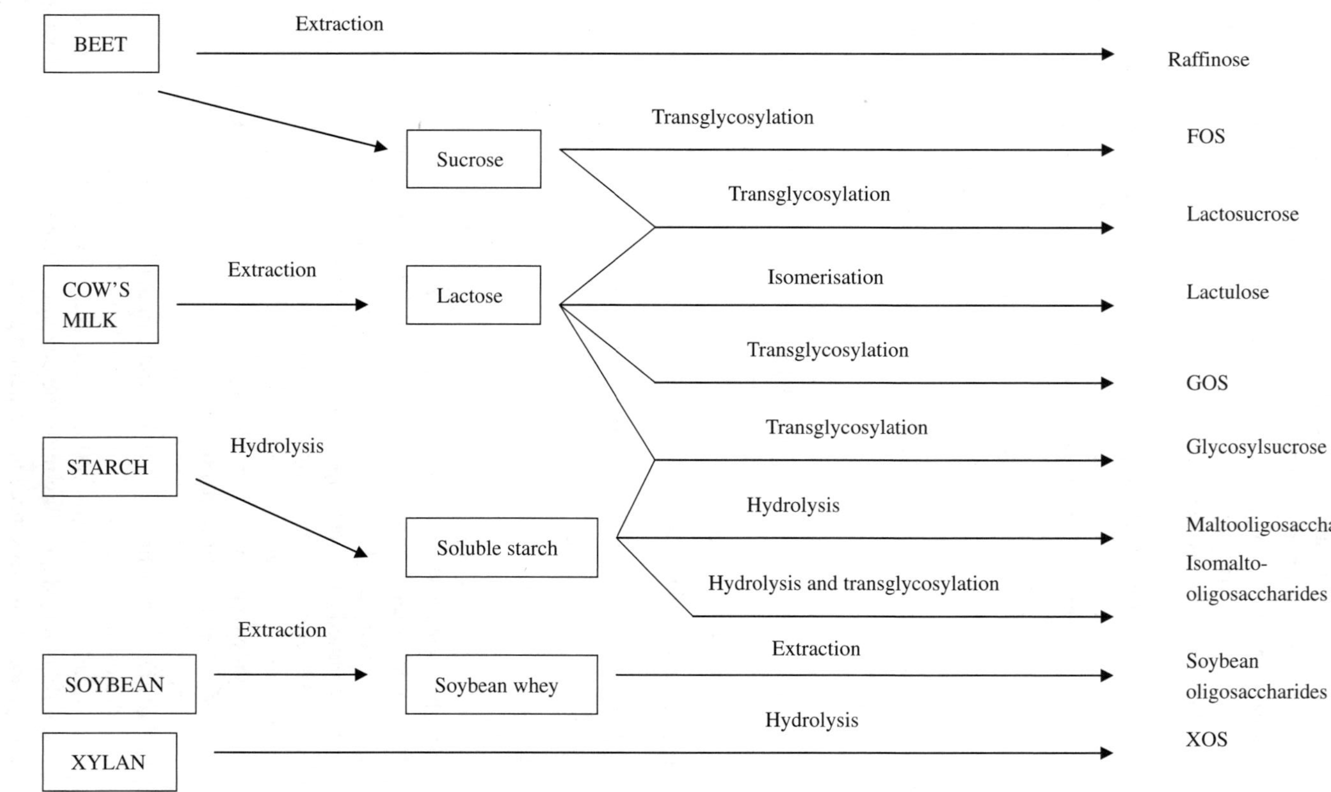

Figure 2.1 *Schematic representation of production processes of nondigestible oligosaccharides. (Adapted from Sako et al., 1999)*

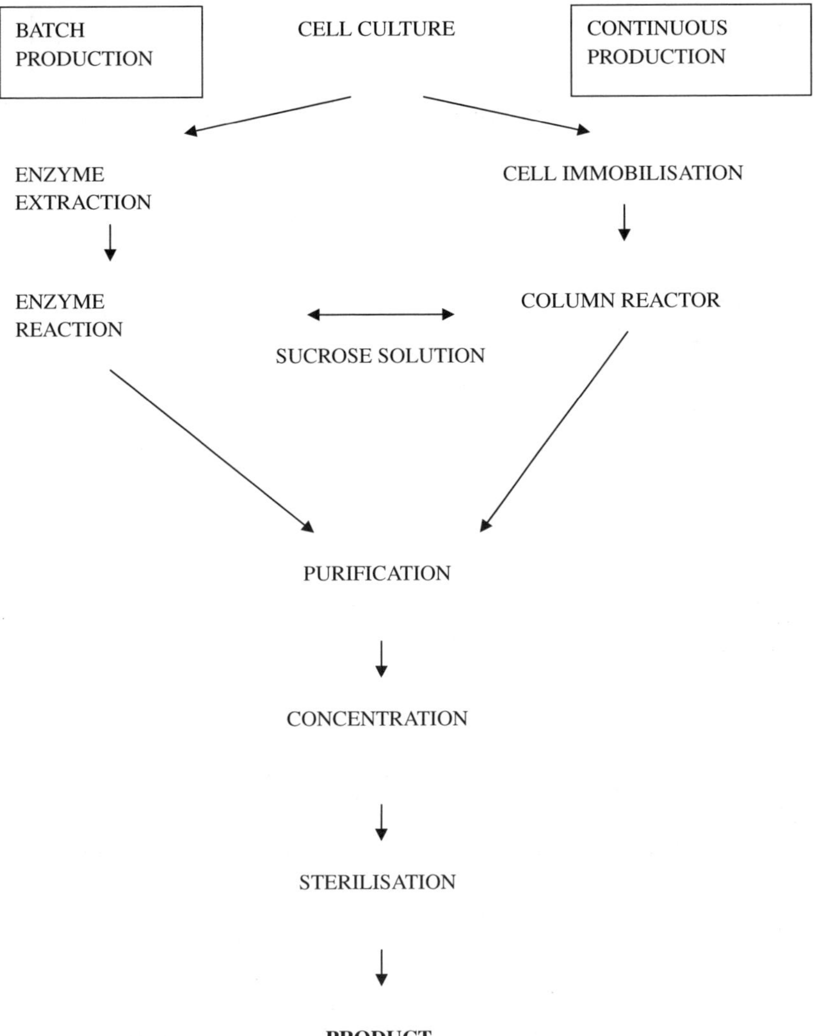

Figure 2.2 *Industrial processes for the production of FOS. (Adapted from Yun, 1996)*

Table 2.2 summarises values of optimum pH, temperature and other conditions that have been reported.

2.2.3 Production Methods

Although a well-established method of producing FOS exists, many researchers have investigated ways to increase the yield and efficiency of the reaction at industrial scales, as well as methods to simplify and accelerate the reaction itself.

Various enzyme sources and reaction conditions have been used in the attempt to optimise enzyme immobilisation techniques. Recently, immobilisation matrices used

Table 2.2 *Summary of fructosyltransferase characteristics*

Reference	Enzyme source	Molecular weight	Optimum pH	Optimum temp.(°C)	Best yield at Sucrose conc. used	Oligosaccharide products	Yield FOS (%)
Cheng *et al.*, 1996	*A. japonicus*	NA	4.0–5.5	NA	NA	NA	NA
Duan *et al.*, 1994a	*A. japonicus*	NA	NA	NA	NA	1-kestose, nystose	NA
Grizard and Bartomeuf, 1999a	Cytolase PCL5	NA	5.5	55	NA	NA	NA
Hayashi *et al.*, 1991	*Aureobasidium sp.*	318 000 346 000	4.5–5.5 4.5–6.0	50–55 50–55	75 % w/v	1-kestose	NA
Hidaka *et al.*, 1988	*A. niger*	NA	NA	NA	50 %	1-kestose	69
Hirayama *et al.*, 1989	*A. niger*	340 000	5.0–6.0	50–60	50 %	1-kestose	NA
Imamura *et al.*, 1994	*B. infantis*	232 000	6.0–6.2	55	NA	NA	NA
Jung *et al.*, 1989	*Aureobasidium pullulans*	NA	5.5	55	NA	NA	NA
Kim *et al.*, 2000	*B. macerans EG-6*	NA	7.0	37	$500\,g\,l^{-1}$	NA	43
L'Hocine *et al.*, 1997		NA	NA	NA	NA	NA	54
Madlova *et al.*, 1999	*Aureobasidium pullulans*	NA	NR	60-65	$700\,g\,dm^{-3}$	1-kestose, nystose	~80
Madlova *et al.*, 1999	*A. niger*	NA	NR	60-65		1-kestose, nystose	~80
Nguyen *et al.*, 1999	*A. niger*	NA	NA	NA	25 % w/v	NA	NA
Park *et al.*, 1999	*B. macerans*	NA	6.0	37	$500\,g\,l^{-1}$	GF_5	33
Tambara *et al.*, 1999	*A. diazotrophicus*	NA	5.5	40	$600–700\,g\,l^{-1}$	1-kestose, nystose, 6-kestose	50
Trujillo *et al.*, 2001	*A. diazotrophicus* (or *Gluconacetobacter diazotrophicus*) recombinant in *Pichia pastoris*	NA	5.0	60	$500\,g\,l^{-1}$	1-kestose, nystose. Liberated D-glucose, fructose and laevan	43

NA: data not available; NR, not relevant.

have included polymeric beads (Lee *et al.*, 1999), gluten (Chien *et al.*, 2001) and ceramic membranes as part of a membrane reactor system (Nishizawa *et al.*, 2000).

Chien *et al.* (2001) achieved the production of FOS through immobilised cells of *A. japonicus* in gluten in a batch reaction. One gram of immobilised mycelia particles was incubated with 100 ml of a 400 g l^{-1} sucrose solution. This gave a 61 % yield of FOS after 5 h, the product consisting mainly of 1-kestose and nystose. Both the sucrose concentration and the cell content in the gluten particles influenced significantly the rate of the enzymatic reaction. According to Chiang *et al.* (1997) the total amount of FOS product reaches a stable value when the concentrations of 1-kestose and nystose are almost identical.

The same enzyme source was used by Lee *et al.* (1999) using oxirane-containing polymeric beads as immobilising material. In a batch system, 1 g of the enzyme particles was added to 100 ml of a sucrose solution ranging between 100 and 750 g l^{-1} in concentration. In the continuous system, the immobilised enzyme particles were packed into a column. A 400 g l^{-1} sucrose solution was passed through the column at an optimal flow rate. It was found, in both systems, that the FOS yield increased with initial sucrose concentration up to 500 g l^{-1} and then decreased due to substrate inhibition.

Nishizawa *et al.* (2000) immobilised the β-FTS from *A. niger* to the inner surface of a ceramic membrane activated by a silane-coupling agent. It was found that the saccharide composition of the product was a function of the permeate flux, which was controlled by transmembrane pressure. By adding enzyme to a 50 % (w/v) sucrose solution of pH 6.0 at 50 °C, sucrose was converted to FOS (GF_2, GF_3 and GF_4) and glucose. After 4 h, the concentration of GF_2 was equal to that of GF_3. The volumetric productivity was found to be 560 times higher than that in a conventional batch system.

An interesting production technique was developed by Sheu *et al.* (2001) in an attempt to increase yields by eliminating the reaction by-product glucose. The reaction, catalysed by β-FTS from *A. japonicus*, was performed in an aerated stirred tank reactor controlled at pH 5.5 by a slurry of $CaCO_3$. The underlying principle was that glucose produced in the reaction would be converted by glucose oxidase to gluconic acid, which would then be precipitated to calcium gluconate thus increasing the reaction rate. The product yield was as high as 90 % (w/w) if the reaction was stopped at the optimum time.

Park *et al.* (1999) found and isolated a new strain, *B. macerans* EG-7, which selectively produced GF_5 only. The reaction mixture consisted of 5 ml of a 500 g l^{-1} sucrose solution, 3 ml of 0.1 M phosphate buffer (pH 6.0) and 2 ml of enzyme solution, incubated at 37 °C and gave a 33 % yield. However, the crude enzyme required a 40–60 h lag period for catalytic activity. It was found that a 10 mM concentration of ammonium nitrate (NH_4NO_3) enhanced the activity of the enzyme 15-fold and made the lag phase disappear completely, giving the same reaction rate.

Kim *et al.* (2000) reported that *B. macerans* EG-6 selectively produced only GF_5 from sucrose. They found that the transfructosylating activity in culture filtrate was higher than in intact cells. The best oligosaccharide yield (43 %) was achieved at 500 g l^{-1} sucrose concentration, at pH 7.0 and at 37 °C. Smaller molecules of oligosaccharides, such as GF_2, GF_3 and GF_4, known to be precursors for GF_5 synthesis in other enzyme systems, were not detected during incubation.

A great deal of effort has gone into the identification and characterisation of laevansucrases (EC 2.4.1.10) from various sources for production of FOS. The enzyme converts sucrose to β-2,6 branched fructans, which correspond to laevan-type fructans, and is found in a number of microorganisms. The size of the enzyme varies, according to source, from 46 kDa (Cha *et al.*, 2001) to 60 kDa (Hernandez *et al.*, 2000), the reported pH and temperature optima for FOS production in native state vary from 30 °C (Sangiliyandi *et al.*, 1999) to 60 °C (Trujillo *et al.*, 2001) and 5.0 (Trujillo *et al.*, 2001), to 7.0 (Cha *et al.*, 2001), respectively. Table 2.2 summarises the enzyme characteristics reported.

2.3 Galactooligosaccharides

2.3.1 Introduction

A number of GOS are currently used as low calorie sweeteners, food ingredients and cosmetic additives (Shin *et al.*, 1998). GOS are more stable than FOS. They remain unchanged after treatment at 120 °C/10 min at pH 3 and at 100 °C/10 min at pH 2. They are bifidogenic and noncariogenic (Sako *et al.*, 1999).

GOS are produced from lactose syrup (Figure 2.3) using the transgalactosylase activity of β-galactosidases (Gibson *et al.*, 2000). β-Galactosidase can be isolated from plant

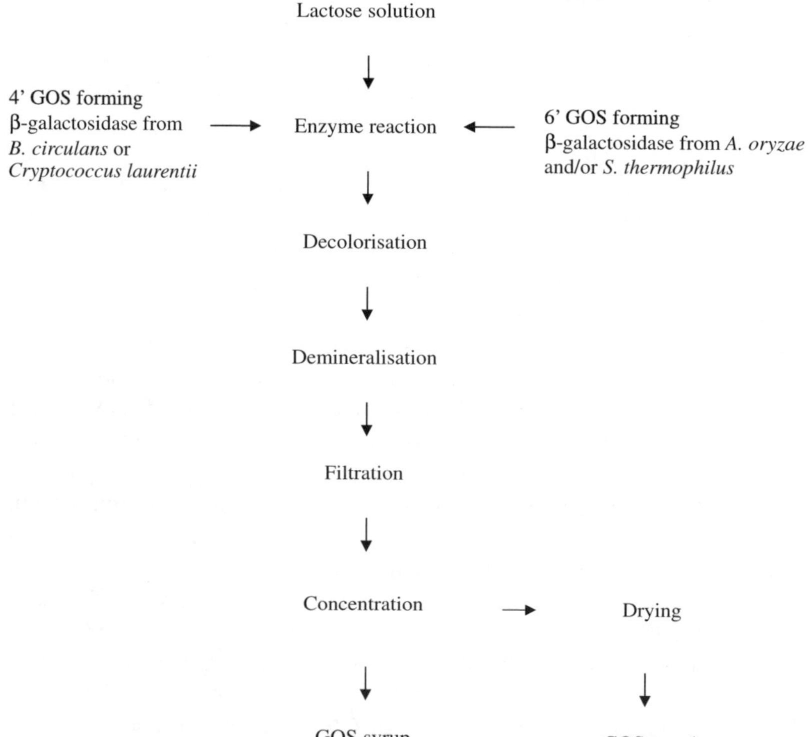

Figure 2.3 *Industrial production process for GOS. (Adapted from Matsumoto, 1990 and Sako et al., 1999)*

(peach, apricot, almond), animal (intestine, brain and skin tissue) and microbial sources (yeast, bacteria, fungi). The microbes offer higher yields for commercial production (Shukla, 1975).

β-Galactosidase (EC 3.2.1.23, β-galactoside galactohydrolase), or lactase, is a hydrolase, which attacks the *O*-galactosyl group of lactose. It is an exoglycosidase. β-Galactosidases from fungal sources, such as *A. niger* and *A. oryzae*, are often used for the hydrolysis of lactose in whey (Sheu *et al.*, 1998). The enzyme catalyses not only the hydrolysis of β-galactosidase linkages of lactose (transferring galactose to a hydroxyl group of water and resulting in the liberation of D-galactose and D-glucose) but also a transgalactosylation reaction (transferring galactose to the hydroxyl groups of the D-galactose or the D-glucose moiety in lactose), the kinetic intermediates of which are GOS (Jorgensen *et al.*, 2001; Petzelbauer *et al.*, 2000; Prenosil *et al.*, 1987a; Sheu *et al.*, 1998; Yang and Tang, 1988; Zarate and Lopez-Leiva, 1990). The production of GOS is according to the following reaction:

$$\text{Lactose} \xrightarrow{\beta\text{-galactosidase}} \text{Galactooligosaccharides} + \text{Glucose} + \text{Galactose}$$

The linkage between the galactose units, the efficiency of transgalatosylation and the components of the final product all depend on the enzyme source and the conditions used in the reactions. It has been observed that β-1,4 bonds are generally formed by enzymes from *B. circulans* (Mozaffar *et al.*, 1984), *Cryptococcus laurentii* (Ozawa *et al.*, 1989) and *Bifidobacterium bifidum* DSM 20082 (Dumortier *et al.*, 1994), whereas enzymes from *A. oryzae* and *S. thermophilus* generate β-1,6 bonds (Matsumoto, 1990; Sako *et al.*, 1999). Depending on the source of the enzyme, the results of GOS production from lactose can be quite different in the final product types and yields (Yang and Bednarcik, 2001). Also, because transgalactosylation products are substrates of β-galacto-catalysed hydrolysis, the composition of the mixture changes quite markedly with progressing reaction time (Boon *et al.*, 1999; Nakayama and Amachi, 1999; Prenosil *et al.*, 1987a,b; Smart, 1991). β-Galactosidase from *A. oryzae*, used in food products, is relatively inexpensive and has optimum pH of 4.5, which is close to the native pH of whey, a common source of lactose.

Usually transglycosylation reactions proceed more efficiently at high substrate concentrations but lactose has low solubility, which imposes a limitation on the efficiency of the reaction. At 25 °C, 25 % (w/v) of lactose can be dissolved in water; at higher temperatures, higher lactose concentrations can be used. Unfortunately, many enzymes denature at high temperatures (Hansson and Aldercreutz, 2001).

Galactose in an inhibitor of β-galactosidase as it competes with lactose for the active site, α-galactose being a significantly stronger inhibitor than the β-form (Flaschel *et al.*, 1982). The inhibition is of a noncompetitive nature (Shukla and Chaplin, 1993).

2.3.2 Enzyme Characteristics and Sources

Many sources of β-galactosidase have been investigated by researchers attempting to improve manufacture and yields (Table 2.3) (Yang and Bednarcik, 2001).

The enzymes can vary in size, reported values ranging from as low as 36 kDa (Tsumura *et al.*, 1991) to 362 kDa (Dumortier *et al.*, 1994), consisting of four subunits. The pH optimum can vary between 4.0 and 9.0 (Tsumura *et al.*, 1991) while the temperature

Table 2.3 *Summary of β-galactosidase characteristics*

Reference	Enzyme source	Molecular Weight	Optimum pH	Optimum temp. (°C)	Best yield at lactose conc. used	Oligosaccharide products	Yield TOS (%)
Akiyama *et al.*, 2001	*Thermus sp.* Z-1	NA	7.0	70	0.88 M lactose solution	3′-Galactosyllactose	40
Dumortier *et al*, 1994	*Bifidobacterium. bifidum*	362 kDa	4.8	45	500 mM lactose	Primarily β-1,3 linkages	29
Fransen *et al.*, 1998	*A. niger*	NA	NA	NA	NA	α-D-Glc*p*-(1,1)-β-D-Gal*p* containing TOS	NA
Hansson and Aldercreutz, 2001	*S. solfataricus* (LacS)	NA	6.5	85	70 % lactose solution	Mostly β-1,3- and β-1, 6-linked oligosaccharides	37
	P. furiosus (CelB)	NA	5.0	95			40
Hashimoto *et al.*, 1995	*Candida guilliermondii* (α)	NA	NA	NA	NA	Mainly α-disaccharides with the 1,6 linkage. Also, 1,1, 1,2, 1,3 linkages	20
Hung and Lee, 2002	*Bifidobacterium infantis HL96* recombinant in *E. coli* β-GAL I subunit	115 kDa	7.5	50	30 %	NA	190 $mg\,ml^{-1}$
Jorgensen *et al.*, 2001	*Bifidobacterium bifidum* C-terminal part of BIF sequence	188 kDa	6.0	37	10, 20 and 40 % w/w	TOS and glucose	39, 44 and 37
Mitsutomi *et al.*, 1991	*Pycnoporus cinnabarinus* (α)	NA	NA	NA	Galactose and sucrose	Raffinose, planteose and 3G-α-galactosylsucrose	15

Onishi and Tanaka, 1995	*Sterigmatomyces elviae*	170 kDa by gel filtration and 86 kDa by SDS gel electrophoresis	4.5–5.0	85	200 mg ml^{-1}	TOS, glucose and little galactose	39
Petzelbauer *et al.*, 2000	*S. solfataricus* (SsβGly)	NA	5.5	70	270 g l^{-1}	β-D-Gal*p*-(1,3)-Glcβ-D-Gal*p*-(1,6)-Glc	26
	P. furiosus (Cel B)					β-D-Gal*p*-(1,3)-lactose β-D-Gal*p*-(1,6)-lactose	32
Reuter *et al.*, 1999	*S. solfataricus*	NA	7.0	65	Lactose and GlcNAc equimolar	β-D-Gal-(1,4)-D-GlcNAc β-D-Gal-(1,6)-D-GlcNAc β-D-Gal-(1,3)-D-GlcNAc	47.7
Shin and Yang, 1996	*P. funicolosum*	NA	NA	40		mainly tri- and tetrasaccharide TOS, glucose and galactose	20
Tsumura *et al.*, 1991	*Bacillus sp.*	S-2, 40 kDa	10	50	Soybean galactan	Gal_3 Gal_4 Gal_2,Gal_3	15 49 31
		S-39, 36 kDa	4.0, 9.0	40			59
Yanahira *et al.*, 1998	*B. circulans*	NA	6.5	40	160 g lactose 40 g NeuNAc (sialic acid) and GlcUA(glucuronic acid) in 200 ml	Acidic TOS N1 N2 G1 G2	(Based on amount of NeuAc and GlcUA used) 17 2.5 23 5

NA: data not available.

optima for GOS production can range from as low as 40 °C (Shin and Yang, 1996) to 95 °C (Hansson and Aldercreutz, 2001) for a thermophilic β-galactosidase from *Pyrococcus furiosus*. In general, fungal lactases have pH optima between 2.5 and 4.5 and yeasts and bacterial lactases in almost neutral region (6–7 and 6.5–7.5, respectively) (Coughlin and Charles, 1980).

To maximise production of GOS many researchers have looked into ways of increasing yields of the reaction or the secretion levels of the enzyme. The total amount of oligosaccharide product depends on the enzyme source and can vary from 1 to 45 % of the total sugar present (Zarate and Lopez-Leiva, 1990). Because the oligosaccharides are constantly being synthesised and degraded, amounts are often expressed as peak levels or maximal amounts. The peak levels usually occur when 40–95 % of the lactose has been hydrolysed as the concentration of donor/acceptor molecules increases. Highest yields are achieved with neutral pH enzymes from bacteria and yeasts rather than from acid ones from moulds. Yields of 40 % have been reported for enzymes from *Kluyveromyces fragilis* (Roberts and Pettinati, 1957), *B. circulans* (Mozaffar *et al.*, 1984), *S. thermophilus* (Smart, 1991) and *Saccharopolyspora rectivergula* (Nakao *et al.*, 1994).

A study by Onishi and Tanaka (1995) indicated that peak oligosaccharide levels increased with removal of glucose. A 37 % yield was obtained by resting cells of *Sterigmatomyces elviae* as opposed to the 64 % yield produced when cells were allowed to grow on glucose (Mahoney, 1998). The highest yield reported was 67 % but this was achieved by using enzymes that are not suitable for food use or not available in sufficient quantities for industrial use (Yanahira *et al.*, 1998). Normal yields range from 5 % with *B. circulans* (Yanahira *et al.*, 1998) and 29 % (Dumortier *et al.*, 1994) with *Bifidobacterium. bifidum* enzyme, when using enzymes from mesophilic bacteria. Yields of up to 48 % (Reuter *et al.*, 1999) can be achieved from the enzyme extracted from the thermostable *S. solfataricus.*

2.3.3 Production Methods

Production techniques are limited and involve either the use of free enzyme in a discontinuous way or an immobilised system, which can be either continuous or discontinuous.

Immobilisation can involve crosslinking, entrapment or binding to a carrier (adsorption, ionic interaction or covalent attachment) (Ekhart and Timmermans, 1996). Factors that affect the rate of reaction are enzyme source, nature and substrate concentration, type of process, process conditions, medium composition and degree of lactose conversion (Ekhart and Timmermans, 1996). Cotton cloth, chitosan beads, glutaraldehyde-treated chitosan beads and agarose beads have been used as immobilisation matrices (Berger *et al.*, 1995).

Albayrak and Yang (2002) produced GOS from lactose by *A. oryzae* β-galactosidase immobilised on cotton cloth. It was found that the total amounts and types of GOS produced depended on the initial lactose concentration in the medium. More and larger GOS are produced by higher initial lactose concentration. The highest yield was 27 %, and the product was 70 % trisaccharides. pH and temperature affected the reaction rate but not the product composition. Galactose and glucose inhibited the reaction. The support matrix of enzyme immobilisation did not affect performance of the enzyme.

Shin *et al.* (1998) produced GOS in a continuous process by immobilising the enzyme from *Bullera singularis* ATCC 24193 in a packed bed reactor by simple adsorption using the chitosan 'Chitopearl BCW 3510' beads (970GU/g resin). The maximum yield obtained was 55 % (w/w) oligosaccharides. The yield of trisaccharide alone was 37 %. Oligosaccharide products were composed mainly of a disaccharide [β-D-Galp-(1,3)-β-D-Glc] and a trisaccharide [β-D-Galp-(1,4)-β-D-Galp–(1,4)-Glc]. The yield and composition of oligosaccharides during the continuous reaction were nearly the same as that of the batch reaction.

Sheu *et al.* (1998) produced GOS using β-galactosidase form *A. oryzae* immobilised on glutaraldehyde-treated chitosan beads in a plug reactor using whey as a substrate containing 5.5 % (w/v) lactose. It was found that the yield of immobilised enzyme activity decreased as the amount of enzyme adsorbed on the chitosan beads increased. It was suggested that this may be caused by diffusion resistance of lactose inside the beads or steric hindrance of the enzyme on the chitosan surface. Maximum yield of GOS was 26 % of the total saccharides on a dry weight basis for an initial concentration of lactose of 300 g l^{-1}. The immobilised enzyme system produced lower GOS than the free enzyme. A 100 g/l^{-1} lactose solution was used for GOS production in a plug reactor for 30 days and the yield was 15 %.

β-Galactosidases from different organisms were found to perform different trans-galactosyl bioconversions. Prenosil *et al.* (1987b) showed that β-galactosidase from *A. oryzae* produced the most GOS compared with *A. niger, K. lactis* and *K. fragilis. B. circulans* enzyme was shown to produce GOS more efficiently than *E. coli* and *S. lactis.* The β-galactosidases isolated from probiotic species have been reported to produce high yields of GOS. Rabiu *et al.* (2001) reported high GOS yields ranging from 27 to 48 % from batch syntheses using β-galactosidases isolated from probiotic species (*Bifidobacterium pseudolongum*, 27 %; *Bifidobacterium bifidum* BB12, 38 %; *Bifidobacterium adolescentis*, 43 %; *Bifidobacterium angulatum*, 44 %; and *Bifidobacterium infantis*, 48 %). Purified *Bifidobacterium bifidum* DSM20082 β-galactosidase gave a GOS yield of 29 % (Dumortier *et al.*, 1994). Hung and Lee (2002) engineered the β-galactosidase gene isolated from *Bifidobacterium infantis* HL96 and reported a GOS yield of 60 %.

Berger *et al.* (1995) studied the production of GOS from immobilised β-galactosidases from *Thermus aquaticus* YT-1. The galactosyl transferase activity of this enzyme with lactose led to the synthesis of two products: a tetrasaccharide (gal/glu 3:1) called OS-1 and a trisaccharide (gal/glu 2:1) called OS-2. It was found that the production of oligosaccharides (OS) was significantly improved when using the immobilised enzyme but the oligosaccharides produced would vary in proportion: the immobilised enzyme produced more OS-2 and less OS-1. The authors suggested that the microenvironment inside the gel matrix may have been responsible for the change in product profile.

Since transgalactosylation is kinetically controlled and is favoured by high lactose concentrations, thermostable enzymes are also of interest. With these thermophilic enzymes, the synthesis can be performed at high temperature (50 °C to 80 °C) at which higher lactose concentrations can be achieved. Studies on GOS synthesis by thermostable enzymes include *Pyrococcus furiosus* and *Sulfolobus solfataricus* β-glycosidases (Petzelbauer *et al.*, 2000), the recombinant β-glycosidase from *Sulfolobus solfataricus* (Splechtna *et al.*, 2001), *Sterigmatomyces elviae* CBS8119 β-galactosidase (Onishi and Tanaka, 1995), *S. elviae* CBS8119 cells (Onishi and Tanaka, 1998),

Sirobasidium magnum β-galactosidase (Onishi and Tanaka, 1997) and *Penicillium simplicissimum* β-galactosidase (Cruz *et al.*, 1999). The amount of GOS formed by these enzymes varied from 30 to 60 %.

Splechtna *et al.* (2001) looked into ways of producing GOS free of lactose and monosaccharides. The approach is based on the selective enzymatic oxidation of lactose into lactobionic acid. The initial sugar mixture was obtained by enzymatic transformation of lactose by recombinant β-galactosidase from *S. solfataricus*, which produces di-, tri- and tetrasaccharides, mainly containing β-1,3 and β-1,6 glycosidic linkages. At 70 °C and at $270\,g\,l^{-1}$ lactose concentration, the yield of GOS was 41 %. The mixture also contained 46 % monosaccharide and 13 % lactose. To further purify the GOS solution, lactose was selectively oxidised into lactobionic acid using fungal cellobiose dehydrogenase (CDH). Oxidation of lactose was coupled to reduction of 2,6-dichloro-indophenol. Ion exchange chromatography was used to remove lactobionic acid, ions and monosaccharides. The final GOS mixture was 97 % pure and contained 1.2 % lactose and 2.1 % monosaccharides.

Goulas *et al.* (2003) found that GOS could be purified by nanofiltration. The choice of membrane proved to be very important in optimising separation. They then went on to purify GOS by continuous diafiltration (Goulas *et al.*, 2002) giving yields of 81–98 % for oligosaccharides, 59–89 % for disaccharides including lactose and only 14–18 % for the monosaccharides.

Foda and Lopez-Leiva (2000) used whey permeate in the continuous production of oligosaccharides by hydrolysis from Maxilact 2000L (*Kluyveromyces lactis* β-D-galactosidase, 135 kDa). The optimum residence time for oligosaccharide production was found to be around 4 h. At pilot plant scale, a UF-hollow fibre membrane reactor gave a yield of 31 % for a whey UF permeate containing 20 % lactose and 0.5 % enzyme. Chockchaisawasdee *et al.* (2005) also studied synthesis of GOS in cross-flow membrane reactors and compared reactor performance with synthesis in batch reactors. They found that enzyme membrane reactors reduced the degree of product inhibition by galactose and thus increased productivity.

A novel approach to GOS production is the immobilisation of *K. lactis* (Lactozym 3000L HP-G) onto Duolite A-568 and the application of focused microwave irradiation (Maugard *et al.*, 2003). GOS selectivity of the β-galactosidase increased when water activity of the medium was reduced. It was found that selectivity for GOS synthesis could be increased 217-fold under microwave irradiation using immobilised enzyme and co-solvents.

2.4 Isomaltooligosaccharides

IMO are composed of glucose monomers linked by α-1,6 (and occasionally α-1,4) glucosidic linkages. Isomalto-900, a commercial product, is produced from cornstarch and consists of isomaltose, isomaltotriose and panose. Starch type dextrans are easily converted to IMO. IMO are the market leaders in the dietary carbohydrate sector of functional foods in Japan (Crittenden and Playne, 1996; Yun *et al.*, 1994).

IMO occur naturally in various fermented foods and sugars such as sake, soybean sauce and honey (Yun *et al.*, 1994). They are a product of an enzymatic transfer reaction

(Nakakuki, 1993). Starch is the starting material and is acted on by three separate enzymes. Firstly, starch is hydrolysed to maltooligosaccharides by α-amylase (EC 3.2.1.1) and pullulanase (EC 3.2.1.41). Then, α-glucosidase (EC 3.2.1.20) is added to catalyse a transfer reaction converting the α-1,4 linked maltooligosaccharide into α-1,6 linked IMO. Glucose is then removed to produce a higher concentration of product (Kohmoto *et al.*, 1998). The maximum concentration of IMO accumulated in the final reaction mixture is around 40 %, and about 40 % of the glucose remains in the mixture (Kuriki *et al.*, 1993). The α-transglucosidase (EC 2.4.1.24) from *A. niger* catalyses the production of panose and higher DP IMO from maltose rich starch hydrolysates. Maltose plays, in this case, both the role of glucosyl donor and acceptor.

IMO are mild in taste and relatively inexpensive to produce (Yun *et al.*, 1994). IMO have desirable physicochemical characteristics such as relatively low sweetness, low viscosity and bulking properties (Lee *et al.*, 2002; Mountzouris *et al.*, 1999; Robyt, 1992). IMOs have been developed to prevent dental caries, as substitute sugars for diabetics, or to improve the intestinal flora (Lee *et al.*, 2002).

Maltogenic amylase from *B. stearothermophilus* (BSMA) and α-GTase (α-glucanotransferase) from *Thermotoga maritima* were used for efficient production of IMOs (Lee *et al.*, 2002). A 58 % yield was obtained from liquefied corn syrup as substrate using BSMA alone. When the two enzymes were used together the resulting yield was 68 % and contained larger IMOs. It was suggested that BSMA hydrolysed maltopentaose and maltohexaose into maltose and maltotriose and transferred the resulting molecules onto acceptor molecules to form IMOs. α-GTase transferred donor sugar molecules onto maltose and maltotriose to form maltopentaose.

Kurimoto *et al.* (1997) synthesised glycosyl-trehaloses (α-isomaltosyl α-glucoside, α-isomaltotriosyl α-glucoside and α-isomaltosyl α-isomaltoside) with an isomaltosyl residue with α-glucosidase from *A. niger*, by using maltotetraose as a glucosyl donor and trehalose as the acceptor. A reaction mixture containing 24 % (w/v) trehalose, 16 % maltotetraose and 400 units of α-glucosidase from *A. niger* was incubated at 60 °C at pH 5.5 for 24 h. The one trisaccharide and two tetrasaccharides formed were isolated by successive column chromatography. In an *in vitro* utilisation test with human intestinal bacteria, these oligosaccharides were predominantly utilised by bifidobacteria.

Duan *et al.* (1994b) investigated the enzymatic synthesis of IMO from maltose using an α-glucosidase from *A. carbonarious* CCRC 30414. Maltose (0.1 g in 1 ml sodium acetate buffer, pH 5.0) was incubated with the enzyme solution at 37 °C for 60 min. The enzyme transferred a glucose unit from the nonreducing end of maltose and other α-linked glucosyl oligosaccharides to glucose and other glucosyl oligosaccharides, which function as accepting co-substrates. The transfer of a glucose unit occurs most frequently to the 6-OH position of the nonreducing end of acceptor, but transfer to 4-OH position also occurs. Treatment of 30 % (w/v) maltose with the enzyme under optimum conditions afforded more than 55 % IMO.

Oguma *et al.* (1994) used a novel enzyme, cyclo-isomaltooligosaccharide glucanotransferase (CITase) purified to homogeneity from the culture filtrate of *Bacillus sp.* T-3040 isolated from soil to catalyse the conversion of dextran to cyclic IMO by intramolecular transglucosylation (cyclisation reaction). The enzyme catalysed the cyclisation reaction and gave three cyclic IMO (cycloisomalto-heptaose, -octaose, and -nonaose) at a total yield of about 20 %.

Sheu *et al.* (1997) immobilised partially purified α-glucosidase from *A. carbonarious* on glutaraldehyde-activated chitosan beads in a packed bed reactor and produced IMO at a yield of 60 % (w/w) using 30 % (w/v) maltose solution. It was noted that the higher concentration of enzyme employed for immobilisation, the less recovery of enzyme activity on the beads. This indicated possible diffusion resistance to substrate inside the porous beads. Immobilised enzyme system produced more IMO than that of intact mycelia system. The reason was possibly due to an enzyme in mycelia with glucosyl hydrolysing activity, but without transglucosylating activity. Relative amounts of tetra-saccharides and glucose are more and panose is less for the IMO solution produced with immobilised enzyme system compared with that with intact mycelia system.

Yun *et al.* (1994) studied the continuous production of IMO from maltose syrup by permeabilised cells of *Aureobasidium pullulans* immobilised into calcium alginate gel using a column reactor. The maltose syrup was fed in the column reactor at a temperature of 50 °C. A maximum productivity of 60 g $l^{-1}h^{-1}$ was obtained when the reactor was operated at a dilution rate of 0.1 h^{-1} at 50 °C using 60 % (w/v) maltose syrup as a substrate.

Kuriki *et al.* (1993) reported a new way of producing IMO syrup from starch, IMO containing one or more α-1,6 glucosidic linkages with or without α-1,4 glucosidic linkages. The new system for production of IMO syrup was based on the strong α-1,6 transglycosylation reaction of neopullulanase. *B. subtilis* saccharifying α-amylase was simultaneously used with neopullulanase to improve yield of the product. The yield was increased from 45.0 to 60.6 %. Neopullulanase and α-amylase from *B. subtilis* were added to a 30 % soluble starch solution and the mixture was incubated at 58 °C for 92 h. The syrup contained isomaltose, a mixture of isopanose and panose, and 6^2-*0*-α-maltosyl-maltose as branched oligosaccharides with DP of 2, 3 and 4, respectively. IMO with DP of 5 and more were also obtained, which had one or more α-1,6 linkages. Almost the same yield of IMO was obtained by using immobilised neopullulanase. The use of neopullulanase immobilised on porous chitosan beads that were crosslinked with an aliphatic or aromatic compound was investigated. The method developed has the advantage of simplicity as only two enzymes are used as opposed to the conventional four.

Oligodextran is a collective term for carbohydrate mixtures derived from dextran. It consists of an IMO part and a relatively low molecular weight (<70 kDa) dextran part (Mountzouris *et al.*, 2002). Oligodextrans, having higher DP, are less digestible than commercially available IMO (Mountzouris *et al.*, 1999). They are obtained commercially either through acidic hydrolysis fractionation of the native dextran from *Leuconostoc mesenteroides* cultivation or by direct *in vitro* synthesis using dextransucrase (EC 2.4.1.5) in the presence of a suitable acceptor (Alsop, 1983; Gasciolli *et al.*, 1991; Robyt, 1992). The dextran produced by *L. mesenteroides* NRRL B512F contains 95 % α-1,6 bonds and 5 % α-1,3 bonds. The backbone of the molecule is made of α-1,6 bonds and side chains are attached to the backbone through α-1,3 bonds (Gasciolli *et al.*, 1991). Dextran is synthesised from sucrose according to the following reaction:

$$n\ \text{sucrose} \longrightarrow \underset{\text{(dextran)}}{[\text{glucose}]_n}\ \text{fructose}$$

Temperature of synthesis affects the branching of dextrans (Gasciolli *et al.*, 1991). Branching seems to appear for number-average molecular weight (M_n) higher than 2000 and closely related to a conformational transition between oligomer and polymer states (Gasciolli *et al.*, 1991).

Mountzouris *et al.* (1999) investigated the types of oligodextrans that would be generated via controlled enzymatic depolymerisation of dextran using an endodextranase (Dextranase 50 L from *Penicillium lilacinum*). The reaction was carried out in an ultrafiltration stirred-cell membrane reactor fitted with a 10 000 molecular weight cut-off (MWCO) membrane. It was found that the dextrose equivalent (DE) increased with increasing enzyme concentration and decreased substrate concentration. DE values ranged from 18 to 38. The oligosaccharide content of the products ranged from 27 to 82 % (w/w), enzyme and substrate concentrations being the most important variables affecting it. It was found that the oligosaccharides consisted mainly of isomaltose and isomaltotriose when enzyme was used in high concentration. At medium and low enzyme concentrations, the oligosaccharide content was lower (30–40 %). DE values of the oligosaccharides produced were in the range 18–22. Anaerobic batch culture fermenters showed that dextran and oligodextrans increased bifidobacterial numbers with high levels of persistence for 48 h. Analysis by high performance liquid chromatography (HPLC) also showed that a high level of butyrate (14.85 mmol l^{-1}) was produced (Olano-Martin *et al.*, 2000).

In a later study (Mountzouris *et al.*, 2002), the continuous production of oligodextrans was investigated by using a recycle continuous stirred tank (CSTR) membrane reactor system. Substrate (industrial grade dextran from *L. mesenteroides* B 512F) and enzyme were the same as those used previously. Oligodextran yields ranged between 84.4 % and 98.7 %. DE values ranged from 22 to 41 and the oligosaccharide content ranged from 55.9 to 93.4 % (w/w), which was higher than in the batch membrane reactor. The reactor produced three types of oligosaccharides with respect to their polysaccharide content: (a) <15 %; (b) 15–30 %; and (c) >30 % and <44 %. The oligosaccharide content was found to be affected not only by substrate and enzyme concentrations but also by residence time. In particular, higher residence time resulted in products with higher oligosaccharide content. The oligosaccharide content (DP 4–10) was fairly constant at 21.7 – 24.8 % in all products, DP 4, 7 and 8 being predominant. Isomaltotriose ranged from 21.6 to 35.2 % (w/w). The novel oligodextrans generated in this work are expected to have lower digestibility than the already commercially available IMO, which have 40 % of the oligosaccharide content of DP 2. The molecular weight distribution is also expected to generate different physicochemical properties. Also, the authors expect the reported oligodextrans to have higher production costs compared with the available IMO, due to the substrate high cost.

Tanriseven and Dogan (2002) produced IMO by using *L. mesenteroides* B 512 FM dextransucrase immobilised in alginate fibres, which according to the authors, is the only successful way to immobilise this enzyme for production of oligosaccharides. The product profile was the following: fructose (20.4 %), glucose (21.2 %), leucrose (3.8 %), DP 2 (4.7 %), DP 3 (5.1 %), DP 4 (6.6 %), DP 5 (6.4 %), DP 6 (5.7 %), DP 7 (4.4 %), DP 8 (2.2 %), DP 9 (2.6 %), DP 10 and bigger (16.9 %). The authors concluded that the use of alginate fibres rather than beads leads to better performance and repetitive use.

Goulas *et al.* (2004a) investigated the synthesis of IMO from sucrose by the combined use of dextransucrase from *L. mesenteroides* and dextranase from *P. lilacinum*. Dextransucrase catalysed the formation of dextran, the molecular weight of which was controlled by action of the dextranase. In this system, a complex of products was formed with DP between 10 and 60. Manipulation of the enzyme ratios allowed some degree of control over the DP profile. This group subsequently (Goulas *et al.*, 2004b) implemented the two enzyme system in a recycle enzyme membrane reactor (EMR). They found that the EMR produced oligosaccharide products with lower DP than did batch synthesis but the productivity was higher.

Lee *et al.* (2002) employed maltogenic amylases and α-glucanitransferase (α-GTase) to develop an efficient process for the production of IMO. For this purpose, *B. stearothermophilus* maltogenic amylase (BSMA) and α-GTase from *Thermotoga maritima* were overexpressed in *E. coli*.

The effect of BSMA and α-GTase activities on the IMO production was investigated by reactions using liquefied corn syrup, in which maltopentaose was the main component. The combined use of the two enzymes considerably improved the speed and the yield of the reaction. It takes about 120 h to produce 53 % IMO by the conventional method. In this study, 68 % IMO were produced in 14 h. Furthermore, the IMO mixture consisting of a high degree of branched oligosaccharides (due to α-GTase) is expected to be particularly effective at promoting bifidobacterial growth in the human intestine, as growth of bifidobacteria is known to be enhanced in proportion to the degree of glucosidic polymerisation of IMO components.

2.5 Lactosucrose

Lactosucrose (lactosylfructoside) is produced from a mixture of lactose and sucrose using the enzyme β-fructosidase. The fructosyl residue is transferred from the sucrose to the 1-position of the glucose moiety in the lactose, producing a nonreducing oligosaccharide (Gibson *et al.*, 2000; Sako et al., 1999) according to the following reaction:

$$\text{Gal}\beta 1 \rightarrow 4\text{Glc} + \text{Glc}\alpha_1 \xleftrightarrow[\beta\text{-fructosidase}]{} 2\beta\text{Fru} \rightarrow \text{Gal}\beta 1 \rightarrow 4\text{Glc}\,\alpha\,1 \leftrightarrow 2\beta\text{Fru} + \text{Glc}$$

This product is bifidogenic in pure culture studies (Fujita *et al.*, 1995; Tamura, 1983). It is an artificial oligosugar that is indigestible and therefore low in calories. It enhances the growth of intestinal bifidobacteria (Pilgrim *et al.*, 2001).

The β-fructofuranosidase (EC 3.2.1.26) from *Arthrobacter sp.* K-1 is known to catalyse three reactions: the transfer reaction producing lactosucrose, sucrose hydrolysis and lactosucrose hydrolysis. The reaction was uncompetitively inhibited by glucose and lactose (Pilgrim *et al.*, 2001).

Kawase *et al.* (2001) produced lactosucrose using a simulated moving bed reactor (SMBR) from sucrose and lactose using β-fructofuranosidase from *Arthrobacter sp.* K-l. Numerical simulation of the batch process showed that the performance of the reaction process is improved by product removal. Because the enzyme has both transferring and hydrolytic activities, selectivity of the reaction depends on composition of the reactants. One way to suppress the backward reaction is to remove one of the products from the reaction. Lactosucrose and glucose were well separated in the SMBR and the lactosucrose yield was increased compared with the conventional batch reaction. The expected optimum lactosucrose yield was not attained due to strong product hydrolysis around the raffinate port. It was found that sucrose conversion varied by changing the enzyme concentration of the enzyme feed. In the case of sucrose conversion higher than 70 %, lactosucrose yield at the same sucrose conversion was higher than in a batch reaction. This resulted from the simultaneous reaction and separation process. The maximum overall lactosucrose yield was 53 % at 70 % sucrose conversion. Simulation of the batch process had shown that product removal would increase the lactosucrose yield from 50 to 80 %.

Takahama et al. (1991) produced lactosucrose from the mixture of sucrose and lactose by *B. natto* levansucrase (EC 2.4.1.10). The enzyme was purified by precipitation with ammonium sulphate and chromatographies on Butyl-Toyopearl 650 M and DEAE-Toyopearl 650 S. The kinetics for lactosucrose formation was explained by the ping-pong mechanism with Michael's constant (K_m) values of 0.21 M for sucrose and 0.42 M for lactose. In the immobilised enzyme, lactosucrose production was reduced, whereas laevan production was promoted compared with those in the native enzyme.

2.6 Xylooligosaccharides

XOS are chains of xylose molecules linked by β-1,4 bonds and mainly consist of xylobiose, xylotriose and xylo-tetraose (Hopkins *et al.*, 1998) and are found naturally in bamboo shoots, fruits, vegetables, milk and honey (Vazquez *et al.*, 2000).

XOS are manufactured by enzymatic hydrolysis of xylan from corn cobs (Crittenden and Playne, 1996). Oat spelt xylan and wheat arabinoxylan can also be used. Typical raw materials for XOS production are hardwoods, corn cobs, straws, bagasses, hulls, malt cakes and bran (Vazquez *et al.*, 2000). The xylan is hydrolysed to xylobiose and smaller quantities of higher oligosaccharides.

Three different approaches have been used for production of XOS from these substrates: (1) enzyme treatment of native, xylan-containing lignicellulosic material (LCM); (2) chemical fractionation of a suitable LCM to isolate xylan, with further enzymatic hydrolysis to produce XOS; and (3) hydrolytic degradation of xylan by steam, water or dilute solutions of mineral acids (Vazquez *et al.*, 2000). For enzymatic production of XOS, enzyme complexes with low exo-xylanase and/or β-xylosidase activity are required, to avoid the production of xylose. For food related applications, a DP of 2–4 is the most desirable (Vazquez *et al.*, 2000).

The purification of crude XOS solutions is a necessary step to remove undesired compounds and to select the required DP range. Membrane techniques and adsorption have been used for this purpose (Vazquez *et al.*, 2000).

Many microorganisms produce xylanases (Table 2.4), the nature of which varies between different organisms: *Trichoderma reesei, T. harzianum, T. viride* and *T. koningii* are well known as producers of both cellulolytic and xylanolytic enzymes (Chen *et al.*, 1997). *T. longibrachiatum* CS-185 has been isolated from a forest soil sample and produces large amounts of xylanases extracellularly.

XOS can be metabolised by bifidobacteria and lactobacilli (Jaskari *et al.*, 1998; Okazaki *et al.*, 1990) in pure culture. In rats, XOS at 6 % (w/w) stimulated caecal and faecal bifidobacteria (Campbell *et al.*, 1997). In relation to human health, XOS selectively enhanced the growth of bifidobacteria thus promoting a favourable intestinal environment (Okasaki *et al.*, 1990, 1991).

Table 2.4 *Summary of xylanase characteristics*

Reference	Enzyme source	Molecular weight	Optimum pH	Optimum temp. (°C)	Best yield conditions	Oligosaccharide products	Yield XOS (%)
Isihara *et al.*, 1997	*P. chyrosporium*, *G. trabeum*, *H. sheweinitzii*, *T. viride*	NA	4.5–5.0	40	Substrate: 1 g of hylglucuronoxylan	Aldotetraouronic acid from *T. viride*	75.7–91.3
Chen *et al.*, 1997	*Trichoderma longibrachiatum*	37.7 kDa	5.0–6.0	45	Oat spelt	Xylobiose, xylotetraose and higher DP XOS	NA
Murakami *et al.*, 1995	*Robillarda sp.*	23.4 kDa	4.5–5.5	55		Xylobiose	
Amano *et al.*, 1994	*Irpex lacteus*				Xyloglycans and aryl-xylooli gosaccharides	β-1,4 glycosidic linkages, or β-1,3 and β-1,4 xylosidic linkages	
Nishimura *et al.*, 1998	*T. viride*				23 g reduced hardwood xylan at 40 °C for 24 h	4-*O*-methyl-α-D -glucopyranosyl-(1-2) -D-xylotriose and 4-*O*-methyl-α-D-glucopyranosyl-(1-2)-D-xylotetraose	Neutral and acidic products were 13.0 g and 2.1 g, respectively
Tokuda *et al.*, 1998	*A. oryzae*	NA	NA	NA	Xylan (3 g)	Xylotriose, xylobiose, and xylose	1 g l^{-1} each
Rydlund and Dahlman, 1997	*T. reesei*	NA	NA	NA	Unbleached birch kraft pulp, with endoxylanase at 40 °C for 24 h at pH 5.5, with β-xylosidase at 40 °C for 24 h at pH 5.5 and then α-glucuronidase at 40 °C for 48 h	Acidic: (4ΔUA)-β-D-xylotetraose, (4ΔUA)-β-D-xylopentaose, (4-*O*-methyl-α-D-glucurono)-β-D-xylotetraose and (4-*O*-methyl-α-D-glucurono)-β-D-xylopentaose Neutral: D-xylose, β-1,4-D-xylobiose and β-1,4-D-xylotriose	NA

NA: data not available.

XOS are stable over a wide range of pH values (2.5–8.0) and temperatures (up to 100 °C). Water activity of xylobiose is reported to be higher than xylose but almost the same as glucose. As food ingredients, XOS have an acceptable odour, are noncariogenic and low-calorie (Vazquez *et al.*, 2000).

2.7 Conclusions

Increasing consumer awareness for health and nutritional issues makes the market for prebiotics very promising; however, because of the high product costs and the still fragmented scientific backup of some of the health claims, it is still a volatile one.

There are at least three areas of scientific research that could provide stability to the prebiotic market: (1) development of cheap and efficient production techniques; (2) studies on structure–function relationships or prebiotics; and (3) elucidation of the *in vivo* mechanisms by which functional oligosaccharides improve host health.

These areas are closely linked, and findings in one often affect the direction of research in the other two.

This chapter has reviewed the production techniques of the most established prebiotics giving special emphasis to the most recent developments and the most promising novel manufacturing approaches.

Enzymatic synthesis of oligosaccharides has the advantage, over chemical synthesis, of low cost, low complexity and high yield processes. Ongoing research into the structure–function relationships of oligosaccharides is attempting to elucidate what makes an oligosaccharide a prebiotic from a structural point of view. However, one of the limitations of the enzymatic synthesis technique is that the product that can be manufactured is defined by the enzyme–substrate combinations available in nature, or that it is possible to obtain by genetic engineering. This means that, if enzymatic techniques are to be the answer to the manufacturing aspect of the prebiotic industry, part of the research is to be dedicated to finding new and interesting enzyme sources and substrates, as well as efficient separation techniques.

References

Akiyama, K., Takase, M., Horikoshi, K., Okonogi, S. (2001) Production of galactooligosaccharides from lactose using a β-glucosidase from *Thermus* sp. Z-1. *Biosci Biotechnol Biochem*, **65**, 438–41.

Albayrak, N., Yang, S. T. (2002) Production of galacto-oligosaccharides from lactose by *Aspergillus oryzae* immobilized on cotton cloth. *Biotechnol Bioeng*, **77**, 8–19.

Alsop, R. M. (1983) Industrial protection of dextrans. In Bushell, M. E., ed. *Progress in Industrial Microbiology.* Amsterdam: Elsevier, pp. 1–44.

Amano, Y., Okazaki, M., Mitani, M., Kanda, T. (1994) Transglycosyl reaction of endo-1,4-β-xylanases from *Irpex lacteus* on aryl-β-xylosides. *Mokuzai Gakkaishi*, **40**, 308–14.

Berger, J. L., Lee, B. H., Lacroix, C. (1995) Oligosaccharides synthesis by free and immobilized β-galactosidases from *Thermus aquaticus* YT-1. *Biotechnol Lett*, **17**, 1077–80.

Boon, M. A., Janssen, A. E. M., van der Padt, A. (1999) Modelling and parameter estimation of the enzymatic synthesis of oligosaccharides by β-galactosidase from *Bacillus circulans*. *Biotechnol Bioeng*, **64**, 559–67.

Campbell, J. M., Fahey, G. C., Wolf, B. W. (1997) Selected indigestible oligosaccharides affect large bowel mass, cecal and fecal short-chain fatty acids, pH and microflora in rats. *J Nutr*, **127**, 130–6.

Cha, J., Park, N. H., Yang, S. J., Lee, T. H. (2001) Molecular and enzymatic characterization of a levan fructotransferase from *Microbacterium* sp. AL-210. *J Biotechnol*, **91**, 49–61.

Chen, C., Chen, J.-L., Lin, T.-Y. (1997) Purification and characterization of a xylanase from *Trichoderma longibrachiatum* for xylooligosaccharide production. *Enzyme Microb Technol*, **21**, 91–6.

Cheng, C. Y., Duan, K. J., Sheu, D. C., Liu, C. T., Li, S. Y. (1996) Production of fructooligosaccharides by immobilized mycelium of *Aspergillus japonicus. J Chem Technol Biotechnol*, **66**, 135–8.

Chiang, C. J., Hsiau, L. T., Lee, W. C. (1997) Immobilization of cell-associated enzymes by entrapment in polymethacrylamide beads. *Biotechnol Tech*, **11**, 121–5.

Chien, C. S., Lee, W. C., Lin, T. J. (2001) Immobilization of *Aspergillus japonicus* by entrapping cells in gluten for production of fructooligosaccharides. *Enzyme Microb Technol*, **29**, 252–7.

Chockchaisawasdee, S., Athanasopoulos, V. I., Niranjan, K., Rastall, R. A. (2005) Process optimisation of galacto-oligosaccharide synthesis by β-galactosidase from *Kluyveromyces lactis* and product characterisation. *Biotechnol Bioeng*, **89**, 434–43.

Coughlin, R. W., Charles, M. (1980) Applications of lactase and immobilised lactase. In Pitcher, W., ed. *Immobilized Enzymes for Food Processing*, Boca Raton: CRC Press, pp. 153–74.

Crittenden, R. G., Playne, M. J. (1996) Production, properties and applications of food-grade oligosaccharides. *Trends Food Sci Technol*, **7**, 353–61.

Cruz, R., Cruz, V. D., Belote, J. G., Khenayfes, M. O., Dorta, C., Oliveira, L. H. S., Ardiles, E., Galli, A. (1999) Production of transgalactosylated oligosaccharides (TOS) by galactosyltransferase activity from *Penicillium simplicissimum. Biores Technol*, **70**, 165–71.

Duan, K. J., Chen, J. S., Sheu, D. C. (1994a) Kinetic-studies and mathematical-model for enzymatic production of fructooligosaccharides from sucrose. *Enzyme Microb Technol*, **16**, 334–9.

Duan, K. J., Sheu, D. C., Lin, M. T., Hsueh, H. C. (1994b) Reaction mechanism of isomaltooligosaccharides synthesis by α-glucosidase from *Aspergillus carbonarious. Biotechnol Lett*, **16**, 1151–6.

Dumortier, V., Brassart, C., Bouquelet, S. (1994) Purification and properties of a β-D-galactosidase from *Bifidobacterium bifidum* exhibiting a transgalactosylation reaction. *Biotechnol Appl Biochem*, **19**, 341–54.

Ekhart, P. F., Timmermans, E. (1996) Techniques for the production of transgalactosylated oligosaccharides (TOS). *Bull, IDF*, **313**, 59–64.

Flaschel, E., Raetz, E., Renken, A. (1982) The kinetics of lactose hydrolysis for the β-galactosidase from *Aspergillus niger. Biotechnol Bioeng*, **24**, 2499–518.

Foda, M. I., Lopez-Leiva, M. (2000) Continuous production of oligosaccharides from whey using a membrane reactor. *Process Biochem*, 35: 581–7.

Fransen, C. T. M., Van Laere, K. M. J., van Wijk, A. A. A., Brull, L. P., Dignum, M., Thomas-Oates, J. E., Haverkamp, J., Schols, H. A., Voragen, A. G. J., Kamerling, J. P., Vliegenthart, J. F. G. (1998) α-D-Glcp-1,1-β-D-Galp-containing oligosaccharides, novel products from lactose by the action of β-galactosidase. *Carbohydr Res*, **314**, 101–14.

Fujita, K., Ogata, Y., Hara, K. (1995) Effect of 4-β-galactosylsucrose (lactosucrose) on intestinal flora. *Seido Gigitsu Kenyukai-shi*, **43**, 83–91.

Fuller, R., Gibson, G. R. (1997) Modification of the intestinal microflora using probiotics and prebiotics. *Gastroenterology*, 32: S28–31.

Gasciolli, V., Choplin, L., Paul, F., Monsan, P. (1991) Viscous properties and molecular characterisation of enzymatically size-controlled oligodextrans in aqueous solutions. *J Biotechnol*, **19**, 193–202.

Gibson, G. R., Beatty, E. B., Wang, X., Cummings, J. H. (1995) Selective stimulation of bifidobacteria in the human colon by oligofructose and inulin. *Gastroenterology*, **108**, 975–82.

Gibson, G. R., Ottaway, P. B., Rastall, R. A. (2000) *Prebiotics*. Oxford: Chandos Publishing (Oxford) Limited.

Goulas, A. K., Kapasakalidis, P. G., Sinclair, H. R., Rastall, R. A., Grandison, A. S. (2002) Purification of oligosaccharides by nanofiltration. *J Membr Sci*, **209**, 321–35.

Goulas, A. K., Grandison, A. G., Rastall, R. A. (2003) Fractionation of oligosccharides by nanofiltration. *J Sci Food Agric*, **83**, 675–80.

Goulas, A. K., Fisher, D. A., Grimble, G. K., Grandison, A. S., Rastall, R. A. (2004a) Synthesis of isomaltooligosaccharides and oligodextrans by the combined use of dextransucrase and dextranase. *Enzyme Microb Technol*, **35**, 327–38.

Goulas, A. K., Cooper, J. M., Grandison, A. S., Rastall, R. A. (2004b) Synthesis of isomaltooligosaccharides and oligodextrans in a recycle membrane bioreactor by the combined use of dextransucrase and dextranase. *Biotechnol Bioeng*, **88**, 778–87.

Grizard, D., Barthomeuf C. (1999a) Enzymatic synthesis and structure determination of NEO-FOS. *Food Biotechnol*, **13**, 93–105.

Grizard, D., Barthomeuf C. (1999b) Non-digestible oligosaccharides used as prebiotic agents: mode of production and beneficial effects on animal and human health. *Repr Nutr Dev*, **39**, 563–88.

Hansson, T., Adlercreutz, P. (2001) Optimization of galactooligosaccharide production from lactose using β-glycosidases from hyperthermophiles. *Food Biotechnol*, **15**, 79–97.

Hashimoto, H., Katayama, C., Goto, M., Okinaga, T., Kitahata, S. (1995) Enzymatic synthesis of α-linked galactooligosaccharides using the reverse reaction of a cell bound α-galactosidase from *Candida guillermondii* H-404. *Biosci Biotechnol Biochem*, **59**, 179–83.

Hayashi, S., Nonoguchi, M., Takasaki, Y., Ueno, H., Imada, K. (1991) Purification and properties of beta-fructofuranosidase from *Aureobasidium* sp. ATCC-20542. *J Ind Microbiol*, **7**, 251–6.

Hernandez, L., Sotolongo, M., Rosabal, Y., Menendez, C., Ramirez, R., Caballer-Mellado, J., Arrieta, J. (2000) Structural levansucrase gene (lsdA) constitutes a functional locus conserved in the species *Gluconacetobacter diazotrophicus*. *Arch Microbiol*, **174**, 120–4.

Hidaka, H., Hirayama, M., Sumi, N. (1988) A fructooligosaccharide-producing enzyme from *Aspergillus niger* ATCC 20611 *Agric Biol Chem*, **52**, 1181–7.

Hirayama, M., Sumi, N., Hidaka H. (1989) Purification and properties of a fructooligosaccharide-producing β-fructofuranosidase from *Aspergillus niger* ATCC 20611. *Agric Biol Chem*, **53**, 667–73.

Hopkins, M. J., Cummings, J. H., Macfarlane, G. T. (1998) Inter-species differences in maximum specific growth rates and cell yields of bifidobacteria cultured on oligosaccharides and other simple carbohydrate sources. *J Appl Microbiol*, **85**, 381–6.

Hung, M.-N., Lee, B. H. (2002) Purification and characterization of a recombinant β-galactosidase with transgalactosylation activity from *Bifidobacterium infantis* HL96. *Appl Microbiol Biotechnol*, **58**, 439–45.

Imamura, L., Hisamitzu, K., Kobashi, K. (1994) Purification and characterization of beta-fructofuranosidase from *Bifidobacterium infantis*. *Biol Pharm Bull*, **17**, 596–602.

Isihara, M., Nojiri, M., Hayashi, N., Nishimura, T., Shimizu, K. (1997) Screening of fungal β-xylanases for production of acidic xylooligosaccharides using *in situ* reduced 4-O-methylglucuronoxylan as substrate. *Enzyme Microb Technol*, **21**, 170–5.

Jaskari, J., Kontula, P., Siitonen, A., Jousimies-Somer, H., Mattila-Sandholm, T., Poutanen, K. (1998) Oat β-glucan and xylan hydrolysates as selective substrates for *Bifidobacterium* and *Lactobacillus* strains. *Appl Microbiol Biotechnol*, **49**, 175–81.

Jorgensen, F., Hansen, O. C., Stougaard, P. (2001) High-efficiency synthesis of oligosaccharides with a truncated β-galactosidase from *Bifidobacterium bifidum*. *Appl MicrobiolBiotechnol*, **57**, 647–52.

Jung, K. H., Yun, J. W., Kang, K. R., Lim, J. Y., Lee, J. H. (1989) Mathematical model for enzymatic production of fructo-oligosaccharides from sucrose. *Enzyme Microb Technol*, **11**, 491–4.

Kawase, M., Pilgrim, A., Araki, T., Hashimoto, K. (2001) Lactosucrose production using a simulated moving bed reactor. *Chem Eng Sci*, **56**, 453–8.

Kim, B. W., Choi, J. W., Yun, J. W. (1998) Selective production of GF_4 fructooligosaccharide from sucrose by a new transfructosylating enzyme. *Biotechnol Lett*, **20**, 1031–4.

Kim, B. W., Kwon, H. J., Park, H. Y., Nam, S. W., Park, J. P., Yun, J. W. (2000) Production of a novel transfructosylating enzyme from *Bacillus macerans* EG-6. *Bioprocess Eng*, **23**, 11–6.

Kohmoto, T., Fukui, F., Takaku, H., Machida, Y., Arai, M., Mitsuoka, T. (1998) Effect of isomaltooligosaccharides on human fecal flora. *Bifidobact Microflora*, **7**, 61–9.

Kosaric, N., Cosentino, G. P., Wieczorek, A., Durnjak, D. (1984) The Jerusalem artichoke as an agricultural crop. *Biomass*, **5**, 1–36.

Kuriki, T., Yanase, M., Takata, H., Takesada, Y., Imanaka, T., Okada, S. (1993) A new way of producing isomalto-oligosaccharide syrup by using the transglycosylation reaction of neopullulanase. *Appl Environ Microbiol*, **59**, 953–9.

Kurimoto, M., Nishimoto, T., Nakada, T., Chaen, H., Fukuda, S., Tsujisaka, Y. (1997) Synthesis by an α-glucosidase of glycosyl-trehaloses with an isomaltosyl residue. *Biosci Biotechnol Biochem*, **61**, 699–703.

Lee, H.-S., Auh, J.-H., Yoon, H.-G., Kim, M.-J., Park, J.-H., Hong, S.-S., Kang, M.-H., Kim, T.-J., Moon, T.-W., Kim, J.-W., Park K.-H. (2002) Cooperative action of α-glucanotransferase and maltogenic amylase for an improved process of isomaltooligosaccharide (IMO) production. *J Agric Food Chem*, **50**, 2812–7.

Lee, W. C., Chiang, C. J., Tsai, P. Y. (1999) Kinetic modelling of fructo-oligosaccharide production catalyzed by immobilized beta-fructofuranosidase. *Ind Eng Chem Res*, **38**, 2564–70.

L'Hocine, L., Jiand, B., Wang, Z. (1997) Production of high content fructooligosaccharides by a multienzyme system. *Food Sci* (China), **18**, 24–7.

L'Hocine, L., Wang, Z., Jiang, B., Xu, S. Y. (2000) Purification and partial characterization of fructosyltransferase and invertase from *Aspergillus niger* AS0023. *J Biotechnol*, **81**, 73–84.

Madlova, A., Antosova, M., Barathova, M., Polakovic, M., Stefuca, V., Bales, V. (1999) Screening of microorganisms for transfructosylating activity and optimization of biotransformation of sucrose to fructooligosaccharides. *Chem Pap – Chem Zvesti*, **53**, 366–9.

Madlova, A., Antosova, M., Polakovic, M., Bales, V. (2000) Thermal stability of fructosyltransferase from *Aureobasidium pullulans. Chem Pap – Chem Zvesti*, **54**, 339–44.

Mahoney, R. R. (1998) Galactosyl-oligosaccharide formation during lactose hydrolysis: a review. *Food Chem*, **63**, 147–54.

Matsumoto K. (1990) Characterisation and utilization of β-galactosidases from lactobacilli and bifidobacteria. *Jpn J Dairy Food Sci*, 39: A239–48.

Maugard, T., Gaunt, D., Legoy, M. D., Besson, T. (2003) Microwave-assisted synthesis of galacto-oligosaccharides from lactose with immobilized β-galactosidase from *Kluyveromyces lactis. Biotechnol Lett*, **25**, 623–9.

Mitsutomi, M., Honda, J., Ohtakara, A. (1991) Enzymatic synthesis of galactooligosaccharides by the condensation action of thermostable α-galactosidase from *Pyroporus cinnabarinus. J Jpn Soc Food Sci Technol*, **38**, 722–8.

Modler, H. W. (1994) Bifidogenic factors–sources, metabolism and applications. *Int Dairy J*, **3**, 383–407.

Monsan, P., Paul, F. (1995) Enzymatic synthesis of oligosaccharides. *FEMS Microbiol Rev*, **16**, 187–92.

Mountzouris, K. C., Gilmour, S. G., Grandison, A. S., Rastall, R. A. (1999) Modelling of oligodextran production in an ultrafiltration stirred-cell membrane reactor. *Enzyme Microb Technol*, **24**, 75–85.

Mountzouris, K. C., Gilmour, S. G., Rastall, R. A. (2002) Continuous production of oligodextrans via controlled hydrolysis of dextran in an enzyme membrane reactor. *J Food Sci*, **67**, 1767–71.

Mozaffar, Z., Nakanishi, K., Matsuno, R., Kamikuro, T. (1984) Purification and properties of β-galactosidases from *Bacillus circulans. Agric Biol Chem*, **48**, 3053–61.

Murakami, T., Kuwahara, T., Sazaki, T., Taniguchi, H. (1995) Purification and characterization of a xylanase from *Rubillarda* sp. Y-20 grown on xylan. *J Jpn Soc Food Sci Technol*, **42**, 531–5.

Nakakuki T. (1993) *Oligosaccharides: Production. Properties and Applications*. Japanese Technology Reviews 3. Yverdon, Switzerland: Gordon & Breach Science Publishers.

Nakao, M., Harada, M., Kodama, Y., Nakayama, T., Shibano, Y., Amachi, T. (1994) Purification and characterization of a thermostable β-galactosidase activity from *Saccharopolyspora rectivirgula. Appl Microbiol Biotechnol*, **40**, 657–63.

Nakayama, T., Amachi, T. (1999) β-Galactosidase. In Flickinger, M. C., Drew, S. W., eds. *Encyclopedia of Bioprocess Technology: Fermentation, Biocatalysis and Bioseparation*, vol. 3. New York: John Wiley & Sons, Ltd, pp. 1291.

Nguyen, Q. D., Mattes, F., Hoschke, A., Rezessy-Szabo, J., Bhat, M. K. (1999) Production, purification and identification of fructooligosaccharides produced by beta-fructofuranosidase from *Aspergillus niger* IMI1303386. *Biotechnol Lett*, **21**, 183–6.

Nishimura, T., Ishihara, M., Ishii, T., Kato, A. (1998) Structure of neutral branched xylooligosaccharides produced by xylanase from *in situ* reduced hardwood xylan. *Carbohydr Res*, **308**, 117–22.

Nishizawa, K., Nakajima, M., Nabetani, H. (2000) A forced-flow membrane reactor for transfructosylation using ceramic membrane. *Biotechnol Bioeng*, **68**, 92–7.

Oguma, T., Tobe, K., Kobayashi, M. (1994) Purification and properties of a novel enzyme from *Bacillus* spp. T-3040, which catalyzes the conversion of dextran to cyclic isomaltooligosaccharides. *FEBS Lett*, **345**, 135–8.

Okazaki, M., Fujikawa, S., Matsumoto, N. (1990) Effect of xylooligosaccharides on growth of bifidobacteria. *J Jpn Soc Nutr Food Sci*, **43**, 395–401.

Okazaki, M., Koda, H., Izumi, R., Fujikawa, S., Matsumoto, N. (1991) Effect of xylooligosaccharide on growth of intestinal bacteria and putrefaction products. *J Jpn Soc Nutr Food Sci*, **44**, 41–4.

Olano-Martin, E., Mountzouris, K. C., Gibson, G. R., Rastall, R. A. (2000) *In vitro* fermentability of dextran, oligodextran and maltodextrin by human gut bacteria. *Br J Nutr*, **83**, 247–55.

Onishi, N., Tanaka, T. (1995) Purification and properties of a novel thermostable galactooligosaccharide producing β-galactosidase from *Sterigmatomyces elviae* CBS8119. *Appl Environ Microbiol*, **61**, 4126–30.

Onishi, N., Tanaka, T. (1997) Purification and characterization of galacto-oligosaccharide-producing β-galactosidase from *Sirobasidium magnum. Lett Appl Microbiol*, **24**, 82–6.

Onishi, N., Tanaka T. (1998) Galacto-oliogosaccharide production using a recycling cell culture of *Sterigmatomyces elviae* CBS8119. *Lett Appl Microbiol*, **26**, 136–9.

O' Sullivan, M. G. (1996) Metabolism of bifidogenic factors by gut flora – an overview. *Bull, IDF*, **313**, 23–30.

Ozawa, O., Ohtsuka, K., Uchida, R. (1989) Production of 4'-galactosyllactose by mixed cells of *Cryptococcus laurentii* and baker's yeast. *Nippon Shokuhin Kogyo Gakkaishi*, **36**, 898–902.

Park, J. P., Bae, J. T., Yun, J. W. (1999) Critical effect of ammonium ions on the enzymatic reaction of a novel transfructosylating enzyme for fructooligosaccharide production from sucrose. *Biotechnol Lett*, **21**, 987–90.

Patel, V., Saunders, G., Bucke, C. (1994) Production of fructooligosaccharides by *Fusarium oxysporum. Biotechnol Lett*, **16**, 1139–44.

Petzelbauer, I., Zeleny, R., Reiter, A., Kulbe, K. D., Nidetzky, B. (2000) Development of an ultra-high-temperature process for the enzymatic hydrolysis of lactose:II. Oligosaccharide formation by two thermostable β-glycosidases. *Biotechnol Bioeng*, **69**, 140–9.

Pilgrim, A., Kawase, M., Ohashi, M., Fujita, K., Murakami, K., Hashimoto, K. (2001) Reaction kinetics and modeling of the enzyme-catalysed production of lactosucrose using β-fructofuranosidase from *Arthrobacter* sp. K-1. *Biosci Biotechnol Biochem*, **65**, 758–65.

Prenosil, J. E., Stuker, E., Bourne, J. R. (1987a) Formation of oligosaccharides during enzymatic lactose hydrolysis. Part I. State of the art. *Biotechnol Bioeng*, **30**, 1019–25.

Prenosil, J. E., Stuker, E., Bourne, J. R. (1987b) Formation of oligosaccharides during enzymatic lactose hydrolysis. Part, I. I., Modeling. *Biotechnol Bioeng*, **30**, 1026–31.

Rabiu, B. A., Jay, A. C., Gibson, G. R., Rastall, R. A. (2001) Synthesis and fermentation properties of novel galactooligosaccharides by β-galactosidases from *Bifidobacterium* spp. *Appl Environ Microbiol*, **67**, 2526–30.

Reuter, S., Nygaard, A. R., Zimmermann, W. (1999) β-galactooligosaccharide synthesis with β-galactosidases from *Sulfolobus solfataricus*, *Aspergillus oryzae* and *Escherichia coli. Enzyme Microb Technol*, **25**, 509–16.

Roberts, C. H., Pettinati, D. J. (1957) Concentration effects in the enzymatic conversion of lactose to oligosaccharides. *J Agric Food Chem*, **5**, 130–4.

Robyt, J. F. (1992) Structure, biosynthesis, and uses of non-starch polysaccharides: Dextran, alternan, pullula and algin. In Alexander, R. J., Zobel, H. F., eds. *Developments in Carbohydrate Chemistry*. St Paul, MN: American Association of Cereal Chemists, pp. 261–92.

Rydlund, A., Dahlman, O. (1997) Oligosaccharides obtained by enzymatic hydrolysis of birch kraft pulp xylan: Analysis by capillary zone electrophoresis and mass spectroscopy. *Carbohydr Res*, **300**, 95–102.

Sako, T., Matsumoto, K., Tanaka, R. (1999) Recent progress on research and applications of non-digestible galacto-oligosaccharides. *Int Dairy J*, **9**, 69–80.

Sangiliyandi, G., Raj, K. C., Gunasekaran, P. (1999) Elevated temperature and chemical modification selectively abolishes levan forming activity of levansucrase of *Zymomonas mobilis*. *Biotechnol Lett*, **21**, 179–82.

Sheu, D. C., Huang, C. I., Duan, K. J. (1997) Production of isomaltooligosaccharides by α-glucosidase immobilized in chitosan beads and by polymethyleneimine-glutaraldehyde treated mycelia of *Aspergillus carbonarious*. *Biotechnol Tech*, **11**, 287–91.

Sheu, D. C., Li, S. Y., Duan, K. J., Chen, C. W. (1998) Production of galactooligosaccharides by β-galactosidase immobilized on glutaraldehyde-treated chitosan beads. *Biotechnol Tech*, **12**, 273–6.

Sheu, D. C., Lio, P. J., Chen, S. T., Lin, C. T., Duan, K. J. (2001) Production of fructooligosaccharides in high yield using a mixed enzyme system of beta-fructofuranosidase and glucose oxidase. *Biotechnol Lett*, **23**, 1499–503.

Shin, H. J., Yang, J. W. (1996) Galactooligosaccharide synthesis from lactose by *Penicillum funiculosum* cellulase. *Biotechnol Lett*, **18**, 143–4.

Shin, H. J., Park, J. M., Yang, J. W. (1998) Continuous production of galacto-oligosaccharides from lactose by *Bullera singularis* β-galactosidase immobilized in chitosan beads. *Process Biochem*, **33**, 787–92.

Shukla, T. P. (1975) Beta-galactosidase technology: a solution to the lactose problem. *CRC Crit Rev Food Technol*, January: 325–56.

Shukla, H., Chaplin, M. (1993) Noncompetitive inhibition of β-galactosidase (*A. oryzae*) by galactose. *Enzyme Microb Technol*, **15**, 297–9.

Smart, J. B. (1991) Transferase reactions of the β-galactosidase from *Streptococcus thermophilus*. *Appl Microbiol Biotechnol*, **34**, 495–501.

Splechtna, B., Petzelbauer, I., Baminger, U., Haltrich, D., Kulbe, K. D., Nidetzky, B. (2001) Production of a lactose-free galacto-oligosaccharide mixture by using selective enzymatic oxidation of lactose into lactobionic acid. *Enzyme Microb Technol*, **29**, 434–40.

Takahama, A., Kuze, J., Okano, S., Akiyama, K., Nakane, T., Takahashi, K., Kobayashi, T. (1991) Production of lactosucrose by *Bacillus natto* levansucrase and some properties of the enzyme. *J Jpn Soc Food Sci Technol*, **38**, 789–96.

Tambara, Y., Hormaza, J. V., Perez, C., Leon, A., Arrieta, J., Hernandez, L. (1999) Structural analysis and optimised production of fructo-oligosaccharides by levansucrase from *Acetobacter diazotrophicus* SRT4. *Biotechnol Lett*, **21**, 117–21.

Tamura Z. (1983) Nutriology of bifidobacteria. *Bifidobact Microflora*, **2**, 3–16.

Tanriseven, A., Dogan, S. (2002) Production of isomalto-oligosaccharides using dextransucrase immobilized in alginate fibres. *Process Biochem*, **37**, 1111–5.

Tokuda, H., Sato, K., Nakanishi, K. (1998) Xylooligosaccharide production by *Aspergillus oryzae* I3 immobilized on a nonwoven fabric. *Biosci Biotechnol Biochem*, **62**, 801–3.

Trujillo, L. E., Arrieta, J. G., Dafhnis, F., Garcia, J., Valdes, J., Tambara, Y., Perez, M., Hernandez, L. (2001) Fructo-oligosaccharides production by the *Gluconacetobacter diazotrophicus* levansucrase expressed in the methylotrophic yeast *Picha pastoris*. *Enzyme Microb Technol*, **28**, 139–44.

Tsumura, K., Hashimoto, Y., Akiba, T., Horikoshi, K. (1991) Purifications and properties of galactanases from alkalophilic bacillus sp. S-2 and S-39. *Agric Biol Chem*, **55**, 1265–71.

Vazquez, M. J., Alonso, J. L., Dominguez, H., Parajo, J. C. (2000) Xylooligosaccharides: manufacture and applications. *Trends Food Sci Technol*, **11**, 387–93.

Yanahira, S., Yabe, Y., Nakakoshi, M., Miura, S., Matsubara, N., Ishikawa, H. (1998) Structures of novel acidic galactooligosaccharides synthesised by *Bacillus circulans* β-galactosidase. *Biosci Biotechnol Biochem*, **62**, 1791–4.

Yanai, K., Nakane, A., Kawate, A., Hirayama, M. (2001) Molecular cloning and characterization of the fructooligosaccharide producing beta-fructofuranosidase gene from *Aspergillus niger* ATGC 20611. *Biosci Biotechnol Biochem*, **65**, 766–73.

Yang, S. T., Bednarcik, J. A. (2001) Production of galacto-oligosaccharides from lactose by immobilized β-galactosidase. In Saha, B. C., Demarjian, D. C., eds. *Applied Biocatalysis in Speciality Chemicals and Pharmaceuticals*. ACS Symposium Series No. 776. Washington, DC: American Chemical Society, pp. 131–54.

Yang, S. T., Tang, I. C. (1988) Lactose hydrolysis and oligosaccharide formation catalysed by β-galactosidase. *Ann NY Acad Sci*, **542**, 417–22.

Yun, J. W. (1996) Fructooligosaccharides – occurrence, preparation, and application. *Enzyme Microb Technol*, **19**, 107–17.

Yun, J. W., Lee, M. G., Song, S. K. (1994) Continuous production of isomalto-oligosaccharides from maltose syrup by immobilized cells of permeabilized *Aureobasidium pullulans. Biotechnol Lett*, **16**, 1145–50.

Zarate, S. and Lopez-Leiva, M. H. (1990) Oligosaccharide formation during enzymatic lactose hydrolysis: a literature review. *J Food Prot*, **53**, 262–8.

3

Inulin-type Fructans as Prebiotics

Jan Van Loo

3.1 Introduction

Inulin is a carbohydrate that is extremely widespread in nature. It occurs in plants mainly as an energy reserve and as a cryoprotectant. The inulin that is stored in plant cells is hydrolysed by enzymes that are induced at low temperatures (around freezing point). This reduces the chain length of the inulin, which can be hydrolysed as far as its composing monosaccharides. The increased osmotic pressure in the cytoplasm, then lowers the freezing point.

Different plant species typically contain inulin with varying chain lengths: wheat, onions and bananas have short-chain inulins (maximal degree of polymerisation $DP_{max}<10$); dahlia tubers, garlic and Jerusalem artichoke have medium-chain inulins ($DP_{max}<40$); and globe artichoke and chicory typically contain long-chain inulin molecules ($DP_{max}<100$) (Van Loo *et al.*, 1995).

Other plants, such as certain types of lily (*Urginea maritima*) and blue agave, and certain bacteria (e.g. *Streptococcus mutans*) produce high DP ($DP_{max}>1000$), highly branched (2–6) inulins.

This chapter will focus upon chicory inulin. The root of this plant contains up to 70 % (dry solids) inulin, and the shape of the root is similar to that of the sugar beet, which makes it a preferred raw material for inulin production. All commercial producers of inulin happen to have 'their roots' in sugar beet processing.

Prebiotics: Development and Application Edited by G. R. Gibson and R. A. Rastall

3.2 Chemical Description and Origin

Inulin is extracted from the chicory root (*Cichorium intybus*) by means of hot water and is called native chicory inulin or standard inulin (ST-inulin). Native chicory inulin is a linear (2-1) fructan presented as GF_n (G, glucose F, fructose; n number of fructose moieties in the chain) composed of oligomers and polymers in which the degree of polymerisation (DP) varies from 2 to 65 units with an average DP (DP_{av}) of 10 (Van Loo *et al.*, 1995) (Figure 3.1).

The partial enzymatic hydrolysis (endo-inulinase EC 3.2.1.7) of chicory inulin produces oligofructose, which is a mixture of both GF_n and F_m molecules, in which the DP varies from 2 to 7 with a DP_{av} of 4. Oligofructose also can be obtained through enzymatic synthesis (transfructosylation) using fructosyltransferases (EC 3.2.1.99 and EC 3.2.1.100). At high concentrations of substrate [e.g. 60 % (w/v) sucrose], these enzymes produce new carbon–carbon bonds without the use of ATP (transferase activity) and molecules with a DP of up to 7 (DP_{av} 4) can be synthesised. The oligofructose obtained via this route contains GF_n-type molecules only.

By applying physical separation techniques (crystallisation and filtration), long-chain inulin such as (HP-inulin) with DP ranging between 10 and 60 (DP_{av} 25) is produced.

Finally, a specific mixture known as Raftilose®Synergy1 is obtained by co-spray-drying oligofructose and HP inulin. There is no scientific name for this second generation product, which distinguishes itself from the other products by its physiological and nutritional properties. The name Synergy1 will therefore be used throughout this text.

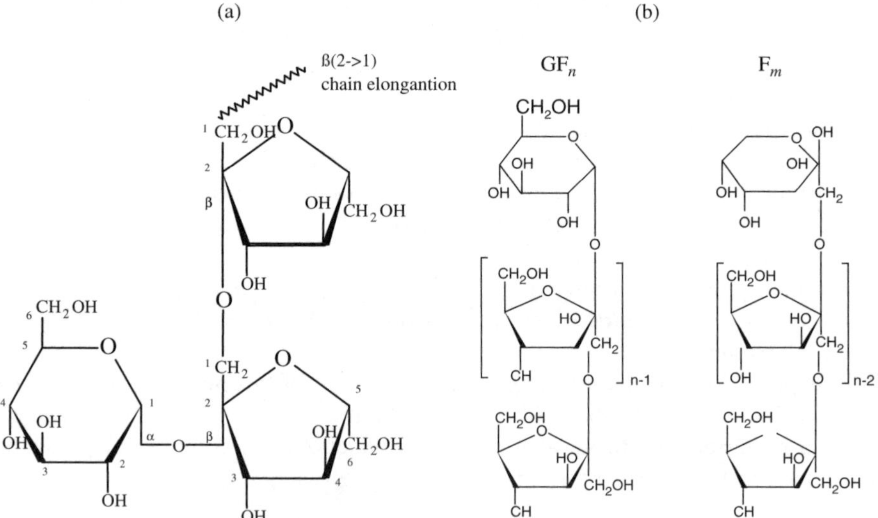

Figure 3.1 *Structure formula of chicory inulin (DP 3–65, DP_{av} = 10) (a) and its partial hydrolysate oligofructose (DP 2–7, DP_{av} = 4) (b). Endo-inulinase activity cuts the inulin molecules into GF_n and F_m molecules. Fructo-oligosaccharides synthesized from sucrose have (a) as general formula (DP 3–7, DP_{av} = 4)*

3.3 Physicochemical Description

Chicory inulin comprises a particular set of molecules. Whereas, in chemical terms, all molecules in inulin differ only in chain length, two groups of molecules that have completely different physicochemical properties can be distinguished. This distinction is made by chain length, whereby DP 10 seems to be a critical limit. It has been observed that molecules with DP < 10 are very soluble in water (up to 85 % w/w) and are fermented rapidly (Figure 3.2). Molecules with DP > 10 are less soluble (up to 5 %, w/w) and hence are less readily accessible for bacterial degradation and, as a consequence, they are fermented more slowly. Fermentation properties are of central importance in nutritional properties of these food ingredients (see below). Both rapid and slow fermentation influence intestinal effects in a particular and complementary way; Synergy1 is designed to make optimal use of this property (Van Loo, 2004a).

3.4 Analysis

In most countries, inulin and oligofructose are included as 'dietary fibre' for food labelling purposes. However, because of their partial solubility in ethanol/water (4/1, v/v) and partial hydrolysis in acidic conditions, the classical methods for analysis of dietary fibre do not include oligofructose, and only partly recover inulin (Quemener, 1994). Recently, AOAC International has adopted method 997.08, known as the Fructan method, that allows the quantitative determination of inulin and oligofructose in foods (Hoebregs, 1997; Prosky and Hoebregs, 1999). The method involves treatment of the sample with amyloglucosidase and inulinase, followed by determination of the released sugars by chromatography. Combined with the standard AOAC 'Total Dietary Fiber' method, it provides values for the total amount of fibre used for food labelling (Quemener, 1994). Another methodology involves: enzymatic hydrolysis of α-glucans to glucose; precipitation of nonstarch polysaccharides in ethanol; enzymatic hydrolysis of fructans to fructose; and chemical ($NaBH_4$) reduction to acid-stable alditols that are quantified by gas-liquid chromatography as acetate derivatives (Quigley *et al.*, 1999). To quantify the individual oligomers, a capillary gas chromatographic method that includes the derivatisation (oximation and silylation) of extracted sugars has been developed (Joye *et al.*, 2000). This method is used to quantify oligofructose in plants and food products. Various enzymatic methods have also been developed and validated (Andersen and Sorensen, 1999; Hofer, 1999; Joye *et al.*, 2000; McCleary *et al.*, 2000; McCleary and Rossiter, 2004).

3.5 Nondigestibility

Inulin and oligofructose are present in food plants; are carbohydrates that are composed of a mixture of either oligosaccharides or oligosaccharides and polysaccharides; resist hydrolysis by acid in the stomach and by human digestive enzymes. Digestive enzymes secreted by the pancreas or brush border of vertebrates, and of mammals in particular, are unable to hydrolyse β-glucosidic bonds. The individual components of

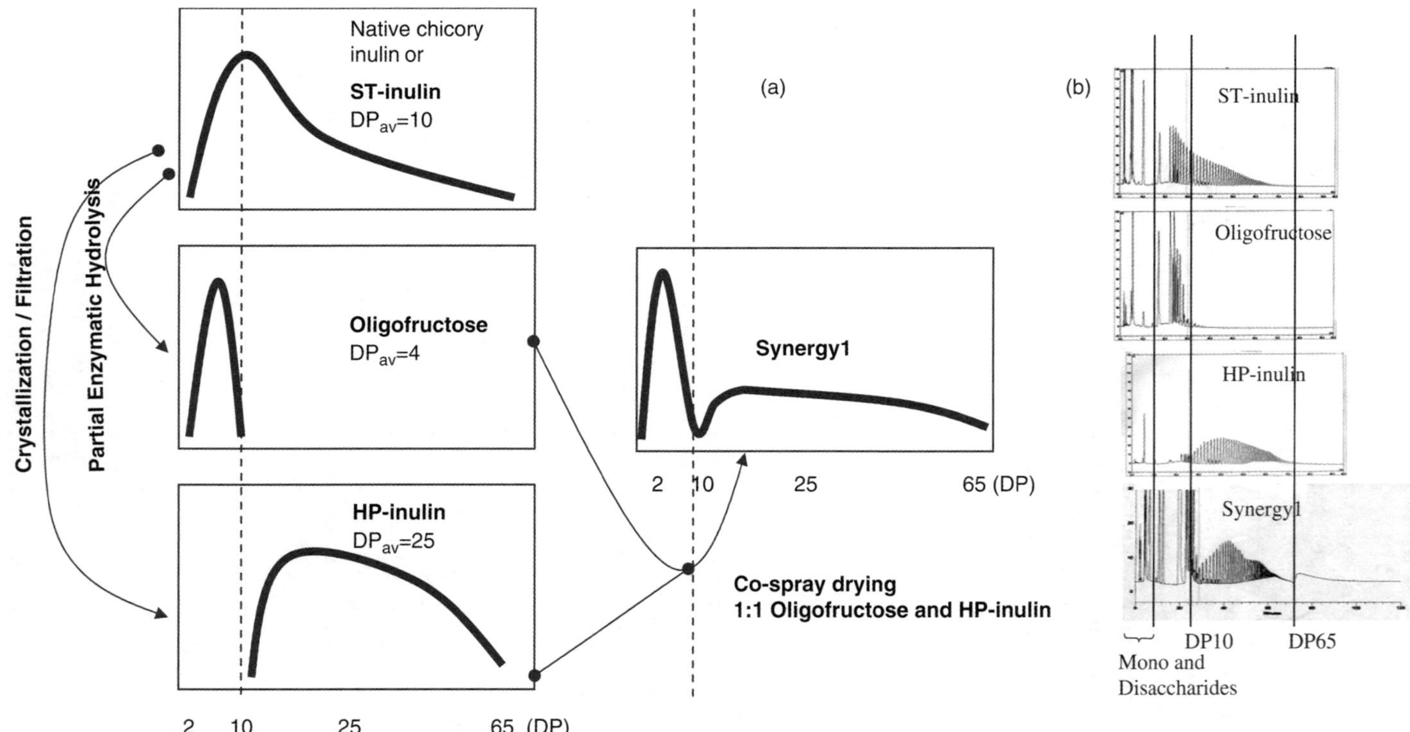

Figure 3.2 *Schematic representation (a) and chromatogram (b) of the different fractions obtained from chicory inulin. ST-inulin (DP 3–65, $DP_{av} = 10$) is extracted from the chicory root by hot water. It is purified by techniques that are used in starch processing. Partial enzymatic hydrolysis of ST-inulin gives oligofructose (DP 2–7, $DP_{av} = 4$). Crystallization and filtration allow isolation of the long chain inulin fraction or HP-inulin (DP 12–65, $DP_{av} = 25$). The second generation prebiotic Synergy1 is obtained by co-spray drying oligofructose and HP-inulin in a ratio of 1:1. This product contains significantly more long chains than ST-inulin*

inulin and oligofructose are linked by such bonds and they are not hydrolysed into the monosaccharides fructose and glucose, i.e. they are nondigestible carbohydrates (Björck, 1991; Molis *et al.*, 1996; Nilsson *et al.*, 1988). The nondigestibility of inulin and oligofructose has been demonstrated in several human volunteer studies (Bach Knudsen and Hessov, 1995; Molis *et al.*, 1996). Ellegård *et al.* (1997) demonstrated that 86–89 % of ingested inulin, as well as oligofructose, was recovered in the ileostomy effluent of colectomised volunteers.

Inulin and oligofructose do not appear to be absorbed significantly in the small intestine, except possibly the very short oligosaccharides (DP 2 or 3) (Ma *et al.*, 1995). However, even if such small oligosaccharides are absorbed, they are not hydrolysed inside the body and are excreted as such in the urine. Indeed, it has been reported that oligofructose that disappears from the small intestine is recovered in urine (Molis *et al.*, 1996). Inulin and oligofructose are hydrolysed and fermented completely by the intestinal microflora, producing gases and short-chain fatty acids (SCFAs).

Thus, inulin and oligofructose are classified as dietary fibre and are labelled as such on foods. Moreover, like other dietary fibres, they have a variety of physiological effects that have been extensively reviewed (Gibson *et al.*, 2004) and are summarised here.

3.6 Interaction with Upper Gastrointestinal Functions

During their passage through the upper intestinal tract, inulin and oligofructose may influence the digestion process and thus affect metabolic responses; especially digestion of disaccharides (Rumessen, 1990), glucose transport, and absorption of leucine, proline and glycyl-sarcosine (Buddington *et al.*, 1999). However, and contrary to what has been reported for some dietary fibre components (Hamberg *et al.*, 1989), inulin does not affect starch absorption in humans nor small intestinal handling of nitrogen, fat nor minerals (Ca, Mg, Zn) in ileostomy volunteers (Ellegård *et al.*, 1997; Rumessen *et al.*, 1990).

Inulin and oligofructose increase the length and weight of the small intestine and small intestinal mucosa and emphasise villus structure in experimental models (Buddington 1999; Ohta *et al.*, 1998; Oku *et al.*, 1984; Yusrizal and Chen, 2003a,b).

Moreover, and probably via their effects in the large bowel, inulin-type fructans may affect the production of gastrointestinal (GI) hormonal peptides and consequently hormonal regulation of gastrointestinal motility as well as systemic metabolic processes (see below).

3.7 Fermentation in the Colon

As inulin and oligofructose resist digestion in the human upper GI tract, they reach the colon almost intact. The substrates are then completely available for fermentation by the colonic microflora (Gibson and Roberfroid, 1995). In the proximal colon (caecum), inulin is hydrolysed and further fermented by the resident flora in a selective way. This results in increased bacterial biomass (caecal and eventual faecal bulking effect), and in increased production of SCFAs such as acetate, propionate (Wang and Gibson,

1993), and butyrate (Djouzi and Andrieux, 1997; Karppinen *et al.*, 2000; Le Blay *et al.*, 1999; Perrin, 2002; Roland *et al.*, 1995; Rycroft, 2001), as well as lactate and gases (H_2, CO_2 and CH_4). The final pattern of production of these molecules is the result of complex metabolic interactions between different genera of bacteria. Novel techniques such as metabonomics to characterise patterns in bacterial metabolite production and fluorescent *in situ* hybridisation (FISH) combined with flow cytometry for high-throughput identification and quantification of bacteria will contribute further to our understanding of this fermentation process and its health implications (Blaut *et al.*, 2002). Indirect observations of SCFA content or inulin in different parts of the intestine of experimental models, or experimental outcomes that typically are related to certain locations in the large intestine (e.g. azoxymethane-induced tumours in the distal part of the colon) provide insights into the fermentation behaviour of inulin and are discussed here.

Different mammals (and fish and birds) have various architectures of the GI system. For any animal, the anatomy of the GI tract varies dramatically with stages in the life cycle (neonate, young, adult, old). GI systems can be characterised by length and volume of the different compartments, by the transit times through and by the bacterial overgrowth in the different segments.

In humans, the small intestine has a very low bacterial content (6–7 logCFU, where CFU stands for colony-forming unit), which explains why about 90 % of the inulin arrives in the colon. In pigs and dogs, which have up to 8 logCFU bacteria in the small intestine, the fraction of inulin that arrives in the caecum typically is 60–70 % of the intake dose. In chickens and rabbits, which have a very rapid (<2 h) oro-ileal transit time, a higher proportion of the ingested inulin as compared with oligofructose arrives in the caecum.

The caecum of mammals typically has a 10000–100000-fold greater amount of bacteria than the upper intestinal tract (Hentges, 1993). When the nondigestible inulin that escapes the mild bacterial fermentation in the upper intestinal tract arrives in the caecum, it is fermented completely in the caecum and the colon. No inulin is excreted in faeces, even when doses of up to 60 have been consumed by human volunteers (Shannon, 1935a,b).

Inulin, however, is fermented selectively: some groups of bacteria (saccharolytic lactic acid-producing bacteria) can metabolise inulin much more efficiently than other groups of bacteria (proteolytic bacteria producing putrefactive compounds). The nondigestibility, combined with a selective fermentation makes inulin a prebiotic food ingredient (see below).

3.8 Colonic Fermentation Characteristics of Different Fractions of Chicory Fructans

Oligofructose (DP_{av} 4; up to 85 % soluble in water) is fermented rapidly. It is likely that this fraction of chicory inulin will be rapidly fermented in the proximal part of the colon (caecum). From various human dietary intervention trials, it has been observed that oligofructose shifts the composition of the faecal flora markedly: *Bifidobacterium* spp. typically are increased 10-fold, whereas the *Clostridium perfringens* group is suppressed;

Bacteroides spp. are often not affected (Buddington, 1996; Gibson *et al.*, 1995; Rao, 2001).

By means of *in vitro* fermentation experiments it has been observed that HP-inulin was fermented more slowly than oligofructose (Baeten, 1999; Roberfroid *et al.*, 1998). Reddy *et al.* (1997) induced aberrant crypt foci with azoxymethane in an anticancer model. This model produces preneoplastic lesions, particularly in the distal part of the intestine. It was observed that HP-inulin was more efficient than oligofructose in suppressing carcinogenesis in the distal colon. These are indications that the more slowly fermented long-chain inulin is fermented in more distal parts of the intestine. HP-inulin modifies the composition of the intestinal flora, but, due to the slower, less intense fermentation, impact on composition of the intestinal flora is less pronounced (Harmsen *et al.*, 2002; Tuohy *et al.*, 2001a).

Characteristics of the colonic fermentation of native chicory inulin are situated between those of oligofructose and HP-inulin. The short-chain fraction is fermented rapidly, whereas the longer chain fractions are fermented more slowly. Significant effects on faecal flora with native inulin have been observed repeatedly (Gibson *et al.*, 1995; Kleessen *et al.*, 1997; Kruse *et al.*, 1999). Interestingly, it was shown that in animal nutrition (pets and livestock), inulin may be the preferred substrate: due to the rather substantial overgrowth of bacteria in the small intestine, a larger proportion of inulin reaches the caecum (70 % of the inulin *versus* 60 % of oligofructose in pigs) (De Kuyper, personal communication, 2003).

Synergy1 is designed to combine the intensive fermentation of oligofructose with the sustained fermentation of the long chains in order to improve the composition of the intestinal flora and maintain the particular metabolic activity for a prolonged period of time (Van Loo, 2004a).

3.9 Impact on Colon Function and Tolerance Aspects

Being fermented in the large bowel, inulin influences colonic integrity and functions as shown by a dose-dependent acidification (on average −0.9 pH unit) of the caecal content (Andrieux, 1991; Campbell *et al.*, 1997; Coudray *et al.*, 2003b; Djouzi and Andrieux, 1997; Fontaine *et al.*, 1996; Levrat *et al.*, 1991; Lopez *et al.*, 2000; Roland *et al.*, 1995; Wolf, 1998;Younes *et al.*, 2001) as well as of the colonic lumen (−1.2 pH unit) (Ohta, 1997). A modulation of several caeco-colonic functions especially those associated with epithelium integrity (Bielecka, 2002; Campbell *et al.*, 1997; Coudray *et al.*, 2003a; Djouzi and Andrieux, 1997; Fontaine *et al.*, 1996; Kleessen *et al.*, 2003; Lopez *et al.*, 2000; Ohta *et al.*, 1997; Sakai *et al.*, 2001; Younes *et al.*, 2001) but no evidence exists in humans at this time. A change in the composition of the mucins in the caecum contents as well as caeco-colonic epithelia is characterised by greater amounts of sulfomucins and smaller amounts of sialomucins (Fontaine *et al.*, 1996; Kleessen *et al.*, 2003). This latter effect could be interesting to consider, since sulphomucins are, in general, associated with a higher level of protection, and their proportion is decreased in several intestinal diseases such as inflammation or certain forms of cancer. Such effects may explain the stimulation of mucosa repair in rats with colitis given oligofructose (Cherbut, 2002). There was an increase in the concentration of putrescine in the caecal contents and of

Table 3.1 *Summary of data demonstrating the stool bulking effect of inulin and oligofructose in rats*

Number of in rats/group	Treatment	Factor increase faecal output	Reference
5	Oligofructose 10 %	1.25–1.40	Roberfroid, 1993
5	12–27 days	1.25–1.40	
15	8–29 days	1.65	
	Oligofructose 20 %		
	17 days		
6	Oligofructose 10 %	1.40	Tokunaga, 1986
6	42 days	2.15	
	Oligofructose 20 %		
	42 days		
10	Oligofructose 10 %	1.30	Delzenne *et al.*, 1995
10	50 days Inulin 10 %	1.40	
	50 days		
6	Oligofructose 5 %	1.20	Kleessen *et al.*, 2001
6	7 days Inulin 5 %	1.15	
6	7 days Oligofructose/inulin	1.20	
	(50/50) 5 % 7 days		
Mean		1.50	

putrescine, spermine and spermidine in the caecal tissue (Delzenne *et al.*, 2000; Propst *et al.*, 2003).

As a consequence of rapid and quantitative fermentation in the large bowel that leads to an increase in bacterial biomass, inulin affects bowel habit. In rats, the weight of faeces is increased, on average 1.5-fold after feeding a diet containing inulin or oligofructose (Table 3.1). The effect is dose-dependent.

A similar increase (1.3- and 1.6-fold) in faecal output has been shown in Beagle dogs fed a diet containing oligofructose (4 %, w/w) or inulin (8 %, w/w) (Diez 1997b; Propst *et al.*, 2003). In humans, the stool-bulking index of inulin (an increase of about 2 g of stool per gram of ingested substrate) and oligofructose (1.2–1.5 g per g) is close to that of other fermentable dietary fibre components, such as pectins (1.3 g per g) or guar gum (1.5 g per g) (Cherbut, 2002; Cummings, 1997; Nyman, 2002).

As expected, the effect on stool weight depends upon the dose and amount of other fibres in the diet (Brighenti *et al.*, 1999; van Dokkum, 1999). In a controlled dietary study, an increased stool weight (1–2 g per gram of oligofructose ingested) was recorded in humans after consuming inulin and oligofructose (Gibson *et al.*, 1995). In postmenopausal women given oligofructose, both wet and dry faecal weights increased significantly, corresponding to an increase of 3.5 g of faeces per gram of oligofructose ingested (Tahiri, 2001).

In addition, inulin and oligofructose intake has consistently been reported to increase faecal water content, another factor that contributes towards increased stool weight (Brighenti *et al.*, 1999; Castiglia-Delavaud *et al.*, 1998; Kleessen *et al.*, 1997; Tahiri 2001; van Dokkum, 1999); to stimulate bowel movements due to the increased production of SCFA (Cherbut *et al.*, 1998, 2003); to normalise stool frequency, especially in slightly constipated subjects (Den Hond *et al.*, 1999; Gibson *et al.*, 1995; Kleessen *et al.*, 1997).

The mechanism of these effects in the large bowel is not physicochemical in nature but rather due to fermentation and subsequent increase in bacterial biomass.

As is the case with any nondigestible and fermented carbohydrate arriving in the colon, the fermentation of oligofructose and inulin results in the production of bacterial biomass and SCFA, and in the production of gases (CO_2, H_2, CH_4). At high doses of inulin or oligofructose, the amount of gas produced becomes too excessive to be evacuated via respiration, and flatulence can occur. The dose of inulin or oligofructose at which this occurs is subject to large interindividual variability. Relying upon the many published human dietary intervention studies, it can be stated that at doses of up to 10 g per day the effects are mild, and not experienced as a point of discomfort in the studies (well-being questionnaires) (Briet *et al.*, 1995; Carabin and Flamm, 1999; Hata, 1985; Juffrie, 2002; Stone-Dorshow, 1987). Most nutritional effects in humans have been observed at doses between 5 and 10 g per day (depending on inulin fraction and experimental outcome).

It should be noted that adaptation to the dose occurs; as the composition of the intestinal flora changes (*Bifidobacterium* spp. do not produce gas), the feeling of discomfort lessens. Coudray (1997) conducted a study in which subjects adapted within 2 weeks to a dose of up to 40 g per day native inulin, and the volunteers reported, at most, more than usual flatulence.

It can be concluded that, on average, the most frequent intestinal side-effects are flatulence and bloating, but these symptoms are only mild or moderate and occur at high intake doses. In terms of acceptability, inulin and oligofructose have the advantage of having a relatively high molecular mass as compared with, for instance, the disaccharide lactulose. Because osmotic pressure and hence osmotic diarrhoea is related inversely to molecular mass, inulin and oligofructose are not laxative.

3.10 The Concept of the Balanced Colonic Microflora

The human intestine has a surface area of about 300 m^2, compared with 2 m^2 for the skin and 100 m^2 for the lungs of an average person. The lining of the GI tract is continuous with the external part of the body. As such, the surface of the intestine is the largest interface between our body and the environment. Through this large interface, the human body is in intimate contact with a very dense microbial ecosystem composed of as many as 10^{12} bacteria ml^{-1}, which is volumetrically about the highest possible microbial population density one can be in contact with (Figure 3.3). The intestinal flora may be the most important environmental factor affecting the appropriate functioning and health of the body, improvement of the GI ecosystem (less pathogens, more saccharolytic activity), and thus represents an important improvement of the 'external environment' in which we live. It is thought that improving the composition of the intestinal ecosystem has a positive impact on physiologic functioning of an organism (Van Loo, 2004b).

It has therefore been proposed that the gut microflora must be a 'balanced microflora', a concept that implies that it must be composed predominantly (in numbers) of bacteria

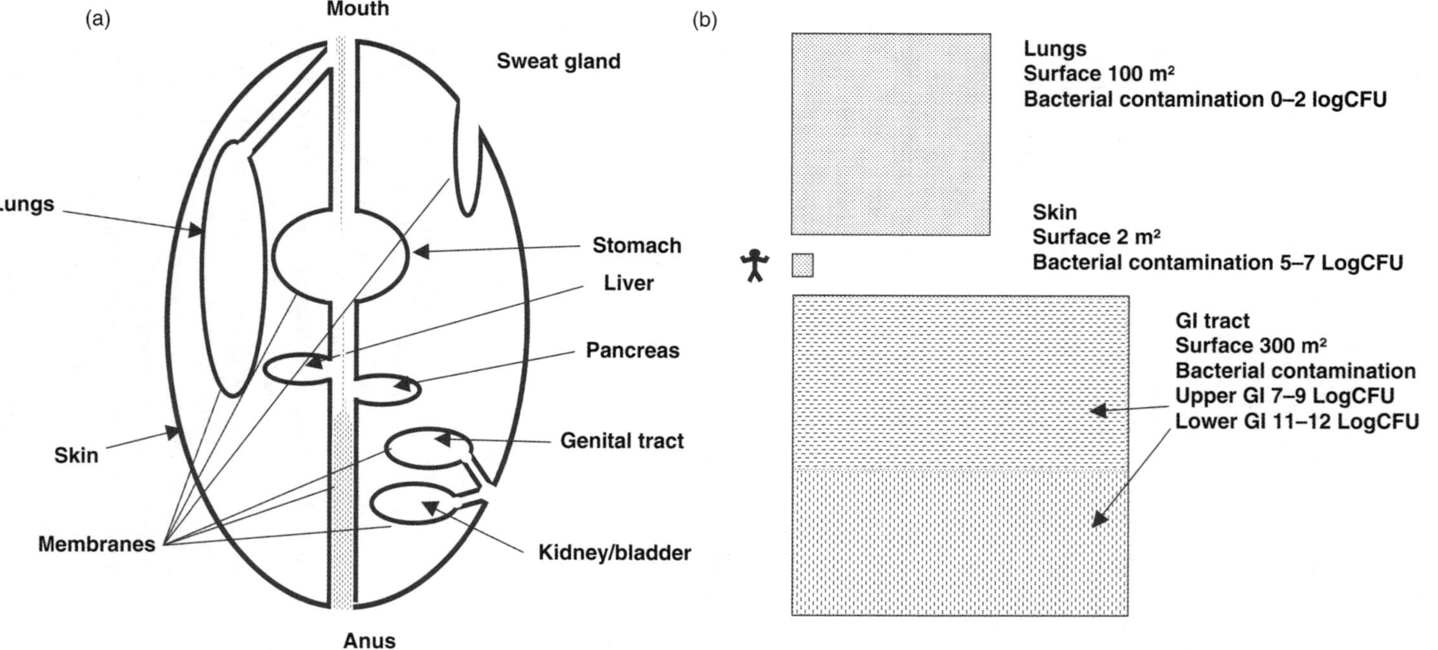

Figure 3.3 *A diagram of the human body and its relation to the bacterial ecosystems (in log CFU* ml^{-1} *for GI content or log CFU* cm^{-2} *for lungs and skin). Skin and membrane systems form a continuous lining of the body, and separate the internal part of the body from the external environment. The lumen of the intestine is 'external environment'. There is a continuous environmental bacterial pressure. It is in the GI tract that a bacterial population is maintained (a). In (b) the relative size of the surface of the skin, the lungs and the GI tract are shown. It is clear that the intestine is by far the largest surface through which the mammalian body is in contact with the external environment. It moreover contains about the densest possible bacterial population (not more than* 10^{12} *bacteria can be volumetrically contained within 1* cm^3*)*

that are associated with good health or that are benign (e.g. *Lactobacillus* and *Bifidobacterium*) and concomitantly low numbers of potentially pathogenic or known harmful microorganisms (e.g. *Escherichia coli*, *Clostridium* spp., *Veillonella* spp., *Candida*, *Salmonella*, etc.) (Gibson and Roberfroid, 1995). Indeed, the gut microflora is a complex ecosystem with a wide variety of potential interactions between the different populations of microorganisms, including interactions between potentially health-promoting and harmful microorganisms.

The concept of a balanced microflora is at the origin of the hypothesis that dietary strategies might be developed to modulate the composition of the colonic microbiota (i.e. by stimulating the growth of health-promoting bacteria and reducing the number of potentially harmful microorganisms) with the objective of improving colonic health and thus, indirectly, the health and well-being of the host as well as the ability to reduce the risk of various diseases (Roberfroid, 1998). Such a strategy includes the consumption of prebiotics in an attempt to stimulate the growth and, perhaps more importantly, the metabolic activity of potentially beneficial bacteria.

3.11 Prebiotic: Definition and Scientific Demonstration

The most recent definition of a prebiotic is: 'A prebiotic is a selectively fermented ingredient that allows specific changes, both in the composition and/or activity in the gastrointestinal microflora that confers benefits upon host well being and health' (Gibson *et al.*, 2004).

Since its introduction, the concept of prebiotics has attracted much attention, stimulating scientific as well as industrial interest. However, a prebiotic effect has been attributed to many food components, sometimes without due consideration to the criteria required. In particular, many food oligosaccharides and polysaccharides (including dietary fibre) have been claimed to have prebiotic activity, but not all dietary carbohydrates are prebiotics (Gibson *et al.*, 2004). Indeed such classification requires a scientific demonstration that the ingredient resists gastric acidity, hydrolysis by mammalian enzymes and gastrointestinal absorption, in order to make sure that most of the ingredient reaches the colon. The ingredient is fermented by the intestinal microflora and more particularly it selectively stimulates the growth and/or metabolic activity of intestinal bacteria associated with health and well-being.

Although each of these criteria is important, the third, concerning the selective stimulation of growth and/or activity of bacteria, is the most contentious and difficult to demonstrate. The prebiotic properties of a candidate prebiotic can be demonstrated only by means of, preferably repeated independently, human dietary intervention trials. The demonstration can be carried out by means of classical plate counting on 'selective' media. This requires anaerobic sampling followed by strictly anaerobic handling, and growing a wide variety of bacterial genera, typically total aerobes and total anaerobes, e.g. *Bacteroides*, *Bifidobacterium*, *Clostridium*, *Enterobacterium*, *Eubacterium* and *Lactobacillus*. Due to the overall poor selectivity of the media it is required that the identity of the individual colonies be confirmed by means of microscopic analysis, membrane triglyceride composition, typical fermentation patterns, or by means of 16 S ribosomal RNA techniques.

Other more modern techniques involve FISH, whereby the selectively hybridised bacteria can be counted with a fluorescent microscope, or in a high-throughput flow cytometer (Blaut *et al.*, 2002; Langendijk *et al.*, 1995).

Simply reporting fermentation in pure cultures of single microbial strains or an increase in a limited number of bacterial genera in complex mixtures of bacteria (e.g. faecal slurries) either *in vitro* or *in vivo* cannot be accepted as demonstrating a prebiotic effect. This is because such a limited approach does not take into account the high level of complexity of the gut microflora and the numerous bacterial interactions that exist therein. Today only two food ingredients meet the requirements for prebiotic classification, i.e. inulin-type fructans and (*trans*)-galactooligosaccharides (Gibson *et al.*, 2004). Moreover, inulin and oligofructose are considered model prebiotics, as it is their nutritional effects that are at the origin of the concept.

3.12 Prebiotic Properties of Different Inulin-type Fructan Fractions ST-inulin, Oligofructose, HP-inulin and Synergy1

Inulin and oligofructose escape digestion in the upper GI tract and reach the large intestine virtually intact, they are thus colonic foods (see above). In addition, they act as prebiotics, as shown in the many short-term and longer-term human dietary intervention studies designed to investigate the effects of inulin on the gut. It has been consistently observed that *Bifidobacterium* spp. and *Lactobacillus* spp., and possibly *Eubacterium* spp., are stimulated. Investigations into the selectivity of these food ingredients has revealed that *Lactobacillus* spp. and *Bifidobacterium* spp. possess selective transporter mechanisms for inulin or oligofructose. As such, inulin and oligofructose are withdrawn from the environment and bacteria having appropriate selective transporters can metabolise and grow on the carbohydrates (Barrangou *et al.*, 2003; Schell *et al.*, 2002). Data on *in vitro* fermentation of inulin and oligofructose in anaerobic batch cultures inoculated with pure genera or strains of different bacteria show that most *Bifidobacterium* spp. ferment inulin and oligofructose as well as, or even better than, glucose whereas, for many other bacterial genera tested, glucose is a more efficient substrate for growth than inulin. Amongst all the *Bifidobacterium* spp. tested for their capacity to ferment inulin and oligofructose, *Bifidobacterium longum* and *Bifidobacterium infantis* prefer both substrates, whereas *Bifidobacterium animalis* grows more efficiently on oligofructose than on inulin and *Bifidobacterium bifidum* ferments neither inulin nor oligofructose (Durieux, 2001; Gibson and Wang, 1994; Hidaka, 1986; Hopkins *et al.*, 1998; Kaplan and Hutkins, 2000; Marx *et al.*, 2000; Mitsuoka *et al.*, 1987; Perrin *et al.*, 2001; Probert and Gibson, 2002; Roberfroid *et al.*, 1998; Wang and Gibson, 1993; Yazawa *et al.*, 1978). Oligofructose and inulin are bifidogenic in the single-stage batch-culture system or a three-stage, continuous-culture model of the colon containing human faecal bacteria. In parallel with an increase in the numbers of *Bifidobacterium* spp., those of *E. coli* and *C. perfringens* were reduced significantly (Dal Bello *et al.*, 2001; Gibson, 1994; Gibson and Wang, 1994; Sghir *et al.*, 1998). A 'measure of the prebiotic effect (MPE)' has been defined. It is based on *in vitro* data and compares measurements of bacterial changes through the determination of maximum growth rates of predominant groups present in faeces, rate of substrate assimilation and the production of lactic acid, acetic acid, propionic acid and butyric acid (Palframan *et al.*, 2003; Vulevic *et al.*, 2004).

Experimental studies *in vivo* have been carried out using animal models, particularly the human microflora-associated rat. In such a model, selective stimulation of growth of bifidobacteria and a reduction in the number of *Clostridium* spp. were induced by oligofructose and inulin (Campbell *et al.*, 1997; Djouzi and Andrieux, 1997; Levrat, *et al.*, 1991; Poulsen, *et al.*, 2002). Extensive studies have compared the effect of inulin with different chain lengths and investigated the changes in microflora composition in the caecum, the colon and the faeces (Kleessen *et al.*, 2001). Besides the bifidogenic effect, the most important finding in these protocols is probably the increase in counts of bacteria belonging to *Clostridium coccoides – Eubacterium rectale*, a cluster that contains most of the nonpathogenic butyrate-producing microorganisms (Duncan *et al.*, 2004). The increase in their counts following inulin feeding might thus explain the increase in the proportion of butyrate in SCFAs resulting from its intestinal fermentation (see above).

There are several human studies that demonstrate the prebiotic effect of oligofructose, inulin and HP-inulin. These studies were carried according to different protocols (controlled diet versus free diet, parallel study design versus cross-over, different doses, different volunteers as what concerns age, sex and colonic bacterial condition) (Bartosh *et al.* 2005; Bouhnik *et al.*, 1999; Buddington *et al.*, 1996; Gibson *et al.*, 1995; Guigoz *et al.*, 2002; Harmsen *et al.*, 2002; Kleessen *et al.*, 1997; Kruse *et al.*, 1999a; Menne *et al.*, 2000; Mitsuoka *et al.*, 1987; Rao, 2001; Tuohy *et al.*, 2001a,b; Williams *et al.*, 1994). In these *in vivo* trials, there were moreover large variations between the initial microflora compositions of the subjects (analysed by applying both classical culture and molecular techniques) as well as in the way they responded to the substrates (Bouhnik *et al.*, 1996; Buddington *et al.*, 1996; Franks *et al.*, 1998; Gibson *et al.*, 1995; Guigoz *et al.*, 2002; Harmsen *et al.*, 2002; Kruse *et al.*, 1999; Rao, 2001; Tuohy *et al.*, 2001a). The efficacy of prebiotics also has been evaluated also in formula-fed neonates (Knol, 2002; Moro, 2001; Rigo, 2001).

Whereas this plethora of studies numerically cannot be directly compared, an increase in the counts of both *Bifidobacterium* spp. and *Lactobacillus* spp. has been consistently reported. In all the human nutrition trials that have tested for the effect of inulin-type fructans, an increase in numbers of *Bifidobacterium* spp. (expressed as logCFU per gram of faeces) becomes significant and reaches its maximum in less than a week and remains as long as the intake of the prebiotic continues (two studies followed-up during a period of 3 months) (Clune *et al.*, 2004; Kruse *et al.*, 1999). When chicory fructan intake stops, the composition of the intestinal flora falls back to the initial values within a period of 3 weeks. The selectivity has been validated in several studies. As *Bifidobacterium* and *Lactobacillus* increase in numbers, pathogenic *Clostridium perfringens* group (plate counts) of *Clostridium histolyticum / C. lituburense* group (FISH) typically decrease in numbers, whereas commensals such as *Bacteroides* or *Fusarium* typically are not affected by the intervention (Gibson *et al.*, 1995; Kruse *et al.*, 1999; Tuohy *et al*, 2001a).

It is to be remarked that the slower fermented HP-inulin induces numerically less important shifts in the composition of the intestinal flora. The HP-inulin however also is completely fermented.

Such observations are well in-line with the hypothesis that chicory fructans are robust prebiotics and act as selective substrates for the fermentation of certain groups of positive gut bacteria.

At present, sampling for microbiological analysis has been limited to faeces. It is generally accepted, and has been demonstrated, that the bacterial composition of faeces is a good representation of the composition of the colonic content (van der Waaij, 2003). Still, the inability to take sample *in situ* certainly limits our understanding of what is happening inside the colon and, more specifically, inside the different segments that are known to differ in their environmental (pH, mineral and water content) conditions and physiological functions (Cummings, 1997).

Another topic of growing interest is the bacterial flora colonising surfaces in the large intestine, especially the mucosa, the mucus layer and the particulate materials in the colonic lumen (Macfarlane *et al.*, 1999). Indeed, a few studies using either biopsy or resected samples have demonstrated the presence in the colonic mucosa of a microflora with a specific composition, different from the luminal colonic microflora (Poxton, 1997). In an *ex vivo* protocol whereby 15 healthy volunteers selected from a colonoscopy waiting list had been asked to supplement their usual diet with Synergy1 for 2 weeks, an increase in both *Bifidobacterium* spp. and *Lactobacillus* spp. counts in the mucosa occurred (Langlands *et al.*, 2004). In a model of rats harbouring a human faecal flora, it was demonstrated that feeding an inulin-supplemented diet significantly increased mucosal bifidobacteria (cells mm^{-2} mucosal surface) (Kleessen *et al.*, 2003). It can thus be hypothesised that the prebiotic effect of inulin concerns both the luminal and the mucosa-associated microflora.

Another observation from these studies is that the prebiotic effect is not especially dose-dependent within the range of 5–40 g per day. The daily dose does not correlate with the absolute increase in bacterial cells. The major factor that controls quantitatively the prebiotic effect is the number of bifidobacteria per gram of faeces the volunteers have before supplementation of the diet with inulin-type fructans begins. To some extent, that parameter correlates inversely with increase in faecal bifidobacteria (Roberfroid *et al.*, 1998).

Finally, it should be recognised that even accurate quantification with modern techniques of a few bacteria (e.g. *Bifidobacterium*, *Bacteroides*, *Clostrium* spp., *Lactobacillus*, *Eubacterium*, *Fusarium*) gives only a small, probably incomplete part of the full diversity picture. The complexity of the colonic bacterial ecosystem is characterised by hundreds, or even thousands, of bacterial species which all interact in particular interdependent ways (Hopkins, 1998; Newton, 1997). It can be supposed that, besides the bacteria that have been studied, several other groups of bacteria and their metabolism (Vanhoutte *et al.*, 2005) are affected.

Inulin and oligofructose are potent prebiotics, and apparently the only direct effect they have is modification of the composition of the intestinal flora and the interaction with bacterial metabolism. The colonic ecosystem changes and, according to the principles illustrated by Figure 3.3, chicory fructan consumption has indirectly an impact on the rest of the body. During the last decade there have been exploratory investigations into the indirect effects of inulin and oligofructose. Through their influences on the colonic microflora, they seem to have the potential to modulate bowel habit, interact with the absorption of minerals, interact with complex processes of cell proliferation in the intestine and in the periferous tissues (anticancer properties), modulate immunological defence mechanisms and interact with the endocrine activities and lipid metabolism (Van Loo *et al.*, 1999). Probably as a consequence of all this, they may have an impact on a general feeling of well-being. These topics are considered in more detail in the following sections.

3.13 Impact on Mineral Absorption and Bone Density in Human Volunteers

The human body contains about 1100 g of calcium (of which over 99 % is in the bone) and 500–800 g of phosphorus. The calcium that circulates in ionised form, or is bound to protein in the blood, is necessary for blood coagulation, muscle contraction and nerve function. Because of its physiological importance, there is a finely regulated calcium homeostasis in the blood, whereby calcium reserves in the bone can be addressed to compensate for possible shortages. As a consequence, it is important to take in sufficient quantities of calcium in the diet. Besides calcium, magnesium is thought to play an important role in bone metabolism and other physiological processes (Lukaski and Nielsen, 2002). A significant proportion of the western population fails to achieve the recommended intakes of these minerals (1200 mg Ca per day; 300–400 mg of Mg per day).

Besides encouraging individuals to meet dietary Ca and Mg requirements, increasing GI absorption is a valid alternative to improve homeostasis. Supplementation of food with inulin influences absorption of these minerals positively (Cashman, 2003; Coxam, 2005).

There is a plethora of data from various types of experimental models (normal healthy, cecectomised or gastrectomised rats; rats receiving Ca and Mg directly by stomach gavage or by caecal intubation; rats fed a Mg- or a Fe-deficient diet; adult virgin ovariectomised female rats and laying hens), which consistently demonstrates that oral intake of oligofructose, inulin, HP-inulin and Synergy1 significantly increase Ca and Mg absorption. The results of these studies are summarised in Table 3.2.

The vast majority of these studies used the metabolic balance method to measure fractional or apparent absorption (FA %). Usually, the animals were rather young (4–6 weeks old) at the start of experiments that lasted for a few weeks (10–31 days). All inulin-type fructans were effective, but Synergy1 was the most active in enhancing absorption of Ca and Mg. HP-inulin and oligofructose restored the Ca balance in ovariectomised rats (model for postmenopausal model).

For both Ca and Mg, the relative increase in intestinal absorption has been shown to correlate inversely with basal absorption capacity, i.e. being higher when basal absorption is lower (Griffin *et al.*, 2003).

In an early study on mineral absorption, oligofructose had no effect on Ca absorption in young men (Van den Heuvel, 1998). One criticism of this study was that urine was collected for only 24 h, potentially missing the late colonic phase of absorption (Van den Heuvel, 1999). Indeed, a subsequent study by the same group, using a 36 h urine collection, showed that oligofructose significantly increased calcium absorption (Van den Heuvel, 1999). In a protocol whereby the volunteers received calcium doses of 1500 mg per day (higher than RDI for Ca), it was shown that Synergy1 still could induce a

Table 3.2 *Effects of inulin-type fructans on Ca and Mg balance in experimental models*

Product tested Animal	Dose (%) Duration	Method	Results Ca	Results Mg	Reference
Oligofructose Sprague-Dawley males Age 5 weeks, N = 10	1, 3, 5 Ca = 0.6 Mg = 0.06 27 days	Balance (ICPS)	FA1(%) 1 % NS 3 % NS 5 % NS	1 % 70→72.9 (+4.1 %) 3 % 70→73.8 (+5.4 %) 5 % 70→77.4 (+ 10.5 %)	Hillman *et al.*, 1988
Oligofructose Fisher 344 males Age 38 weeks, N = 8	5 (Ca = 0.5) 1	Ratio 47Ca/47Sc	FA(%) 12.8→21. (+67 %)		Miller *et al.*, 1989
Inulin Wistar males Age 8 weeks, N = 8	10 Ca = 0.7 Mg = 0.09 21 days	Balance Atomic abs. spectrom.	FA1(%) 23.→37 (+61 %)	35.→57 (+63 %)	Pallarés *et al.*, 1993
Oligofructose Sprague–Dawley males Age 4 weeks, *N* = 7	5 Ca = 0.5 Mg = 0.05 28 days	Balance At days: 4, 10, 17, 24 Atomic abs. spectrom.	FA1(%) Days 4–8: 64.5→80.5 (+24 %) Days 10–14: 64.5→76 (+18 %) Days 17–21: 59.5→71 (+19 %) Days 24–28: 53.5→60 NS	Days 4–8: 62.8→87.7 (+41 %) Days 10–14: 62.8→87.6 (+41 %) Days 17–21: 62.8→87.4 (+40 %) Days 24–28: 62.8→82.8 (+ 33 %)	Ohta *et al.*, 1994
Oligofructose Sprague–Dawley males Age 8 4 weeks, N = 7	1–15 Ca = 0.5 Mg = 0.05 10–31 days	Balance Ion plasma spectral analysis	FA1(%) 1 % NS 3 % 56.5→61 (+ 8 %) 5 % 56.5→65 (+ 15 %)	1 % NS 3 % 69→75 (+ 8.7 %) 5 % 69→83 (+20.5 %)	Delzenne, 1995

			15 % 56.5→82 (+ 15 % 45 %) Days 3–7: 58.6→75 (+ 28 %) Days 14–18: 55.3→64 (+ 15.7 %) Days 27–31: 38.3→51.4 (+ 34.2 %)	69→89 (+ 28 %) day 3–7: 61→83 (+ 36.2 %) Days 14–18: 55.3→78 (+ 41.6 %) Days 27–31: 49.8→73 (+ 47.2 %)	
Inulin Wistar males Age 6 weeks, N = 8	10 Ca = 0.5 Mg = 0.05 21 days	Balance Atomic abs. spectrom.	FA1(%) 33→42 (+ 27 %)	35→69 (+ 97 %)	Ohta *et al.*, 1995
Oligofructose Sprague–Dawley males Age: 6 weeks, N = 7	5 Ca = 0.5 Mg = 0.05 14 days	Balance (ICPS)	FA1(%) Days 3–6: 66.8→75.2 (+ 12.6 %) 63.2→75 (+ 11.9 %) Days 10–13: 58→65 (+ 12 %) 57.9→69.5 (+ 20 %)	 Days 3–6: 58.9→76 (+ 29.2 %) 54→79.5 (+ 45.9 %) Days 10–13: 43→74.9 (+ 72.2 %) 49.3→76 (+ 54.4 %)	 Coudray, 1997
Oligofructose Sprague–Dawley males Age 5 weeks, N = 11	5 Ca = 0.5 Mg = 0.05 15days	Balance (ICPS)	FA1(%) Days 3–7: 45.4→60.9 (+ 34.1 %) Days 10–14: 47.7→52.9 (+ 11 %) Days 3–7: 47.9→59.6 (+ 25 %) Days 10–14: 44→66.8 (+ 51.8 %)	 Days 3–7: 58.8→76.3 (+ 30 %) Days 10–14: 52→68.1 (+ 31 %) Days 3–7: 56→87.9 (+ 57.2 %) Days 10–14: 52→83.6 (+ 60.8 %)	Lopez, *et al.*, 2000
Oligofructose Sprague–Dawley males Age 5 weeks, N = 6	1 and 5 Ca = 1.0 Mg = 0.025	Balance (ICPS)	FA1(%) 1 %NS	1 % Days 7–11: NS Days 21–25:	 Younes *et al.*, 2001

Table 3.2 *(Continued)*

Product tested Animal	Dose (%) Duration	Method	Results Ca	Results Mg	Reference
	25 days		5 % NS	34→51.6 (+ 51.5 %) 5 % Days 7–11: 54.4→71.8 (+ 32 %) Days 21–25: (34→62.5) (+ 83 %)	
Oligofructose Inulin Dogs. Age 4 weeks, N= 10	10 Ca = 0.7 Mg = 0.1 24 days	Balance Plasma atomic abs. spectrom.	FA1(%) Oligofructose: 25.4→43.5 (= 71 %) Inulin: 25.4→40.3 (+ 59 %)	Oligofructose: 27→64.7 (+ 140 %) Inulin: 27→64.7 (+ 140 %)	Beynen *et al.*, 2002
Oligofructose Male wistars Age 6 weeks, N= 8	5 Ca = 0.5 18 days	Balance Atomic abs. spectrom.	FA1(%) 43.7→53.2 (+ 22 %)		(Richardson *et al.*, 2002
Oligofructose Wistar males Age 8 weeks, N = 8	5 Ca = 0.75 Mg = 0.025 28 days	Balance Atomic abs. spectrom.	FA1(%) 52.9.→41 (− 23 %)	84.6.→90.5 (+ 7 %)	Baba, 1996
Oligofructose Ovariectomized rats	2.5, 5.0 and 10 (8 and 16 weeks) Ca = 0.5 and 1.0 %	Balance Trabecular bone structure (mirroradiographs)	Increased Ca absorption especially in diets with high Ca content Effect of ovari-ectomization was compensated Strengthened trabecular bone structure		Scholz Ahrens *et al.*, 2002

Oligofructose HP-inulin Synergy1 Wistar males Body wt 170 g, N = 10	2.5 (1 week) 5.0 (1 week) 10 (2 weeks) Ca = 0.5 Mg = 0.05 28 days	Balance Atomic abs. spectrom.	FA1(%) Oligofructose +10 % (NS) HP inulin +13 % (NS) Synergy1 47.9 → 58.1 (+ 21 %)	Oligofructose 48.8→71.3 (+ 46 %) HP inulin 4976.4 (+ 56.5 %) Synergy1 48.8 →76.7 (+57 %)	Coudray *et al.*, 2003b
Synergy1 Ovariectomized Sprague–Dawley	5.5 (3 weeks)	^{45}Ca (AAS) Bone strength and bone resorption rate	Increased Ca absorption Improved bone strength Reduced resorption rate		Zafar *et al.*, 2004

FA, Fractional absorption; ICPS, inductive coupled plasma emission spectrometry.

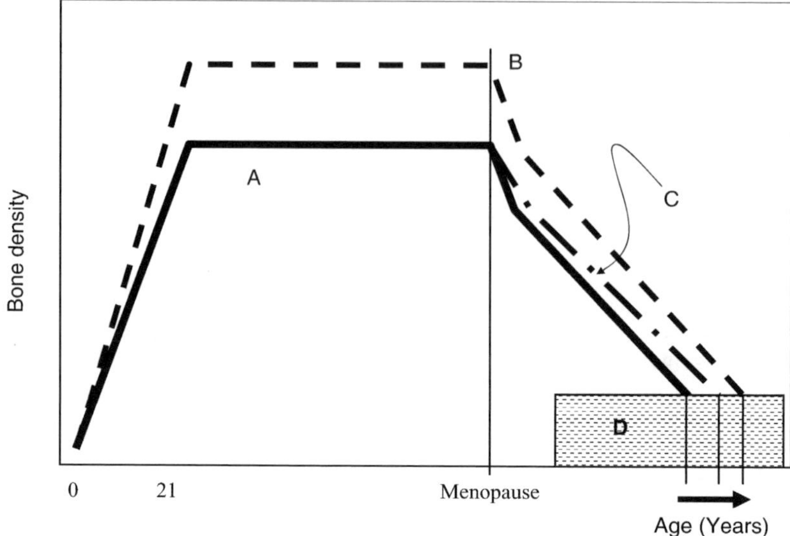

Figure 3.4 Demonstration of the idea behind the possible reduction of risk for osteoporosis via dietary intake of inulin-type fructans. Curve (A) presents evolution of bone mineral content during life. Increasing peak bone density during adolescence (B) postpones the moment that postmenopausal decalcification increases risk for spontaneous fractures (D). Moreover, increasing mineral absorption in postmenopausal women (C) slows down the decalcification rate, also postponing the process of the bone becoming fragile (D)

significant increase in calcium absorption (+18 % compared with control) in adolescent girls. Under these restrictive conditions of high Ca supplementation, oligofructose alone did not show the stimulation of Ca absorption that occurred with the other protocol using normal levels of Ca intake (Griffin *et al.*, 2002). This corresponds with data from experimental models (Coudray *et al.*, 2003b) and suggests that this 1:1 co-spray-dried mixture of the long-chain HP-inulin and the short-chain oligofructose (Figure 3.1) may be more potent than oligofructose in promoting Ca absorption. Synergy1 increased Ca absorption by approximately 90 mg per day. This may be physiologically relevant; if only part of the additional Ca absorbed was used for bone mineral production, it could lead to a significant increase in peak bone mineral density during this critical stage of the life cycle (see below). Another interesting result of studies with adolescents is the inverse correlation between the relative increase in absorption caused by Synergy1 and basal absorption capacity before the intervention, which again seems to be in agreement with animal data. This indicates that, with regard to mineral absorption, consuming inulin-type fructans would be of most benefit to adolescents, who have a low basal Ca absorption level, a situation that could correlate with genetic polymorphisms known to account for differences in Ca absorption (Griffin *et al.*, 2002). More recently, in a long-term trial, in which 50 children (25 girls and 25 boys) were given Synergy1 (8 g per day) using maltodextrins as a control in a double-blind parallel study design for a period of up to 1 year, it was observed that the enhanced Ca absorption persisted during the whole 12 months of the study (Abrams *et al.*, 2005b). Furthermore, and very interestingly, at the same time, whole-body bone mineral density and whole-body bone mineral content were increased significantly in the adolescents who had consumed Synergy1, thus demonstrating that the excess of absorbed Ca was well utilised by the bones (Abrams *et al.*, 2005a).

In postmenopausal women, studies have confirmed the effect of Synergy1 but not oligofructose on Ca absorption, whereas both ingredients were active in enhancing Mg absorption (Holloway *et al.*, 2003; Tahiri, 2001, 2003). These data again support the observation that Synergy1 is more effective than other inulin-derivatives in enhancing mineral (especially Ca) absorption.

All these findings show that regular intake of even modest amounts of Synergy1 increases the absorption of Ca in girls, and in prepubertal and adolescent boys, and that Synergy1 and oligofructose enhances Mg absorption in postmenopausal women, with adequate or high intakes of Ca. As is the case with experimental models (Roberfroid *et al.*, 2002), the extra Ca absorbed is used to improve bone density. In adolescents, such physiological benefits, which persist for at least 1 year (Abrams *et al.*, 2005a), are particularly interesting in view of the importance of peak bone mass in the growing years. This is a parameter that is thought to influence the risk of osteoporosis in later life, especially after menopause (Weaver *et al.*, 2000, Figure 3.4).

3.14 Reduced Risk for Cancer in Human Volunteers

A reduced risk for cancer has been demonstrated repeatedly by means of various anticancer-related experimental models, and recently was confirmed in a first exploratory human trial.

In chemoprevention models, the animals are challenged by a chemical that induces carcinogenesis in specific organs. Azoxymethane (AOM) and dimethylhydrazine (DMH) target the intestine, more particularly the colon, whilst 4-methyl nitrosourea (4-MNU) targets the breast in a female model.

The animals receive either a control diet or a diet containing a certain chicory fructan fraction (2.5–10 %, w/w). The reduction in number of either preneoplastic lesions, aberrant crypt foci (ACF), mucin-depleted foci (MDF) (Femia *et al.*, 2004) (after 8 weeks) or tumours (after 35 weeks and up to 52 weeks) in the test group versus a control group is a measure of suppression of carcinogenic processes or reduced risk for cancer.

The chemopreventive effects of chicory fructans are dose-dependent (Figure 3.5) and become more pronounced with increasing chain length of the prebiotic inulin-type fructans, with Synergy1 being the most efficient (Figure 3.6).

This observation again seems to support the Synergy1 hypothesis, i.e. it is the combined action of improving the composition of the intestinal microflora (by the rapidly fermented oligofructose part) and subsequent maintenance of the metabolism of the improved flora (by the slowly but still selectively fermented HP-inulin).

A lowered risk for carcinogenesis was observed when prebiotics are fed before injection of the carcinogens, i.e. during the initiation phase. The effect of prebiotics upon the promotion phase, tested by feeding the prebiotic after injection of the

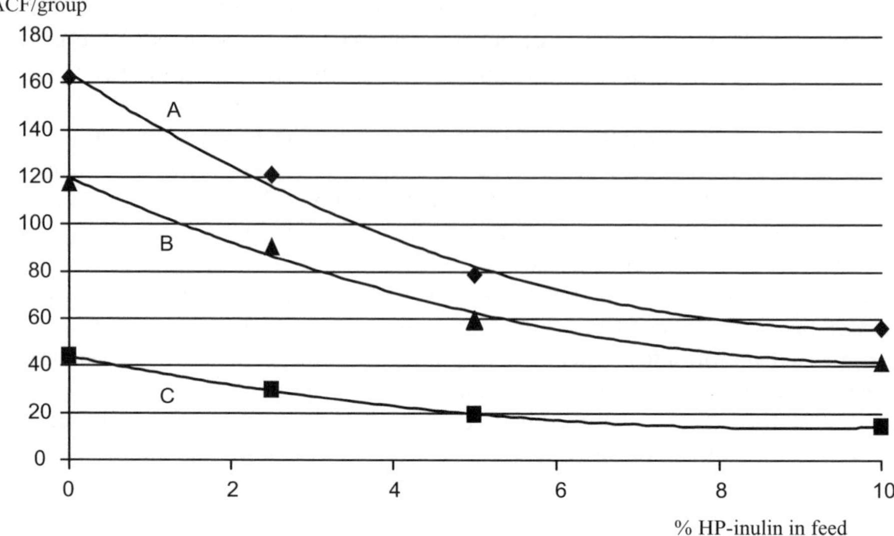

Figure 3.5 *Demonstration of the dose–effect relationship of HP-inulin in a chemoprevention model (AOM injection, count ACF after 8 weeks). The ACF in the whole intestine decrease with increasing dose of HP-inulin (A). It is observed that for a certain inulin fraction (HP-inulin) the anticarcinogenic capacity is proportional in the proximal (C) and distal colon (B). (After Verghese et al., 2002a)*

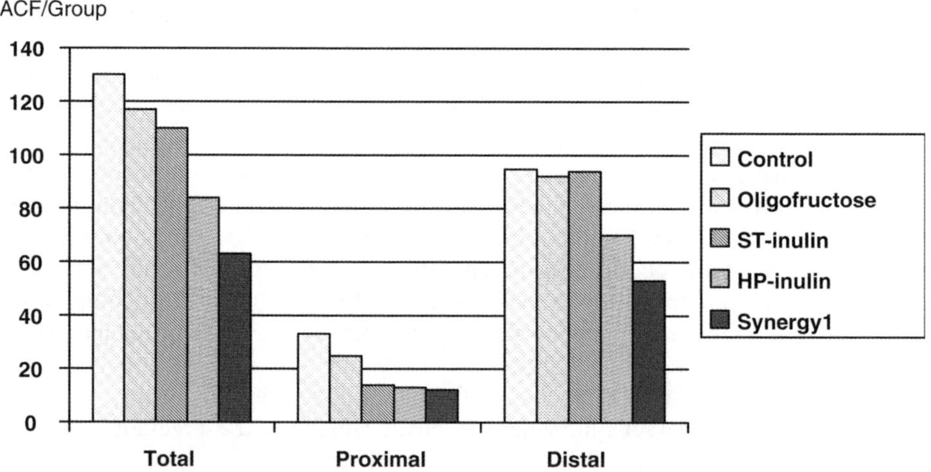

Figure 3.6 *Impact of the chicory fructan fraction on suppression of carcinogenesis. In a chemoprevention model (AOM injection, count ACF after 8 weeks) it is observed that short chain oligofructose exerts beneficial effects, particularly in the proximal colon. The higher the proportion of long chains, the more important the anticarcinogenic effect in the distal part of the colon (series ST – HP). There is a smaller proportion of long chains in Synergy1 than in HP, but the additional effect on the composition of the flora in the proximal colon induced by the short-chain fraction of Synergy1 may explain the difference. (After Verghese et al., 2005)*

carcinogen, is even more pronounced (Reddy *et al.*, 1997; Verghese, 1999; Verghese *et al.*, 2002a,b, 2005).

The development of chemically induced breast cancer was suppressed by feeding the animals a diet containing chicory fructan versus a control diet (Taper and Roberfroid, 1999). This indicates that systemic effects were involved and that the cancer-preventing action was not simply a local phenomenon.

The underlying mechanisms are not known. It should be noted that growth of the prebiotic-fed animals did not differ from that of control animals at any time; therefore, the reduction in tumour development was not due to energy restriction.

Animals fed prebiotic had a smaller amount of 'high-risk' compounds in the colon. There was less β-glucuronidase and less free ammonium present (Rowland *et al.*, 1998). The apoptotic index of prebiotic-fed rats was significantly higher than that of the control group (Hughes and Rowland, 2001). The genotoxic potential of faecal water was shown to be lower in inulin-fed animals (Burns and Rowland, 2004). All observations that might indicate possible mechanisms are related directly to the impact of the chicory fructan fractions on caecal, or colonic, bacterial fermentation. Any difference in effects between the fractions seems to be a direct consequence of their fermentation characteristics (Van Loo, 2004a). Targeted interaction with colonic fermentation would be the basis of the anticarcinogenic properties of the prebiotic fructans.

In tumour implantation models, aggressive and invasive tumour cells are implanted into either muscle (solid tumour growth) or the peritoneum (ascitic tumour growth) of mice. The animals are fed either a control diet or foods containing chicory fructan for 1–2 weeks before tumour implantation. Mortality curves, expressed as increase in life span or mean survival time, are then monitored (Taper *et al.*, 1997, 1998). Microscopic histological evaluation allows conclusions to be drawn on metastatic aspects of size and type of tumour (Taper and Roberfroid, 2000). These data show consistently that prebiotics retard the the growth of transplanted liver tumour cells and slow down mortality curves. These observations again confirm that systemic factors are involved.

When the tumour cells were allowed to metastasise (intraperitoneal application of the tumour cells), less lung metastases were observed in animals fed HP-inulin as compared with the control animals. The metastasis that still developed in the chicory fructan fed rats were smaller than metastases in the control group (Taper and Roberfroid, 2000).

It has been demonstrated in chemoprevention models and in tumour implantation models, that chicory fructan fractions potentiate various types of pharmaceutical products that are used in fighting cancer [nonsteroidal anti-inflammatory drugs such as Sulindac (Verghese *et al.*, 2005) and celecoxib (Buecher *et al.*, 2003)] and also various types of cytostatic compounds such as antimetabolites and 5-fluorouracyl (Taper and Roberfroid, 2002a,b).

In genetically predetermined mice models, such as the Apc Min mice that develop tumours spontaneously mainly in the small intestine but a few in the colon, it was shown that consumption of oligofructose or Synergy1 reduce the incidence of tumours in the small intestine as well as in the colon (Pierre *et al.*, 1997, 1999; Pool-Zobel *et al.*, 2002). These results are relevant for humans, as it is clear that whereas various life-style factors explain over 75 % of cancer incidence [World Cancer Research Fund (WCRF) and American Institute for Cancer Research (AICR)], hereditary genetic cancer risk factors (for which the Min mice are a model) play an important role in an individual's risk for cancer.

Recently, in the EU-sponsored research project SYNCAN (www.syncan.be), a group of 40 polypectomised patients and a group of 40 surgically treated colon cancer patients were given a placebo (maltodextrins) or Synergy1 in combination with commercial *Bifidobacterium animalis* and *Lactobacillus rhamnosus* strains for a period of 3 months (Van Loo and Jonkers, 2001) in a double-blind parallel study design ($N=20$ per test group). Because of the exploratory nature of the project, it was decided to sample biopsies, blood, faeces and urine at the start and at the end of the intervention period, and to quantify a quasi-exhaustive list of anticancer-related biomarkers in these samples. A very distinctive prebiotic effect (increase in *Bifidobacterium* and *Lactobacillus*; decrease in coliforms and *Clostridium perfringens* group and no effect on *Bacteroides*) was observed in both the polypectomised and the cancer patient groups, confirming compliance of the volunteers, and the efficacy of the administered dose of Synergy1 of 12 g/per day (Clune *et al.*, 2004; Van Loo *et al.*, 2005). It was demonstrated that DNA damage was reduced in the test group (Comet assay) (Klinder *et al.*, 2004a,b). Faecal water from polypectomised patients improved tight junction integrity (a marker for reduced metastatic potential of tumour cells) and genotoxic potential of the volunteers administered Synergy1 and the probiotics was reduced. Perhaps the most important finding of the project was the reduction of cell proliferation rate as observed in biopsies from polypectomised patients (Clune *et al.*, 2004; Rafter *et al.*, 2005). This first trial with human volunteers confirmed that, as seen in experimental models, Synergy1 may help reduce the risk for colon cancer in human volunteers. It was decided to administer a combination of Synergy1 and bacteria because of the observation by Rowland *et al.* (1998) in a chemoprevention model that such a combination may be more effective than either of the components alone. This synergistic effect was not confirmed in the SYNCAN project, however, where a reduction in tumour formation in an experimental model was observed only where the prebiotic was administered (Caderni *et al.*, 2003; Femia *et al.*, 2002). The antitumourigenic effect correlated with increased production of SCFAs (particularly with butyrate production) and with stimulation of the gut-associated immune system (Roller *et al.*, 2004a,b).

3.15 Modulation of Immune Function in Human Volunteers

From experimental models, it appears that functioning of the gut-associated immune system (GALT) may be modulated by prebiotics (Kelly-Quagliana *et al.*, 2003; Roller *et al.*, 2004a,b). These organs (mesenteric lymph nodes, Peyer's patches) cannot be sampled in humans, and the protocol of the SYNCAN project did not show changes in the immune markers that can be measured in the blood. In a recent interpretation of the biochemical and clinical relevance of various immunology-related biomarkers it was concluded that to show the effects of a food ingredient in human dietary intervention trials, it is necessary to follow the reaction against a vaccine (challenge model) or to look at populations at risk (Albers *et al.*, 2005). Double-blind, randomised, placebo-controlled studies focusing specifically on several clinical parameters related to common acute paediatric illnesses or the immunological response to measles vaccine in infants and toddlers were carried out with oligofructose (Firmansyah, 2000; Saavedra, 1999; Saavedra and Tschernia, 2002; Tschernia, 1999). The intervention group that received oligofructose, showed reduced occurrence of concurrent fever as an indicator of severity during episodes of diarrhoea, medical intervention during such an event, day-care absenteeism rate, and reduced use of antibiotics during respiratory illness.

In children who were immunised with a standard live attenuated measles vaccination, feeding Synergy1 stimulated the production of anti-measles IgG (Firmansyah *et al.*, 2001). These observations were confirmed in a recent study with toddlers given oligofructose for a period of 4 weeks. Besides the distinct prebiotic effect resulting in an improved bacterial intestinal colonisation (characterised by more *Bifidobacterium* spp. and less pathogens), it was observed that the test group had less fever, less flatulence, less diarrhoea and less regurgitation (Waligora-Dupriet *et al.*, 2005).

In combination with *Bifidobacterium longum*, Synergy1 was shown to initiate resolution of inflammation in patients with active ulcerative colitis. Even a short-term treatment (1 month) was able to improve the clinical appearance of chronic inflammation (Furrie *et al.*, 2005). It is known that dysfunction of the immune system is at the basis of inflammatory bowel disease (ulcerative colitis and Crohn's disease).

3.16 Interaction with Lipid Metabolism in Human Volunteers

Lowering of serum triglycerides by inulin-type fructans in experimental models has been reported (Delzenne, 1993; Hata, 1983; Kok, 1995). Cholesterol-lowering effects were observed only with prolonged intervention periods (12 weeks) (Fiordaliso *et al.*, 1995). Studies with isolated hepatocytes showed reduced hepatic lipogenic activity in rats fed nondigestible oligosaccharides (NDO), and led to the observation that the activity of all hepatic lipogenic enzymes was significantly down-regulated (−50 %) in the animals fed chicory fructan (Kok *et al.*, 1996). This may be due to modified genetic regulation (interaction with gene expression) in the liver. Further research revealed that this might be induced by hormonal changes (inulin, glucagon-like protein of type 1 or GLP1 and glucose-dependent insulinotropic polypeptide or GIP) (Kok *et al.*, 1998a). It is hypothesised that bacterial metabolites (particularly propionate) from the fermentation of NDO were involved (Alamowitch, 1993).

Furthermore, it was demonstrated that the increase in postprandial triglyceride levels that is induced by a fat-rich diet (Western-type diet) is decreased (−50 %) in rats upon feeding oligofructose, and that the increase in liver triacylglycerol induced by a large fructose load was suppressed in oligofructose-fed rats (Kok *et al.*, 1998b).

When these findings were checked in human volunteer studies, it appeared that there were some studies showing a cholesterol-lowering effect in hyperlipidemic volunteers (Yamashita *et al.*, 1984), other experiments showed a triglyceride-lowering effect (Davidson, 1998; Jackson *et al.*, 1999). There were studies in which both a cholesterol-lowering and a triglyceride-lowering effect were reported (Brighenti *et al.*, 1999) and other studies showed no effect (Alles *et al.*, 1997; van Dokkum *et al.*, 1999). Given the apparent inconsistency of these data, which may be due to experimental conditions in general and of the condition of the volunteers in particular, it cannot be claimed unequivocally that inulin-type fructans could be efficient in reducing risk for cardiovascular disease.

The mechanisms that are at the basis of lipid and cholesterol metabolism in the human body are complex, and include various interdependent biochemical mechanisms involving hormones (insulin, GLP1, GIP) and their regulation mechanisms. These processes take place in the liver and pancreas, intestine, but also in periferous tissues (such as fat tissue). Moreover, they depend on various external environmental factors, including the training that volunteers take, dietary habits, etc.

Research in this field of interest, however, evolved and focused upon the endocrine activity of the gut.

3.17 Interaction with the Endocrine Activity of the Gut

Inulin, HP-inulin, oligofructose as well as Synergy1 are prebiotics that improve Ca and Mg bioavailability, and exert beneficial systemic effects (Delzenne, 2003; Roberfroid and Delzenne, 1998) as shown by a decrease in hepatic lipogenesis and in postprandial triglyceridemia (Delzenne and Kok, 2001) leading to a reduced accumulation of triacylglycerol in the liver and the epididymal fat pad in both normal and obese Zucker Fa/Fa rats (Daubioul *et al.*, 2002). These effects correlate with a decrease in food-derived energy intake, suggesting that inulin-type fructans could help reduce food intake and, subsequently, fat mass development (Yusrizal and Chen, 2003a). However, a fundamental question remains unanswered: how can, from a mechanistic point of view, chicory fructans affect satiety?

The intestinal, mostly caeco-colonic, mucosa contains endocrine L-cells that secrete miscellaneous peptides, especially GLP-1 and PYY are important modulators of appetite, and/or pancreatic functions (Druce *et al.*, 2004; Drucker, 2002; Stanley *et al.*, 2004). It has been demonstrated that SCFAs (the main fermentation products of dietary fibre) modulate the expression of proglucagon, a precursor of these peptides (Drozdowski *et al.*, 2002; Tappenden *et al.*, 1998), and since some dietary fibres were shown to increase proglucagon expression in dogs and rats (Massimo *et al.*, 1998; Reimer and McBurney, 1996), experiments were carried out to test the hypothesis that chicory fructans modulate the production of these gut peptides in the different segments of the intestine, and that this effect correlates with the modulation of food intake, fat mass development, and pancreatic functions.

In rats, oligofructose and Synergy1 increased significantly the total caecal pool as well as the portal blood concentration of both GLP-1 and PYY (Cani *et al.*, 2004). In addition, oligofructose decreases the plasma level of ghrelin, a peptide secreted by stomach endocrine cells whose production is regulated by GLP-1 (Lippl *et al.*, 2004). Ghrelin stimulates feeding behaviour, lowers energy expenditure and drives body weight increase (Kojima *et al.*, 1999; Tschop *et al.*, 2000).

Nutritional research with the chicory fructans initially focused on human health aspects. To date, research results can justify various functional claims. The effects that the chicory fructans may exert, suggest health promotion through prevention of various diseased states. In order to demonstrate authentic health effects however, more-long term large scale human dietary intervention trials need to be carriied out. A pan-European group of scientists evaluated criteria (type and number of studies) that need to be fulfilled in order to be able to claim certain health effects (EU Directorate Research sponsored PASSCLAIM project (HTTP://EUROPE.ILSI.ORG/PASSCLAIM).

3.18 Animal Nutrition

In livestock (pigs, poultry, calves, rabbits, fish, and others) as well as in pets (cats and dogs) the nutritional benefits of inulin and oligofructose can be used to improve animal well-being (Flickinger *et al.*, 2003). Depending on the animal species, the nutritional benefits of the chicory fructans have a different focus.

In livestock, the animals are not allowed to grow old and hence prevention of chronic disease is not relevant. The short-term direct impact on intestinal flora and the concomitant prevention of intestinal infection, however, is one of the main applications. These problems occur mainly in young animals. It was observed that inulin and oligofructose could help the prevention of intestinal problems that often occur after weaning the young animal. A change of diet from breast-feeding to solid food disturbs the intestinal flora, which often results in diarrhoea and other intestinal discomfort, which in turn may retard growth and hence reduced productivity. As antibiotics are being banned as growth promotors in animal feed, prebiotics such as inulin offers a good alternative. Young animals are typically fed relatively high doses of inulin [up to 2 % (w/w) in feed]. The follow-up diets can contain lower preventive doses (typically around 0.05–0.3 %, depending on the type of livestock). Improved zootechnical performance (faster growth, improved feed conversion ratio, less mortality) has been observed in piglets, calves, and poultry (Houdijk, 1998; Mikkelsen, 2001; Naughton, 2001; Nemcova, 1999; Xu *et al.*, 2002a,b; Yusrizal and Chen 2003a,b).

In livestock production, 'price issues' compete with 'nutritional benefits'. In animal production, both can be measured accurately, and brought into balance. As the young animal only ingests small amounts of food compared with what it eats until it is slaughtered, this strategy can be implemented successfully and economically in animal production facilities.

When the use of antibiotics as growth promotor is forbidden, it may be expected that common infections such as *Clostridium* (pigs, calves, poultry, rabbits) pathogenic *E. coli* (pigs) or *Campylobacter* and *Salmonella* (broilers) will become an important cause of mortality (Chambers *et al.*, 1997; van der Sluis, 2000). The use of 'preventative maintenance doses' of inulin in animal nutrition could help to prevent this problem of impaired animal well-being and jeopardised productivity.

In addition to protection against bacterial infection, inulin was seen to be efficient in removing intestinal parasites (*Oesophagostomum*) in pigs (Petkevicius, 2003, 2004).

The shift of nitrogen excretion from urine to faeces is an additional interesting environmental factor (reduced odour) (Rideout *et al.*, 2004; Russell, 1995). This is linked to colonic fermentation. Increased bacterial biomass fixes nitrogen as bacterial proteins and reduced pH makes ammonia less soluble. Both factors increase faecal nitrogen excretion. Faecal nitrogen is released more slowly and can be absorbed by soil bacteria. Urinary nitrogen is liberated rapidly and causes additional odour problems.

There are some interesting applications for inulin or oligofructose in laying hens. It has been observed that in older laying hens whose productivity is declining, the egg production rate can be increased significantly. The improvement of mineral absorption results in stronger egg shells, resulting in less broken eggs during processing. The effect on lipid metabolism results in over 20 % reduced cholesterol content in the egg yolk (Chen and Chen, 2005a,b).

In adult rabbits, caecal fermentation is important. If there is reduced SCFA producing saccharolytic fermentation, caecal pH has a tendency to increase, giving an opportunity for *Clostridium* to develop, with increased mortality as a consequence. Inulin was shown to arrive in the caecum efficiently, where it can maintain the saccharolytic fermentation (Maertens *et al.*, 2004), thus presumably reducing the risk for lethal clostridiosis.

Nutritional research into fish farming focuses on the intestinal flora and on immune stimulation, again with the ban on the use of antibiotics in the overcrowded cultivation vessels as a goal. There are also applications in horses, where reduced incidence of colics can increase the availability of the animal (Mahious *et al.*, 2005).

In pets, the issue of feeding inulin or oligofructose is different. As is the case in humans, it is desirable for pets to have a long and healthy life. To date, however, there has been little research into the prevention of chronic disease in dogs and cats. There have been a few studies on prevention of diabetes in dogs (Diez, 1997a; Diez *et al.*, 1997, 1998; Jeusette *et al.*, 2004) or prevention of small intestinal bacterial overgrowth (SIBO) (Willard *et al.*, 1994), and the shift of the nitrogen flow from urine to faeces (Younes *et al.*, 1995) might help in reducing the risk for kidney disease in cats. Another important issue in pets is consistency of stools. Companion animals live in houses. In cities, dog owners are supposed to collect dog waste. Improved stool consistency and reduced odour as a result of feeding inulin may be relevant here (Hesta, 2001; Hesta *et al.*, 2003a,b; Propst *et al.*, 2003; Swanson *et al.*, 2002).

In animal nutrition, knowledge of the differences in intestinal architecture of the various animal species is important. The gut is characterised by the length and volume of the different compartments, transit times, and bacterial overgrowth. Some animals have very short transit times (chickens) and low bacterial overgrowth in the upper intestinal tract. The use of oligofructose thus may be advisable in poultry or laying hens. In most livestock and domestic pet animals, however, there is considerable bacterial overgrowth in the upper intestinal tract. Also, it must be taken into consideration that prolonged selective stimulation of bacterial fermentation may be more important than rapidly changing the composition of the intestinal flora.

3.19 Physicochemical and Technological Properties

Inulin is present in many plants (Carpita, 1989), and also in many fruits and roots that humans have been eating for millions of years. Evolutionary development of the digestive system is orders of magnitude slower than the very ‘sudden’ change in dietary habits that started to take place a few thousand years ago when people settled in agricultural units. As such, bringing inulin back to the human diet seems to be plausible. People however do not want to compromise taste (fat rich) and palatable nutrition. Therefore it is important to investigate whether inulin, HP-inulin, oligofructose and Synergy1 can be applied in food without jeopardising the so much appreciated organoleptic properties of the diet.

Native chicory inulin, HP-inulin, Synergy1 and oligofructose are white, odourless powders. Oligofructose is available also as a viscous syrup (75 %, w/w). All products have a high level of purity and have a well-known chemical composition. Their physicochemical and technological properties are summarised in Table 3.3.

Native inulin has a bland neutral taste, without any off-flavour or aftertaste. Because it contains fructose, glucose and sucrose, native inulin is slightly sweet [10 % sweetness (w/w) in comparison with sugar], whereas HP-inulin, which contains no sugars with DP<10, is not sweet. Solubility in water is moderate [maximum 10 % (w/v) at room temperature for native inulin] and viscosity is rather low [less than 2 mPa·s for a 5 % (w/w) solution in water]. When thoroughly mixed with water or another aqueous liquid, submicrometre crystalline inulin (especially HP-inulin) particles form a three-dimensional gel network resulting in a white creamy structure with a short spreadable texture, a smooth fatty mouth feel, as well as a glossy aspect and a well-balanced flavour release. Such a gel can be incorporated easily into foods to replace up to 100 % of the fat (Franck, 1993). Inulin works in synergy with most gelling agents. Inulin improves the stability of foams and emulsions, such as aerated desserts, ice creams, table spreads and sauces (Franck and Coussement, 1997). Oligofructose is much more soluble than inulin [up to 85 % (w/w) in water at room temperature] and has a sweetness

Table 3.3 *Physicochemical and technological properties of chicory inulin and oligofructose and derivatives in powder form*

	Inulin	Oligofructose	HP-inulin	Synergy1
Chemistry	GF_nDP 2–65	GF_n and F_mDP 2–7	GF_nDP 10–65	GF_n and F_nDP 2–7 and 10–65 (1:1)
DP_{av}	12	4	25	4 and 25 (two peaks)
Oligosaccharide content[% (w/w) dry matter]	92	95	99.5	90–94
Dry matter (%, w/w)	95	95	95	95
Sugars[% (w/w) dry matter]	8	5	< 0.5	6–10
pH(10 % in H_2O)	5–7	5-7	5-7	5–7
Relative fermentation times[a/7]	1.2	1	3	Diauxic fermentation 1 + 5

[a/7] Relative value of the time required to ferment 1 % initial product concentration by 5 % faecal slurry as inoculum at 37 °C and controlled pH 6.7. Fermentation time of oligofructose is set at 1.0.

of about 35 % in comparison (w/w) with sucrose. Its sweetening profile closely approaches that of sugar, with a very clean taste without any lingering effect and it enhances fruit flavours. Oligofructose combines with intense sweeteners such as aspartame and acesulpham K, providing mixtures with a rounder mouth feel and improved sustained flavour with reduced aftertaste, as well as stability. Oligofructose contributes towards improved mouth feel, shows humectant properties, reduces water activity ensuring high microbiological stability, and affects boiling point and freezing point (Crittenden, 1996)

3.20 Food Applications

Inulin and oligofructose can be used for either their nutritional advantages or technological properties, but they are often applied to offer a dual benefit: an improved organoleptic quality and a better-balanced nutritional composition. Table 3.4 gives an overview of their applications in foods and drinks.

The use of inulin or oligofructose in bakery products and breakfast cereals often leads to improved taste and texture, and gives more crispiness (Coussement, 1998; Franck and Coussement, 1997). These ingredients keep breads and cakes moist and fresh for longer. Their solubility allows incorporation into watery systems such as drinks, dairy products and table spreads (Coussement, 1996; Coussement and Franck, 1998). Oligofructose is often included in low-calorie dairy products, frozen desserts and meal replacers.

Because of its specific gelling characteristics, inulin is used to develop low-fat foods without compromising taste or texture; in table spreads, it allows the replacement of significant amounts of fat and the stabilisation of the emulsion, while providing a short spreadable texture. Inulin can be used as a fat replacer in frozen desserts, providing an easy processing, a fatty mouth feel, excellent melting properties, as well as freeze–thaw stability, without any unwanted off-flavour. Fat replacement can further be applied in meal replacer, meat products, sauces and soups, e.g. to produce sausages and pâtés with a creamier and juicier mouth feel and improved stability due to better water immobilisation

Table 3.4 *Typical examples of food technology applications of inulin and inulin derivatives. (Adapted from Franck and Coussement, 1997)*

Food products	Applications
Dairy products	Body and mouth feel Sugar and fat replacement Synergy with sweeteners Foam stability
Frozen desserts	Melting and texture Sugar and fat replacement Synergy with sweeteners
Table spreads	Emulsion stability Fat replacement Spreadability and textue
Baked goods and breads	Retention of moisture Sugar replacement
Breakfast cereals	Crispness and expansion
Fruit preparations	Body and mouth feel Sugar replacement Synergy with sweeteners
Meat products	Texture and stability Fat replacement
Chocolate	Heat resistance Sugar replacement

(Coussement, 1996; Franck, 1995, 1997a). The incorporation of inulin (1–3 %) in fruit yoghurts improves the mouth feel, and offers a synergistic taste effect in combination with aspartame and/or acesulfam K (Franck, 1997b).

Inulin has found an interesting application as a low-calorie bulk ingredient in chocolate without added sugar, often in combination with a polyol, and is used as a fibre or sugar replacer in tablets (De Soete, 1994).

Inulin has thus become a key ingredient for the food industry, offering new opportunities for the development of well-balanced and yet better-tasting products (Franck, 2002).

3.21 Summary and Conclusion

β(2-1) linked fructans or inulin-type fructans are nondigestible carbohydrates that are widespread in nature, and are also present in several food plants. As such they have always been part of the human diet.

The nondigestibility of the β-glucosidic bonds is a particularity in mammal digestive system. It is of the same type as the clear 'preference' of anabolic and catabolic enzymes for L-sugars or for D-amino acids.

In terms of fermentation, inulin-type fructans are a selective substrate for some groups of bacteria. From a taxonomic point of view this preferential use of a certain source of carbohydrates is a well-known and even characteristic property of bacteria. In the interaction with mammal GI bacterial ecosystems, however, it is observed that inulin-type fructans stimulate groups of bacteria that seem to contribute directly (interaction with immune system, impact on bowel habit) and indirectly (various systemic effects attributed to the impact of inulin on bacterial metabolism and metabolite production) towards improved health. This again may be related to co-evolution of intestinal ecosystem and the host in an environment of inulin-rich foods.

Chicory roots are a known crop in the north of France, Belgium and southern regions of The Netherlands, where it is grown and roasted as a coffee substitute. The roots contain high amounts of inulin and their shape is similar to the sugar beet root. It therefore is a preferred source of inulin in countries with a tradition of sugar beet processing.

It has been observed that the inulin which is extracted from chicory with a chain length of 3 to 65 fructose moieties ($DP_{av} = 10$), contains two physicochemically distinct fractions. Short chain oligofructose ($DP_{av} = 4$) which is highly soluble (up to 80 % in water) and which is rapidly fermented and and long chain HP-inulin ($DP_{av} = 25$) which is poorly soluble (up to 5 % in water) and which is slowly fermented.

It was observed in experimental models and confirmed in human dietary intervention trials, that dietary inulin-type fructans have various nutritional properties. The different fractions and their combination modify composition of the intestinal flora (oligofructose = Synergy1>ST-inulin>HP-inulin). The modification typically increases *Bifidobacterium* and *Lactobacillus* and butyrate producers of the *Cl. coccoides-Eu ractale* group, leaves commensals as *Bacteroides* and *Fusarium* unaffected and suppresses pathogens such as *Cl. perfringens* group or *Cl. histolyticum/lituburense*. This has a direct impact on improved bowel habit (stool bulking, stool frequency, reduced infectious and inflammatory stress).

Most likely as a consequence of the modified colonic fermentative environment (composition and metabolic activity) several other physiological functions are improved. There is an increased absorption of minerals from the diet, which moreover are used to strengthen bone structure (experimental models and confirmed in human volunteers) (Synergy1>HP-inulin>ST-inulin = oligofructose). There is a modulation of the immune system which results in improved resistance to infection (models and confirmed in human volunteers) (Synergy1>HP-inulin>ST-inulin = oligofructose). There is interaction with the gut endocrine system (experimental models) (oligofructose>HP-inulin), whose complex regulation also might be at the basis of effects observed on lipid metabolism (experimental models but not unequivocally confirmed in human volunteers) (no data with Synergy1, HP-inulin>ST-inulin = oligofructose). Dietary intake of the β(2-1) fructans moreover has the most interesting potential to reduce the risk of intestinal cancer (experimental models confirmed in human volunteers) and various types of systemic cancer (experimental models) (Synergy1>HP-inulin>ST-inulin = oligofructose). Initiation of carcinogenesis but also promotion of tumour cells, and even their metastatic potential are suppressed.

In vitro, long chains have been shown to be slower fermented, and there are indirect indications of beneficial effects in more distal parts of the intestine. Interestingly, whereas they do not change the composition of the intestinal flora to the same extent as their short chain counterparts, HP-inulin seems to be physiologically more active (mineral absorption, anticancer and lipid metabolism models). As HP-inulin eventually also is completely fermented it seems that the impact on bacterial metabolism may be at least as important as modifying the composition of the intestinal flora. Results in models and human volunteers with the combination product of oligofructose and HP-inulin (Synergy1) seem to support this hypothesis: maintaining metabolic activity (long chain) and an improved bacterial ecosystem (short chain) has a more pronounced impact than of its compounds (effect of flora composition-only by oligofructose or effect on metabolism-only by HP-inulin) on physiological processes such as mineral absoption, anticancer properties and lipid metabolism.

In animal nutrition, ST-inulin but also oligofructose, is able to improve zootechnical performance of livestock (pigs, broilers and laying hens, calves). The impact of the prebiotics seems to be maximal in the young animal (post-weaning). Continuous addition of a relatively low dose of inulin or oligofructose may protect the animal against intestinal infection as improved flora composition suppresses the development of pathogens. This might be an important asset in livestock production, especially when antibiotics will be banned from the diets as a growth promotor.

In pets, the use of inulin could contribute towards the healthy condition of the animal, and also a less odorous stool.

Extensive nutritional research with the chicory fructans has revealed that the different fractions (oligofructose, HP-inulin) influence the same physiological processes (mineral absorption, anticancer, lipid metabolism) but at different levels. The second generation prebiotic Synergy1 makes optimal use of these observations. This observation only can be caused by, and hence explained by colonic fermentation properties of these food ingredients. Fermentation properties can be grouped into two aspects: impact on composition of the bacterial flora and impact on bacterial metabolism. Progress has been made in identifying and quantifying colonic bacteria. The newly developed techniques led to the discovery of many hitherto unrecognised genera. Research on the metabolic activity and metabolite production by the colonic bacterial ecosystem however seems to be as important as change in composition of that flora (e.g. effects of HP-inulin interaction with metabolism versus oligofructose interaction with composition of flora, and additive effects of the two as observed with Synergy1).

Future research in the field of prebiotic Synergy1, HP-inulin, ST-inulin and oligofructose should focus on larger-scale long-term nutritional studies, in order to justify the use of health claims. These studies however should be paralleled, or supported, by more intensive investigations into the composition of the colonic flora but perhaps more importantly into metabolite production and mutual interactions.

References

Abrams, S. A., Griffin, I., Hawthorne, K. M., Liang, L., Gunn, S. K., Darlington, G., Ellis, K. 2005a. An inulin-type fructan enhances calcium absorption and bone mineral accretion in adolescents. *Am. J. Clin. Nutr.*, **82**, 471–476.

Abrams, S. A., Griffin, I. J., Hawthorne, K. M., Chen, Z., Gunn, S. K., Wilde, M., Darlington, G., Shypailo, R., Ellis, K. 2005b. Vitamin D receptor FOK1 polymorphisms affect calcium absorption, kinetics and bone mineralization rates during puberty. *Am. J. Clin. Nutr.*, in press.

Alamowitch, C. 1993. Effects of products of dietary fibre fermentation, short-chain fatty acids on sugar and lipid metabolism *J. Annu. Diabetol.*, 245–259.

Albers, R., Antoine, J., Bourdet-Sicard, R., Calder, P., Gleeson, M., Lesourd, B., Samartin, S., Sandersen, I., Van Loo, J., Vas Dias, F., Watzl, B. 2005. Markers to measure immunomodulation in human nutrition intervention studies. *Br. J. Nutr.*, **93**, 1–31.

Alles, M. S., De Roos, N. M., Backx, J. C., van de Lisdonk, E., Hautvast, J. 1997. Consumption of fructo-oligosaccharides does not affect blood glucose and serum lipids in patients with type 2 diabetes, *Am. J. Clin. Nutr.*, vol. **69**, 64-69.

Andersen, R., Sorensen, A. 1999. An enzymatic method for the determination of fructans in foods and food products. Comparison of the results by HPAEC-PAD. *Eur. J. Food Res. Technol.*, **210**, 148–152.

Andrieux, C. 1991. Physiological effects of inulin in germ free rats and in heteroxenic rats inoculated with a human flora. *Food Hydrocolloids*, **5**, 49–56.

Baba, S. O. 1996. Fructooligosaccharides stimulate the absorption of magnesium from the hindgut in rats, *Nutr. Res.*, **16**, 657–666.

Bach Knudsen, K. E., Hessov, I. 1995. Recovery of inulin from Jerusalem artichoke (*Helianthus tuberosus* L.) in the small intestine of man. *Br. J. Nutr.*, **74**, 101–113.

Baeten, J. 1999. Invloed van inuline-type fructanen op de darmflora: simulaties door middel van *in vitro* batchfermentaties met fecale slurry. (Influence of inulin-type fructans on intestinal flora: simulations by means of *in vitro* batch fermentation with faecal slurry. Masters thesis. Vrije Universiteit Brussel, Belgium.

Barrangou, R., Altermann, E., Hutkins, R., Cano, R., Klaenhammer, T. R. 2003. Functional and comparative genomic analyses of an operon involved in fructooligosaccharide utilization by *Lactobacillus acidophilus. Proc Natl Acad Sci USA*, **100**, 8957–8962.

Bartosch, S., Woodmansey, E. J., Paterson, J. C. M., McMurdo, M. T., MacFarlane, G. T. 2005. Microbiological effects of consuming a synbiotic containing *Bifidobacterium bifidum, Bifidobacterium lactis*, and oligofructose in elderly persons, determined by real-time PCR and counting of viable bacteria. *Clin. Infect. Dis.*, **40**, 28–37.

Beynen, A. C., Baas, J. C., Hoekemeijer, P. E., Kappert, H. J., Bakker, M. H., Koopman, J. P., Lemmens, A. G. 2002. Faecal bacterial profile, nitrogen excretion and mineral absorption in healthy dogs fed supplemental oligofructose. *J. Animal Physiol. Animal Nutr.*, **86**, 298–305.

Bielecka, M. 2002. Effect of non-digestible oligosaccharides on gut microecosystem in rats. *Food Res. Int.*, **35**, 139–144.

Björck, I. 1991. On the possibility of acid hydrolysis of inulin in the rat stomach. *Food Chem.*, **41**, 243–250.

Blaut, M., Collins, M. D., Welling, G. W., Doré, J., Van Loo, J., de Vos, W. 2002. Molecular biological methods for studying the gut microbiota: the EU human gut flora project. *Br. J. Nutr.*, **87**, 203–211.

Bouhnik, Y., Flourie, B., Andrieux, C., Bisetti, N., Briet, F., Rambaud, J. C. 1996. Effects of *Bifidobacterium* sp. fermented milk ingested with or without inulin on colonic bifidobacteria and enzymatic activities in healthy humans. *Eur. J. Clin. Nutr.*, **50**, 269–273.

Bouhnik, Y., Vahedi, K., Achour, L., Attar, A., Salfati, J., Pochart, P., Marteau, P., Flourie, B., Bornet, F., Rambaud, J. C. 1999. Short-chain fructo-oligosaccharide administration dose-dependently increases fecal bifidobacteria in healthy humans. *J. Nutr.*, **129**, 113–116.

Briet, F., Achour, L., Flourie, B., Beaugerie, L., Pellier, P., Franchisseur, C., Bornet, F., Rambaud, J. C. 1995. Symptomatic response to varying levels of fructo-oligosaccharides consumed occasionally or regularly. *Eur. J. Clin. Nutr.*, **49**, 501–507.

Brighenti, F., Casiraghi, M. C., Canzi, E., Ferrari, A. 1999. Effect of consumption of a ready-to-eat breakfast cereal containing inulin on the intestinal milieu and blood lipids in healthy male volunteers. *Eur. J. Clin. Nutr.*, **53**, 726–733.

Buecher, B., Thouminot, C., Menanteau, J., Bonnet, C., Jarry, A., Heymann, M.-F., Cherbut, C., Galmiche, J.-P., Blottiere, H. 2003. Fructooligosaccharide associated with celecoxib reduces the number of aberrant crypt foci in the colon of rats. *Reprod. Nutr. Dev.*, vol. **43**, 347–356.

Buddington, R. K. 1996. Dietary supplement of neosugar alters the fecal flora and decreases activities of some reductive enzymes in human subjects. *Am. J. Clin. Nutr.*, **63**, 709–716.

Buddington, R. K. 1999. Influence of fermentable fibre on small intestinal dimensions and transport of glucose and proline in dogs. *Am. J. Vet. Res.*, **60**, pp. 354–358.

Buddington, R. K., Buddington, K. K., Sunvold, G. D. 1999. Influence of fermentable fiber on small intestinal dimensions and transport of glucose and proline in dogs. *Am. J. Vet. Res.*, **60**, 354–358.

Buddington, R. K., Williams, C. H., Chen, S. C., Witherly, S. A. 1996. Dietary supplement of neosugar alters the fecal flora and decreases activities of some reductive enzymes in human subjects. *Am. J. Clin. Nutr.*, **63**, 709–716.

Burns, A., Rowland, I. R. 2004. Antigenotoxicity of probiotics and prebiotics on fecal water-induced DNA damage in human colon adenocarcinoma cells. *Mutat. Res.*, **551**, 233–243.

Caderni, G., Femia, A. P., Giannini, A., Favuzza, A., Luceri, C., Salvadori, M., Dolara, P. 2003. Identification of mucin-depleted foci in the unsectioned colon of azoxymethane-treated rats: correlation with carcinogenesis. *Cancer Res.*, **63**, 2388–2392.

Campbell, J. M., Fahey Jr, G. C., Wolf, B. W. 1997. Selected indigestible oligosaccharides affect large bowel mass, cecal and fecal short-chain fatty acids, pH and microflora in rats, *J. Nutr.*, **127**, 130–136.

Cani, P. D., Dewever, C., Delzenne, N. M. 2004. Inulin-type fructans modulate gastrointestinal peptides involved in appetite regulation—Glucagon-like Peptide-1 and Ghrelin—in rats. *Br. J. Nutr.*, **92**, 521–526.

Carabin, I. G., Flamm, W. G. 1999. Evaluation of safety of inulin and oligofructose as dietary fiber. *Regul. Toxicol. Pharmacol.*, **30**, 268–282.

Carpita, N. C., Kanabus, J., Housley, T. L. 1989. Linkage structure of fructans and fructan oligomers from *Triticum aestivum* and *Festuca arundacea* leaves. *J. Plant Physiol.*, **134**, 162–168.

Cashman, K. 2003. Prebiotics and calcium bioavailability. *Curr. Issues Intest. Microbiol.*, **4**, 21–32.

Castiglia-Delavaud, C., Verdier, E., Besle, J. M., Vernet, J., Boirie, Y., Beaufrere, B., De Baynast, R., Vermorel, M. 1998. Net energy value of non-starch polysaccharide isolates (sugarbeet fibre and commercial inulin) and their impact on nutrient digestive utilization in healthy human subjects. *Br. J. Nutr.*, **80**, 343–352.

Chambers, J. R., Spencer, J. L., Modler, H. W. 1997. The influence of complex carbohydrates on *Salmonella typhimurium* colonization, pH, and density of broiler ceca. *Poult. Sci.*, **76**, 445–451.

Chen, Y., Chen, T. 2005a. Effects of chicory fructans on egg cholesterol in commercial layers. *Int. J. Poultry Sci.*, **4**, 109–114.

Chen, Y., Chen, T. 2005b. Improvement of layer performance by dietary prebiotic chicory oligofructose and inulin. *Int. J. Poultry Sci.*, **4**, 103–108.

Cherbut, C. 2002. Inulin and oligofructose in the dietary fibre concept. *Br. J. Nutr.*, **87**, S159–S162.

Cherbut, C., Ferrier, L., Roze, C., Anini, Y., Blottiere, H., Lecannu, G., Galmiche, J. P. 1998. Short-chain fatty acids modify colonic motility through nerves and polypeptide YY release in the rat. *Am. J. Physiol*, **275**, G1415–G1422.

Cherbut, C., Salvador, V., Barry, J.-L., Doulay, F., Delort-Laval, J. 2003. Dietary fibre effects on intestinal transit in man: involvement of their physicochemical and fermentative properties. *Food Hydrocolloids*, **1**, 15–22.

Clune, Y., Klinder, A., May, M., Roller, M., Karlsson, P., Van Loo, J., Rechkemmer, G., Rafter, J., Watzl, B., Caderni, G., Rowland, I., Pool-Zobel, B., Collins, K. 2004. Synbiotic consumption modulates cancer risk biomarkers in colon cancer and polypectomised human subjects. *Proc. 3rd Frontiers Cancer Prev. AACR*, 70–71.

Coudray, C. 1997. Effect of soluble or partly soluble dietary fibres supplementation on absorption and balance of calcium, magnesium, iron and zinc in healthy young men. *Eur J. Clin. Nutr.*, **51**, 375–380.

Coudray, C., Demigne, C., Rayssiguier, Y. 2003a. Effects of dietary fibers on magnesium absorption in animals and humans. *J. Nutr.*, **133**, 1–4.

Coudray, C., Tressol, J. C., Gueux, E., Rayssiguier, Y. 2003b. Effects of inulin-type fructans of different chain length and type of branching on intestinal absorption and balance of calcium and magnesium in rats. *Eur. J. Nutr.*, **42**, 91–98.

Coussement, P. 1996. Neue Chancen fuer die Suesswarenindustrie. *Zucker- und Suesswarenwirtsch.*, **49**, 43–44.

Coussement, P. 1998. The use of inulin and oligofructose in breakfast cereals. *China Food Ind.*, **5**, 16–18.

Coussement, P., Franck, A. 1998, New food applications for inulin. *Agro Food Ind. Hi-Tech*, **9**, 26–28.

Coxam, V. 2005. Inulin-type fructans and bone health: state of the art and perspectives in the management of osteoporosis. *Br. J. Nutr.*, **93** (Suppl. 1), S111–S123.

Crittenden, R. G. 1996. Production, properties and applications of food-grade oligosaccharides. *Trends Food Sci. Technol.*, **7**, 353–361.

Cummings, J. H. 1997. *The Large Intestine in Health and Disease*. Danone Chair Monograph, Danone, Paris.

Dal Bello, F., Walter, J., Hertel, C., Hammes, W. P. 2001. *In vitro* study of prebiotic properties of levan-type exopolysaccharides from lactobacilli and non-digestible carbohydrates using denaturing gradient gel electrophoresis, *System. Appl. Microbiol.*, **24**, 232–237.

Daubioul, C., Rousseau, N., Demeure, R., Gallez, B., Taper, H., Declerck, B., Delzenne, N. 2002. Dietary fructans, but not cellulose, decrease triglyceride accumulation in the liver of obese Zucker fa/fa rats. *J. Nutr.*, **132**, 967–973.

Davidson, M. H. 1998. Evaluation of the influence of dietary inulin on serum lipids in adults with hypercholesterolemia. *Nutr. Res.*, **18**, 503–517.

Delzenne, N. M. 1993. Dietary fructooligosaccharides modify lipid metabolism in rats. *Am. J. Clin. Nutr.*, **57**, 820S.

Delzenne, N. 1995. Effect of fermentable fructo-oligosaccharides on energy and nutrients absorption in the rat. *Life Sci.*, **57**, 1579–1587.

Delzenne, N. 2003. Fructo-oligosaccharides: state of the art. *Proc. Nutr. Soc.*, **62**, 177–182.

Delzenne, N., M. Kok, N. 2001. Effects of fructans-type prebiotics on lipid metabolism. *Am. J. Clin. Nutr.*, **73** (Suppl.), 458S.

Delzenne, N., Aertssens, J., Verplaetse, H., Roccaro, M., Roberfroid, M. 1995. Effect of fermentable fructo-oligosaccharides on mineral, nitrogen and energy digestive balance in the rat. *Life Sci.*, **57**, 1579–1587.

Delzenne, N. M., Kok, N., Deloyer, P., Dandrifosse, G. 2000. Dietary fructans modulate polyamine concentration in the cecum of rats. *J. Nutr.*, **130**, 2456–2460.

Den Hond, E., Geypens, B., Ghoos, Y. 1999. Effect of high performance chicory inulin on constipation. *Nutr. Res.*, **20**, 731–736.

De Soete, J. 1994. Inulin in schokolade—neue wege der kalorienreduzierung. (Inulin in chocolate—new methods of fat reduction.) *Suesswaren*, 16–17.

Diez, M. 1997a. Dietary fibre in the dog: influence of diets containing sugarbeet or chicory pulps on nutrient digestibility and on plasma metabolites. *Rev. Med. Vet.*, **148**, 991–998.

Diez, M. 1997b. Influence of a blend of fructo-oligosaccharides and sugar beet fiber on nutrient digestibility and plasma metabolite concentrations in healthy Beagles. *Am. J. Vet. Res.*, **58**, 1238–1242.

Diez, M., Hornick, J., Baldwin, P., Istasse, L. 1997. Influence of a blend of fructo - oligosaccharides and sugar beet fiber on nutrient digestibility and plasma metabolite. *Am. J. Vet. Res.*, **58**, 1238–1242.

Diez, M., Hornick, J. L., Baldwin, P., Van Eenaeme, C., Istasse, L. 1998. The influence of sugar-beet fibre, guar gum and inulin on nutrient digestibility, water consumption and plasma metabolites in healthy Beagle dogs. *Res. Vet. Sci.*, **64**, 91–96.

Djouzi, Z., Andrieux, c. 1997. Compared effects of three oligosaccharides on metabolism of intestinal microflora in rats inoculated with a human faecal flora. *Br. J. Nutr.*, **78**, 313–324.

Drozdowski, L. A., Dixon, W. T., McBurney, M. I., Thomson, A. B. 2002. Short-chain fatty acids and total parenteral nutrition affect intestinal gene expression. *J. Parent. Ent. Nutr.*, **26**, 145–150.

Druce, M. R., Small, C. J., Bloom, S. R. 2004. Peptides regulating satiety. *Endocrinology,* **145**, 2660–2665.

Drucker, G. 2002. The growing years and prevention of osteoporosis in later life. *Gastroenterology*, **122**, 531–544.

Duncan, S., Holtrop, G., Lobley, G. E., Calder, A. G., Stewart, C., Flint, H. 2004. Contribution of acetate to butyrate formation by human faecal bacteria. *Br. J. Nutr.*, **91**, 915–923.

Durieux, A. 2001. Metabolism of chicory fructooligosaccharides by bifidobacteria. *Biotechnol. Lett.*, **23**, 1523–1527.

Ellegård, L., Andersson, H., Bosaeus, I. 1997. Inulin and oligofructose do not influence the absorption of cholesterol, or the excretion of cholesterol, Ca, Mg, Zn, Fe, or bile acids but increases energy excretion in ileostomy subjects. *Eur. J. Clin. Nutr.*, **51**, 1–5.

Femia, A., Dolara, P., Caderni, G. 2004. Mucin-depleted foci (MDF) in the colon of rars treated with azoxymethane (AOM) are useful biomarkers for colon carcinogenesis. *Carcinogenesis*, **25**, 277–281.

Femia, A. P., Luceri, C., Dolara, P., Giannini, A., Biggeri, A., Salvadori, M., Clune, Y., Collins, K. J., Paglierani, M., Caderni, G. 2002. Antitumorigenic activity of the prebiotic inulin enriched with oligofructose in combination with the probiotics *Lactobacillus rhamnosus* and *Bifidobacterium lactis* on azoxymethane-induced colon carcinogenesis in rats. *Carcinogenesis*, **23**, 1953–1960.

Fiordaliso, M., Kok, N., Desager, J. P., Goethals, F., Deboyser, D., Roberfroid, M., Delzenne, N. 1995. Dietary oligofructose lowers triglycerides, phospholipids and cholesterol in serum and very low density lipoproteins of rats. *Lipids*, **30**, 163–167.

Firmansyah, A., Pramita, G., Fassler, C., Haschke, F., Link-Amster, H. 2001. Improved humoral immune response to measles vaccine in infants receiving infant cereal with fructooligosaccharides. *J. Paediatr. Gastroenterol. Nutr.*, **31**, A521.

Flickinger, E. A., Van Loo, J., Fahey Jr, G. C. 2003. Nutritional responses to the presence of inulin and oligofructose in the diets of domesticated animals: a review. *Crit Rev. Food Sci. Nutr.*, **43**, 19–60.

Fontaine, N., Meslin, J. C., Lory, S., Andrieux, C. 1996. Intestinal mucin distribution in the germ-free rat and in the heteroxenic rat harbouring a human bacterial flora: effect of inulin in the diet. *Br. J. Nutr.*, **75**, 881–892.

Franck, A. 1993. Rafticreming: The new process allowing to turn fat into dietary fibre. *FIE Conf. Proc. 1992*, 193–197.

Franck, A. 1995. Raftiline: more than only a fat replacer. (Raftiline—Mehr als nur Fettersatz.) *Lebensmitt. Milchwirtschaft*, **21**, 1038–1042.

Franck, A. 1997a. Inulin improves spreadable food products. (Inulin, verbessert streichfähige Lebensmittel-Produkte.) *Food Technol. Mag.*, **4**, 4–6.

Franck, A. 1997b. Oligofructose—an ingredient without equal in fruit preparations and milk products. *Lebensmitt. Milchwirtschaft*, **118**, 414–416.

Franck, A. 2002. Technological functionality of inulin and oligofructose. *Br. J. Nutr.*, **87** (Suppl. 2), S287–S291.

Franck, A., Coussement, P. A. A. 1997. Multifunctional inulin. *Food Ingred. Anal. Int.*, **October**, 8–10.

Franks, A. H., Harmsen, H. J. M., Raangs, G. C., Jansen, G. J., Schut, F., Welling, G. W. 1998. Variations of bacterial populations in human feces measured by fluorescent *in situ* hybridization with group-specific 16S rRNA-targeted oligonucleotide probes. *Appl. Environ. Microbiol.*, **64**, 3336–3345.

Furrie, E., Macfarlane, S., Kennedy, A., Cummings, J. H., Walsh, S. V., O'Neil, D. A., Macfarlane, G. T. 2005. Synbiotic therapy (*Bifidobacterium longum* / Synergy1) initiates resolution of inflammation in patients with active ulcerative colitis: a randomised controlled pilot trial. *Gut*, **54**, 242–249.

Gibson, G. R. 1994. Bifidogenic properties of different types of fructose containing oligosaccharides. *Food Microbiol.*, **11**, 491–498.

Gibson, G. R., Roberfroid, M. B. 1995. Dietary modulation of the human colonic microbiota: introducing the concept of prebiotics. *J. Nutr.*, **125**, 1401–1412.

Gibson, G. R., Wang, X. 1994. Enrichment of bifidobacteria from human gut contents by oligofructose using continuous culture. *FEMS Microbiol. Lett.*, **118**, 121–127.

Gibson, G. R., Beatty, E. R., Wang, X., Cummings, J. H. 1995. Selective stimulation of bifidobacteria in the human colon by oligofructose and inulin. *Gastroenterology*, **108**, 975–982.

Gibson, G. R., Probert, H. M., Van Loo J., Rastall, R. A., Roberfroid, M. 2004. Dietary modulation of the human colonic microbiota: updating the concept of prebiotics. *Nutr. Res. Rev.*, **17**, 259–275.

Griffin, I. J., Davila, P. M., Abrams, S. A. 2002. Non-digestible oligosaccharides and calcium absorption in girls with adequate calcium intakes. *Br. J. Nutr.*, **87**, S187–S191.

Griffin, I. J., Hicks P. M. D., Heaney R. P., Abrams, S. A. 2003, Enriched chicory inulin increases calcium absorption mainly in girls with lower calcium absorption. *Nutr. Res.*, **23**, 901–909.

Guigoz, Y., Rochat, F., Perruisseau-Carrier, G., Rocat, F., Schiffrin, E. J. 2002. Effects of oligosaccharide on the faecal flora and non-specific immune system in elderly people. *Nutr. Res.*, **22**, 13–25.

Hamberg, O., Rumessen, J. J., Gudmand-Hoyer, E. 1989. Inhibition of starch absorption by dietary fibre. *Scand. J. Gastroenterol.*, **24**, 102–111.

Harmsen, H. J., Raangs, G. C., Franks, A., Wildeboer-Veloo, A. C., Welling, G. W. 2002. The effect of the prebiotic inulin and the probiotic *Bifidobacterium longum* on the fecal microflora of healthy volunteers measured by FISH and DGGE. *Microbial. Ecol. Health Dis.*, **14**, 219.

Hata, Y. 1983. The effect of fructo-oligosaccharides (Neosugar(r)) on lipidemia. *Geriat. Med.*, **21**, 156–167.

Hata, Y. 1985. The relationship between fructooligosaccharides intake and intestinal symptoms: observations on the maximum non-effective dose and 50 % effective dose on diarrhea. *Geriat. Med.*, **23**, 817–828.

Hentges, D. J. 1993. The anaerobic microflora of the human body. *Clin. Infect. Dis. (USA)*, **16**, S175–S180.

Hesta, M. 2001. The effect of oligofructose and inulin on faecal characteristics and nutrient digestibility in healthy cats. *J. Anim. Physiol. Anim. Nutr.*, **85**, 135–141.

Hesta, M., Janssens, G., Debraekeleer, J., Millet, S., De Wilde, R. 2003a. Fecal odor components in dogs: nondigestible oligosaccharides and resistant starch do not decrease fecal H_2S emission. *J. Appl. Res. Vet. Med.*, **1**, 225–232.

Hesta, M., Roosen, W., Janssens, G., Millet, S., De Wilde, R. 2003b. Prebiotics affect nutrient digestibility but not faecal ammonia in dogs fed increased dietary protein levels. *Br. J. Nutr.*, **90**, 1007–1014.

Hidaka, H. 1986. Effects of fructooligosaccharides on intestinal flora and human health. *Bifidobacteria Microflora*, **5**, 37–50.

Hillman, L. S., Tack, E., Covell, D. G., Vieira, N. E., Yergey, A. L. 1988. Measurement of true calcium absorption in premature infants using intravenous 46Ca and oral 44Ca. *Pediatr. Res.*, **23**, 589–594.

Hoebregs, H. 1997. Fructans in foods and food products, ion-exchange chromatographic method: collaborative study. *J. of AOAC Int.*, **80**, 1029–1037.

Hofer, K. 1999. Enzymatic determination of inulin in food and dietary supplements. *Eur. Food Res. Technol.*, **209**, 423–427.

Holloway, L., Moynihan, S., Friedlander, A. L. 2003. Effects of Synergy1 on mineral absorption and markers of bone turnover. Report.

Hopkins, M., Cummings, J., Macfarlane, G. 1998. Inter-species differences in maximum specific growth rates and cell yield of bifidobacteria cultured on oligosaccharides and other simple carbohydrate sources. *J. Appl. Microbiol.*, **85**, 381–386.

Hopkins, M. J. 1998. Inter-species differences in maximum specific growth rates and cell yields of bifidobacteria cultured on oligosaccharides and other simple carbohydrate sources. *J. Appl. Microbiol.*, **85**, 381–386.

Houdijk, J. G. M. 1998. Effects of dietary oligosaccharides on the growth performance and fecal characteristics of young growing pigs. *Anim. Feed Sci. Technol.*, **71**, 35–48.

Hughes, R., Rowland, I. R. 2001. Stimulation of apoptosis by two prebiotic chicory fructans in the rat colon. *Carcinogenesis*, **22**, 43–47.

Jackson, K. G., Taylor, G. R. J., Clohessy, A. M., Williams, C. M. 1999. The effect of the daily intake of inulin on fasting lipid, insulin and glucose concentrations in middle-aged men and women. *Br. J. Nutr.*, **82**, 23–30.

Jeusette, I., Grauwels, M., Cuvelier, C., Tonglet, C., Istasse, L., Diez, M. 2004. Hypercholesterolemia in a family of rough collie dogs. *J. Small Anim. Pract.*, **45**, 319–324.

Joye, D., Hoebregs, H., Joye, D., Hoebregs, H. 2000. Determination of oligofructose, a soluble dietary fiber, by high-temperature capillary gas chromatography. *J. AOAC Int.*, **83**, 1020–1025.

Juffrie, M. 2002. Fructooligosaccharide and diarrhea. *Biosci. Microflora*, **21**, 31–34.

Kaplan, H., Hutkins, R. W. 2000. Fermentation of fructooligosaccharides by lactic acid bacteria and bifidobacteria. *Appl. Environ. Microbiol.*, **66**, 2682–2684.

Karppinen, S., Liukkonen, K., Aura, A., Forssell, P., Poutanen, K. 2000. In vitro fermentation of polysaccharides of rye, wheat and oat brans and inulin by human fecal bacteria. *J. Sci. Food Agric.*, **80**, 1469–1476.

Kelly-Quagliana, K. A., Nelson, P. D., Buddington, R. K. 2003. Dietary oligofructose and inulin modulate immune functions in mice. *Nutr. Res.*, **23**, 257–267.

Kleessen, B., Hartmann, L., Blaut, M. 2001. Oligofructose and long-chain inulin: influence on the gut microbial ecology of rats associated with a human faecal flora. *Br. J. Nutr.*, **86**, 291–300.

Kleessen, B., Hartmann, L., Blaut, M. 2003. Fructans in the diet cause alterations of intestinal mucosal architecture, released mucins and mucosa-associated bifidobacteria in gnotobiotic rats. *Br. J. Nutr.*, **89**, 597–606.

Kleessen, B., Sykura, B., Zunft, H. J., Blaut, M. 1997. Effects of inulin and lactose on fecal microflora, microbial activity, and bowel habit in elderly constipated persons. *Am. J. Clin. Nutr.*, **65**, 1397–1402.

Klinder, A., Foerster, A., Caderni, G., Femia, A., Pool-Zobel, B. 2004a. Fecal water genotoxocity is predictive of tumor preventive activities by inulin like oligofructoses, probiotics (*L. rhamnosus* and *B. lactis*), and their synbiotic combination. *Nutr. Cancer*, **49**, 144–155.

Klinder, A., Gietl, E., Hughes, R., Jonkers, N., Karlsson, P., McGlyn, H., Pistoli, S., Tuohy, K., Rafter, J., Rowland, I., Van Loo, J., Pool-Zobel, B. 2004b. Gut fermentation products of chicory inulin-derived prebiotics inhibit markers of tumour progression in human colon tumour cells. *J. Cancer Prevent.*, **1**, 19–32.

Knol, J. 2002. Bifidobacterial species that are present in breast-fed infant are stimulated in formula fed infants by changing to a formula containing prebiotics. *J. Periatr. Gastroenterol. Nutr.*, **34**, 477.

Kojima, M., Hosoda, H., Date, Y., Nakazato, M., Matsuo, H., Kangawa, K. 1999. Ghrelin is a growth-hormone-releasing acylated peptide from stomach. *Nature*, **402**, 656–660.

Kok, N. 1995. Regulation of lipogenesis and triglyceride secretion in oligofructose-fed rats. *Arch. physiol. biochem.*, **103**, B80.

Kok, N., Roberfroid, M., Delzenne, N. 1996. Dietary oligofructose modifies the impact of fructose on hepatic triacylglycerol metabolism. *Metabolism*, **45**, 1547–1550.

Kok, N. N., Morgan, L. M., Williams, C. M., Roberfroid, M. B., Thissen, J. P., Delzenne, N. M. 1998a. Insulin, glucagon-like peptide 1, glucose-dependent insulinotropic polypeptide and insulin-like growth factor I as putative mediators of the hypolipidemic effect of oligofructose in rats. *J. Nutr.*, **128**, 1099–1103.

Kok, N. N., Taper, H. S., Delzenne, N. M. 1998b. Oligofructose modulates lipid metabolism alterations induced by a fat-rich diet in rats. *J. Appl. Toxicol.*, **18**, 47–53.

Kruse, H. P., Kleessen, B., Blaut, M. 1999. Effects of inulin on faecal bifidobacteria in human subjects. *Br. J. Nutr.*, **82**, 375–382.

Langendijk, P. S., Schut, F., Jansen, G. J., Raangs, G. C., Kamphuis, G. R., Wilkinson, M. H. F., Welling, G. W. 1995. Quantitative fluorescent in situ hybridisation of *Bifidobacterium* with genus-specific 16S rRNA-targeted probes and its application in fecal samples. *Appl. Environ. Microbiol.*, **61**, 3069–3075.

Langlands, S. J., Hopkins, M. J., Coleman, N., Cummings, J. H. 2004. Prebiotic carbohydrates modify the mucosa associated microflora of the human large bowel. *Gut*, **53**, 1610–1616.

Le Blay, G., Michel, C., Blottiere, H. M., Cherbut, C. 1999. Prolonged intake of fructo-oligosaccharides induces a short-term elevation of lactic acid-producing bacteria and a persistent increase in cecal butyrate in rats. *J. Nutr.*, **129**, 2231–2235.

Levrat, M. A., Remesy, C., Demigne, C. 1991. High propionic acid fermentations and mineral accumulation in the cecum of rats adapted to different levels of inulin. *J. Nutr.*, **121**, 1730–1737.

Lippl, F., Kircher, F., Erdmann, J., Allescher, H. D., Schusdziarra, V. 2004. Effect of GIP, GLP-1, insulin and gastrin on ghrelin release in the isolated rat stomach. *Regul. Pept.*, **119**, 93–98.

Lopez, H. W., Coudray, C., Levrat-Verny, M., Feillet-Coudray, C., Demigne, C., Remesy, C. 2000. Fructooligosaccharides enhance mineral apparent absorption and counteract the deleterious effects of phytic acid on mineral homeostasis in rats. *J. Nutr. Biochem.*, **11**, 500–508.

Lukaski, H. C., Nielsen, F. H. 2002. Dietary magnesium depletion affects metabolic responses during submaximal exercise in postmenopausal women. *J. Nutr.*, **132**, 930–935.

Ma, T. Y., Hollander, D., Erickson, R. A., Truong, H., Nguyen, H., Krugliak, P. 1995. Mechanism of colonic permeation of inulin: is rat colon more permeable than small intestine? *Gastroenterology*, **108**, 12–20.

Macfarlane, S., Cummings, J. H., Macfarlane, G. 1999. Bacterial colonisation of surfaces in the large intestine. In *Colonic Microbiota, Nutrition and Health*, Gibson, G. and Roberfroid, M., eds, pp. 71–88, Springer-Verlag, New York.

Maertens, L., Aerts, J., De Boever, J. 2004. Degradation of dietary oligofructose and inulin in the gastro-intestinal tract of the rabbit and the effects on caecal pH and volatile fatty acids. *World Rabbit Sci.*, **12**, 235–246.

Mahious, A. S., Gatesoupe, F. J., Hervé, M., Metailler, R., Ollevier, F. 2005. Effect of dietary inulin and oligosaccharides as prebiotics for weaning turbot *Psetta maxima*. Aquaculture Int., in press.

Marx, S. P., Winkler, S., Hartmeier, W. 2000. Metabolization of beta-(2,6)-linked fructose-oligosaccharides by different bifidobacteria. *FEMS Microbiol. Lett.*, **182**, 163–169.

Massimo, S. P., McBurney, M. I., Field, C. J. 1998. Fermentable dietary fiber increases GLP-1 secretion and improves glucose homeostasis despite increased intestinal glucose transport capacity in healthy dogs. *J. Nutr.*, **128**, 1786–1793.

McCleary, B. V., Rossiter, P. 2004. Measurement of novel dietary fibers. *J. AOAC Int.*, **87**, 707–717.

McCleary, B. V., Murphy, A., Mugford, D. C. 2000. Measurement of total fructan in foods by enzymatic/spectrophotometric method: collaborative study. *J. AOAC Int.*, **83**, 356–364.

Menne, E., Guggenbuhl, N., Roberfroid, M. 2000. Fn-type chicory inulin hydrolysate has a prebiotic effect in humans. *J. Nutr.*, **130**, 1197–1199.

Mikkelsen, L. 2001. Fructooligosaccharides and galactooligosaccharides as prebiotics for piglets at weaning: effect on the intestinal microbial ecosystem with emphasis on the population of bifidobacteria. PhD thesis, Danish Institute of Agricultural Sciences, Research Centre Foulum and The Royal Veterinary and Agricultural University.

Miller, J. Z., Smith, D. L., Flora, L., Peacock, M., Johnston, C. 1989. Calcium absorption in children estimated from single and double stable calcium isotope techniques. *Clin. Chim. Acta*, **183**, 107–114.

Mitsuoka, T., Hidaka, H., Eida, T. 1987. Effect of fructo-oligosaccharides on intestinal microflora. *Nahrung*, **31**, 427–436.

Molis, C., Flourie, B., Ouarne, F., Gailing, M. F., Lartigue, S., Guibert, A., Bornet, F., Galmiche, J. P. 1996. Digestion, excretion, and energy value of fructooligosaccharides in healthy humans. *Am. J. Clin. Nutr.*, **64**, 324–328.

Moro, G. 2001. Dosage effect of oligosaccharides on faecal flora and stool characteristics in term infants. *J. Pediatr. Gastroenterol. Nutr.*, **32**, 401.

Naughton, P. J. 2001. Effects of nondigestible oligosaccharides on *Salmonella enterica* serovar *typhimurium* and nonpathogenic *Escherichia coli* in the pig. *Appl. Environ. Microbiol.*, **67**, 3391–3395.

Nemcova, R. 1999. Study of the effect of *Lactobacillus paracasei* and fructooligosaccharides on the faecal microflora in weanling piglets. *Berl. Münch. Tierärztl. Wschr.*, **112**, 225–228.

Newton, D. F., Cummings, J. H., Macfarlane, S., Macfarlane, G. 1998. Development of reproductive model ecosystems in the chemostat based on defined populations of human intestinal bacteria. *J. Appl. Microbiol.,* **85**, 372–380.

Nilsson, U., Oste, R., Jagerstad, M., Birkhed, D. 1988. Cereal fructans: *in vitro* and *in vivo* studies on availability in rats and humans. *J. Nutr.*, **118**, 1325–1330.

Nyman, M. 2002. Fermentation and bulking capacity of indigestible carbohydrates: the case of inulin and oligofructose. *Br. J. Nutr.*, **87** (Suppl. 2), S163–S168.

Ohta, A., Baba, S., Ohtsuki, M., Takizawa, T., Adachi, T., Hara, H. 1997. *In vivo* absorption of calcium carbonate and magnesium oxide from the large intestine in rats, *J. Nutr. Sci. Vitaminol. (Tokyo)*, **43**, 35–46.

Ohta, A., Baba, S., Takizawa, T., Adachi, T. 1994. Effects of fructooligosaccharides on the absorption of magnesium in the magnesium-deficient rat model, *J. Nutr. Sci. Vitaminol.(Tokyo)*, **40**, 171–180.

Ohta, A., Motohashi, Y., Ohtsuki, M., Hirayama, M., Adachi, T., Sakuma, K. 1998. Dietary fructooligosaccharides change the concentration of calbindin-D9k differently in the mucosa of the small and large intestine of rats. *J. Nutr.*, **128**, 934–939.

Ohta, A., Ohtsuki, M., Baba, S., Adachi, T., Sakata, T., Sakaguchi, E. 1995. Calcium and magnesium absorption from the colon and rectum are increased in rats fed fructooligosaccharides. *J. Nutr.*, **125**, 2417–2424.

Oku, T., Tokunaga, T., Hosoya, N. 1984. Nondigestibility of a new sweetener, Neosugar in the rat. *J. Nutr.*, **114**, 1574–1581.

Pallarés, I., Lisbona, F., Alaga, I. L., Barrionuevo, M., Alfrez, M. J. M., Campos, M. S. 1993. Effect of iron deficiency on digestive utilization of iron, phosphorus, calcium and magnesium in rats. *Br. J. Nutr.*, **70**, 609–620.

Palframan, R., Gibson, G., Rastall, R. A. 2003. Development of a quantitative tool for the comparison of the prebiotic effect of dietary oligosaccharides. *Lett. Appl. Microbiol.*, **37**, 281–284.

Perrin, S. F. 2002. Fermentation of chicory fructo-oligosaccharides in mixtures of different degrees of polymerization by three strains of bifidobacteria. *Can. J. Microbiol.*, **48**, 759–763.

Perrin, S., Warchol, M., Grill, J. P., Schneider, F. 2001. Fermentations of fructo-oligosaccharides and their components by *Bifidobacterium infantis* ATCC 15697 on batch culture in semi-synthetic medium. *J. Appl. Microbiol.*, **90**, 859–865.

Petkevicius, S., Knudsen, K. E. B., Murrell, K. D., Wachmann, H. 2003. The effect of inulin and sugar beet fibre on *Oesophagostomum dentatum* infection in pigs. *Parasitology*, **127**, 61–68.

Petkevicius, S., Murrell, K. D., Knudsen, K. E. B., Jorgensen, H., Roepstorff, A., Laue,. A., Wachmann, H. 2004. Effects of short chain fatty acids on survival of *Oesophagostomum dentatum* in pigs. *Vet. Parasitol.*, **122**, 293–301.

Pierre, F., Perrin, P., Bassonga, E., Bornet, F., Meflah, K., Menanteau, J. 1999. T cell status influences colon tumor occurrence in Min mice fed short chain fructo-oligosaccharides as a diet supplement. *Carcinogenesis*, **20**, 1953–1956.

Pierre, F., Perrin, P., Champ, M., Bornet, F., Meflah, K., Menanteau, J. 1997. Short-chain fructo-oligosaccharides reduce the occurrence of colon tumors and develop gut-associated lymphoid tissue in Min mice. *Cancer Res.*, **57**, 225–228.

Pool-Zobel, B., Van Loo, J., Rowland, I., Roberfroid, M. B. 2002. Experimental evidence on the potential of prebiotic fructans to reduce the risk of colon cancer. *Br. J. Nutr.*, **87** (Suppl. 2), S273–S281.

Poulsen, M., Molck, A. M., Jacobsen, B. L. 2002. Different effects of short- and long-chained fructans on large intestinal physiology and carcinogen-induced aberrant crypt foci in rats. *Nutr. Cancer*, **42**, 194–205.

Poxton, I. R. 1997. Mucosa-associated bacterial flora of the human colon. *J. Med. Microbiol.*, **33**, 85–91.

Probert, H. M., Gibson, G. R. 2002. Investigating the prebiotic and gas-generating effects of selected carbohydrates on the human colonic microflora. *Lett. Appl. Microbiol.*, **35**, 473–480.

Propst, E. L., Flickinger, E. A., Bauer, L. L., Merchen, N. R., Fahey Jr, G. C. 2003. A dose–response experiment evaluating the effects of oligofructose and inulin on nutrient digestibility, stool quality, and fecal protein catabolites in healthy adult dogs. *J. Anim. Sci.*, **81**, 3057–3066.

Prosky, L., Hoebregs, H. 1999. Methods to determine food inulin and oligofructose. *J. Nutr.*, **129** (Suppl.), 1418S–1423S.

Quemener, B. 1994. Determination of inulin and oligofructose in food products, and integration in the AOAC method for measurment of total dietary fibre. *Lebensm. Technol.*, **27**, 125–132.

Quemener, B., Thibault, J. F., Coussement, P. 1997. Integration of inulin determination in the AOAC method for measurement of total dietary fibre. *Int. J. Biol. Macromol.*, **21**, 175–178.

Quigley, M. E., Hudson, G., Englyst, H. 1999. Determination of resistant short-chain carbohydrates (non-digestible oligosaccharides) using gas–liquid chromatography. *Food Chem.*, **65**, 381–390.

Rafter, J., Bennet, M., Caderni, G., Clune, Y., Collins, K., Hughes, R., Karlsson, P., Klinder, A., Pool-Zobel, B. L., Rechkemmer, G., Roller, M., Rowland, I., Salvadori, M., Van Loo, J., Watzl, B. 2005. The probiotics *Lactobacillus rhamnosus* and *Bifidobacterium lactis* in combination with the prebtioc inulin enriched with oligofructose (Synergy1) alter colorectal cancer biomarkers in patients with resected polyps and colon cancer. *Am. J. Clin. Nutr.*, in press.

Rao, V. A. 2001. The prebiotic properties of oligofructose at low intake levels. *Nutr. Res.*, **21**, 843–848.

Reddy, B. S., Hamid, R., Rao, C. V. 1997. Effect of dietary oligofructose and inulin on colonic preneoplastic aberrant crypt foci inhibition. *Carcinogenesis*, **18**, 1371–1374.

Reimer, R. A. McBurney, M. I. 1996. Dietary fiber modulates intestinal proglucagon messenger ribonucleic acid and postprandial secretion of glucagon-like peptide-1 and insulin in rats. *Endocrinology*, **137**, 3948–3956.

Richardson, J. E., Verghese, M., Walker, I. A., Bonsi, I. A., Howard, C., Shackelford, L., Chawan, C. B. 2002. Effects of prebiotics on bone mineralization in Fisher 344 male weanling rats. *Institute of Food Technology*, Annual Meeting, Annaheim, CA, Poster.

Rideout, T. C., Fan M. Z., Cant, J. P., Wagner-Riddle, C. 2004. Excretion of major odour-causing and acidifying compounds in response to dietary supplementation of chicory inulin in growing pigs. *J. Anim. Sci.*, **82**, 1678–1684.

Rigo, J. 2001. Clinical evaluation in term infants of a new formula based on prebiotics beta palmitate and hydrolysed proteins. *J. Pediatr. Gastroenterol. Nutr.*, **32**, 402.

Roberfroid, M. 1993. Dietary fiber, inulin, and oligofructose: a review comparing their physiological effects. *Crit Rev. Food Sci. Nutr.*, **33**, 103–148.

Roberfroid, M. B. 1998. Prebiotics and synbiotics: concepts and nutritional properties. *Br. J. Nutr.*, **80**, S197–S202.

Roberfroid, M. B., Cumps, J., Devogelaer, J. P. 2002. Dietary chicory inulin increases whole-body bone mineral density in growing male rats. *J. Nutr.*, **132**, 3599–3602.

Roberfroid, M. B., Delzenne, N. M. 1998. Dietary fructans. *Annu. Rev. Nutr.*, **18**, 117–143.

Roberfroid, M. B., Van Loo, J. A., Gibson, G. R. 1998. The bifidogenic nature of chicory inulin and its hydrolysis products. *J. Nutr.*, **128**, 11–19.

Roland, N., Nugon-Baudon, L., Andrieux, C. 1995. Comparative study of the fermentative characteristics of inulin and different types of fiber in rats inoculated with a human whole fecal flora. *Br. J. Nutr.*, **74**, 239–249.

Roller, M., Femia, A., Caderni, G., Rechkemmer, G., Watzl, B. 2004a. Intestinal immunity of rats with colon cancer is modulated by oligofructose-enriched inulin combined with *Lactobacillus rhamnosus* and *Bifidobacterium lactis*. *Br. J. Nutr.*, **92**, 931–938.

Roller, M., Rechkemmer, G., Watzl, B. 2004b. Prebiotic inulin enriched with oligofructose in combination with the probiotics *Lactobacillus rhamnous* and bifidobacterium immune functions in rats. *J. Nutr.*, **134**, 153–156.

Rowland, I. R., Rumney, C. J., Coutts, J. T., Lievense, L. C. 1998. Effect of *Bifidobacterium longum* and inulin on gut bacterial metabolism and carcinogen-induced aberrant crypt foci in rats. *Carcinogenesis*, **19**, 281–285.

Rumessen, J. J., Bode, S., Hamberg, O., Gudmand-Hoyer, E. 1990. Fructans from Jerusalem artichokes: intestinal transport, absorption, fermentation and influence on blood glucose, insulin and C-peptide response in healthy subjects. *Am. J. Clin. Nutr.*, **52**, 675–681.

Russell, T. J. 1995. Feeding fructooligosaccharide to the weaned pig improves nitrogen metabolism and reduces odor of metabolite excretion. *J. Anim. Sci.*, **73**, 70.

Rycroft, C. E. 2001. A comparative *in vitro* evaluation of the fermentation properties of prebiotic oligosaccharides. *J. Appl. Microbiol.*, **91**, 878–887.

Saavedra, J. M. 1999. Gastro-intestinal function in infants consuming a weaning food supplemented with oligo-fructose, a prebiotic. *J. Pediatr. Gastroenterol. Nutr.*, **29**, A95.

Saavedra, J., Tschernia, A. 2002. Human studies with probiotics and prebiotics: clinical implications. *Br. J. Nutr.*, **87**, 241–246.

Sakai, K., Aramaki, K., Takasaki, M., Inaba, H., Tokunaga, T., Ohta, A. 2001. Effect of dietary short-chain fructooligosaccharides on the cecal microflora in gastrectomized rats. *Biosci. Biotechnol. Biochem.*, **65**, 264–269.

Schell, M. A., Karmirantzou, M., Snel, B., Vilanova, D., Berger, B., Pessi, G., Zwahlen, M. C., Desiere, F., Bork, P., Delley, M., Pridmore, R. D., Arigoni, F. 2002. The genome sequence of *Bifidobacterium longum* reflects its adaptation to the human gastrointestinal tract. *Proc. Natl. Acad. Sci. USA*, **99**, 14422–14427.

Scholz Ahrens, C., Yahya, A., Schrezenmeir, J. 2002. Effect of oligofructose or dietary calcium on repeated calcium and phosphorus balances, bone mineralization and trabecular structure in ovariectomized rats. *Br. J. Nutr.*, **88**, 365–377.

Sghir, A., Chow, J. M., Mackie, R. I. 1998. Continuous culture selection of bifidobacteria and lactobacilli from human faecal samples using fructooligosaccharide as selective substrate. *J. Appl. Microbiol.*, **85**, 769–777.

Shannon, J. A. 1935a. The excretion of inulin by the dog. *Am. J. Physiol.*, **112**, 405–413.

Shannon, J. A. 1935b. The excretion of inulin, xylose and urea by normal and phlorizinised man. *J.Clin.Invest.*, **14**, 393–401.

Stanley, S., Wynne, K., Bloom, S. 2004. Gastrointestinal satiety signals III. Glucagon-like peptide 1, oxyntomodulin, peptide YY, and pancreatic polypeptide. *Am. J. Physiol. (Gastrointestinal Liver Physiol.)*, **286**, G693–G697.

Stone-Dorshow, T. 1987. Gaseous response to ingestion of a poorly absorbed fructo-oligosaccharide sweetener. *Am. J. Clin. Nutr.*, **46**, 61–65.

Swanson, K. S., Grieshop, C. M., Flickinger, E. A., Merchen, N. R., Fahey Jr, G. C. 2002. Effects of supplemental fructooligosaccharides and mannanoligosaccharides on colonic microbial populations, immune function and fecal odor components in the canine. *J. Nutr.*, **132**, 1717S–1719S.

Tahiri, M. 2001. Five-week intake of short-chain fructo-oligosaccharides increases intestinal absorption and status of magnesium in postmenopausal women. *J. Bone Mineral Res.*, **16**, 2152–2160.

Tahiri, M. 2003. Effect of short-chain fructooligosaccharides on intestinal calcium absorption and calcium status in postmenopausal women: a stable-isotope study. *Am. J. Clin. Nutr.*, **77**, 449–457.

Taper, H. S., Delzenne, N. M., Roberfroid, M. B. 1997. Growth inhibition of transplantable mouse tumors by non-digestible carbohydrates. *Int. J. Cancer*, **71**, 1109–1112.

Taper, H. S., Lemort, C., Roberfroid, M. B. 1998. Inhibition effect of dietary inulin and oligofructose on the growth of transplantable mouse tumor. *Anticancer Res.*, **18**, 4123–4126.

Taper, H. S., Roberfroid, M. 1999. Influence of inulin and oligofructose on breast cancer and tumor growth. *J. Nutr.*, **129** (Suppl.), 1488S–1491S.

Taper, H. S., Roberfroid, M. B. 2000. Inhibitory effect of dietary inulin or oligofructose on the development of cancer metastases. *Anticancer Res.*, **20**, 4291–4294.

Taper, H. S., Roberfroid, M. B. 2002a. Inulin/oligofructose and anticancer therapy. *Br. J. Nutr.*, **87** (Suppl. 2), S283–S286.

Taper, H. S., Roberfroid, M. B. 2002b. Non-toxic potentiation of cancer radiotherapy by dietary oligofructose or inulin. *Anticancer Res.*, **22**, 3319–3323.

Tappenden, K. A., Drozdowski, L. A., Thomson, A. B., McBurney, M. I. 1998. Short-chain fatty acid-supplemented total parenteral nutrition alters intestinal structure, glucose transporter 2 (GLUT2) mRNA and protein, and proglucagon mRNA abundance in normal rats. *Am. J. Clin. Nutr.*, **68**, 118–125.

Tokunaga, T. 1986, Influence of chronic intake of new sweetener fructo-oligosaccharide (Neosugar) on growth and gastrointestinal function of the rat. *J. Nutr. Sci. Vitaminol.*, **32**, 111–121.

Tschernia, A. 1999. Effects of long-term consumption of a weaning food supplemented with oligofructose, a prebiotic, on general infant health status. *J. Pediatr. Gastroenterol. Nutr.*, **29**, A58.

Tschop, M., Smiley, D. L., Heiman, M. L. 2000. Ghrelin induces adiposity in rodents. *Nature*, **407**, 908–913.

Tuohy, K., Finlay R. K., Wynne A. G., Gibson, G. R. 2001a. A human volunteer study on the prebtioc effects of HP inulin – faecal bacteria enumerated using fluorescent *in situ* hybridisation (FISH). *Anaerobe*, **7**, 113–118.

Tuohy, K., Kolida, S., Lusteberger, A., Gibson, G. R. 2001b. The prebiotic effects of biscuits containing partially hydrolysed guar gum and fructo-oligosaccharides—a human volunteer study. *Br. J. Nutr.*, **86**, 341–348.

Van den Heuvel, E. G. H. 1998. Nondigestible oligosaccharides do not interfere with calcium and nonheme-iron absorption in young, healthy men. *Am. J. Clin. Nutr.*, **67**, 445–451.

Van den Heuvel, E. G. H. 1999. Oligofructose stimulates calcium absorption in adolescents. *Am. J. Clin. Nutr.*, **69**, 544–548.

van der Sluis, W. 2000. Clostridial enteritis—a syndrome emerging worldwide. *World Poultry*, **16**, 56–59.

van der Waaij, L. A., Harmsen, H. J. M., Madjipour, M., Kroese, F. G. M., Zwiers, M., van Dullemen, H. M., de Boer, N. K., Welling, G. W., Jansen, P. L. M. 2005. Bacterial population analysis of human colon and terminal ileum biopsies with 16S rRNA-based fluorescent probes: commensal bacteria live in suspension and have no direct contact with epithelial cells. *Inflamm. Bowel Dis.*, **11**, 865–871.

van Dokkum, W., Wezendonk, B., Srikumar, T. S., van den Heuvel, E. G. 1999. Effect of nondigestible oligosaccharides on large-bowel functions, blood lipid concentrations and glucose absorption in young healthy male subjects. *Eur. J. Clin. Nutr.*, **53**, 1–7.

Van Loo, J. 2004a. The specificity of the interaction with intestinal fermentation by prebiotics determines their physiological efficacy. *Nutr. Res. Rev.*, **17**, 89–98.

Van Loo, J. 2004b. Prebiotics promote good health. The basis, the potential and the emerging evidence. *J. Clin. Gastroenterol.*, **38**, S70–S75.

Van Loo, J., Jonkers, N. 2001. Evaluation in human volunteers of the potential anticarcinogenic activities of novel nutritional concepts: prebiotics probiotics and synbiotics (the SYNCAN project). *Nutr., Metab. Cardiovasc. Dis.*, **11**, 87–93.

Van Loo, J., Clune, Y., Bennett, M., Collins, K. 2005. The SYNCAN project: goals, set-up, first results and settings of the human intervention study. *Br. J. Nutr.*, **92**, 91–98.

Van Loo, J., Coussement, P., De Leenheer, L., Hoebregs, H., Smits, G. 1995. On the presence of inulin and oligofructose as natural ingredients in the Western diet. *Crit. Rev. Food Sci. Nutr.*, **35**, 525–552.

Van Loo, J., Cummings, J., Delzenne, N., Englyst, H., Franck, A., Hopkins, M., Kok, N., Macfarlane, G., Newton, D., Quigley, M., Roberfroid, M., Vam Vliet, T., Van Den Heuvel, E. 1999. Functional food properties of non-digestible oligosaccharides: a consensus report from the ENDO project (DGXII AIRII-CT94-1095). *Br. J. Nutr.*, **81**, 121–132.

Vanhoutte, T., Huys, G., De Brandt, E., Fahey, G., Swings, J. 2005. Molecular monitoring and characterisation of the faecal microbiota of healthy dogs during a prebiotic feeding study. *FEMS Microbiol. Lett.*, **249**, 65–71.

Verghese, M., Rao, D. R., Chawan, C. B., Shackelford, L. 2002a. Dietary inulin suppresses azoxymethane-induced preneoplastic aberrant crypt foci in mature Fisher 344 rats. *J. Nutr.*, **132**, 2804–2808.

Verghese, M., Rao, D. R., Chawan, C. B., Williams, L. L., Shackelford, L. 2002b. Dietary inulin suppresses azoxymethane-induced aberrant crypt foci and colon tumors at the promotion stage in young Fisher 344 rats. *J. Nutr.*, **132**, 2809–2813.

Verghese, M., Walker, L. T., Shackelford, L. A., Chawan, C. D., Van Loo, J. 2005. Inhibitory effects of nondigestible carbohydrates of different chain lengths on azoxymethane-induced aberrant crypt foci in Fisher 344 rats. *Nutr. Res.* **25**, 859–868.

Vulevic, J., Rastall, R. A., Gibson, G. 2004. Developing a quantitative approach for determining the *in vitro* prebiotic potential of dietary oligosaccharides. *FEMS Microbiol. Lett.*, **236**, 153–159.

Waligora-Dupriet, A.-J., Campeotto, F., Nicolis, I., Bonet, A., Dupont, C., Butel, M.-J. 2005. Effects of oliofructose supplementation in infants 6 to 24 months of age on gut microflora and well-being: a double-blind, placebo controlled randomised trial. *Br. J. Nutr.*, in press.

Wang, X., Gibson, G. R. 1993. Effects of the *in vitro* fermentation of oligofructose in comparison to other carbohydrates by bacteria growing in the human large intestine. *J. Appl. Bacteriol.*, **75**, 373–380.

Weaver, C., Heineman, E., Thunissen, F., van den Bogaard, A. 2000. The growing years and prevention of osteoporosis in later life. *Proc. Nutr. Soc.*, **59**, 303–306.

Willard, M. D., Simpson, R. B., Delles, E. K., Cohen, N. D., Fossum, T. W., Kolp, D., Reinhart, G. 1994. Effects of dietary supplementation of fructo-oligosaccharides on small intestinal bacterial overgrowth in dogs. *Am. J. Vet. Res.*, **55**, 654–659.

Williams, C. H., Witherly, S. A., Buddington, R. K. 1994. Influence of dietary fructo-oligosaccharides on selected bacterial groups of the human fecal microbiota. *Microbial* Ecol. *Health* Dis., **7**, 91–97.

Wolf, B. W. 1998. Varying dietary concentrations of fructooligosaccharides affect apparent absorption and balance of minerals in growing rats. *Nutr. Res.*, **18**, 1791–1805.

Xu, Z. R., Hu, C. H., Wang, M. Q. 2002a, Effects of fructooligosaccharide on conversion of L-tryptophan to skatole and indole by mixed populations of pig fecal bacteria. *J. Gen. Appl. Microbiol.*, **48**, 83–89.

Xu, Z. R., Zou, X. T., Hu, C. H., Xia, M. S., Zhan, X. A., Wang, M. Q. 2002b. Effects of dietary fructooligosaccharide on digestive enzyme activities, intestinal microflora and morphology of growing pigs. *Asian-Austr. J. Anim. Sci.*, **15**, 1784–1789.

Yamashita, K., Kawai, K., Itakura, M. 1984. Effects of fructo-oligosaccharides on blood glucose and serum lipids in diabetic subjects. *Nutr. Res.,* **4**, 961–966.

Yazawa, K., Imai, K., Tamura, Z. 1978. Oligosaccharides and polysaccharides specifically utilizable by bifidobacteria. *Chem. Pharm. Bull*, **26**, 3306–3311.

Younes, H., Coudray, C., Bellanger, J., Demigne, C., Rayssiguier, Y., Remesy, C. 2001. Effects of two fermentable carbohydrates (inulin and resistant starch) and their combination on calcium and magnesium balance in rats. *Br. J. Nutr.*, **86**, 479–485.

Younes, H., Garleb, K., Behr, S., Remesy, C., Demigne, C. 1995. Fermentable fibers or oligosaccharides reduce urinary nitrogen excretion by increasing urea disposal in the rat cecum. *J. Nutr.*, **125**, 1010–1016.

Yusrizal, Chen, T. C. 2003a. Effect of adding chicory fructans in feed on broiler growth performance, serum cholesterol and intestinal length. *Int. J. Poultry Sci.*, **2**, 214–219.

Yusrizal, Chen, T. C. 2003b. Effect of adding chicory fructans in feed on fecal and intestinal flora. *Int. J. Poultry Sci.*, **2**, 188–194.

Zafar, T. A., Weaver, C., Zhao, Y., Martin, B. R., Wastney, M. E. 2004. Nondigestible oligosaccharides increase calcium absorption and suppress bone resorption in ovariectomized rats. *J. Nutr.*, **134**, 399–402.

4

Galacto-oligosaccharides as Prebiotic Food Ingredients

Robert A. Rastall

4.1 Introduction

Galacto-oligosaccharides (GOS) are manufactured from lactose by Friesland Foods as Vivinal GOS® in Europe whilst Yakult manufacture it as Oligomate® in Japan. It is manufactured from lactose in whey using β-galactosidase (E.C. 3.2.1.23). At high lactose concentrations this enzyme will catalyse glycosyl transfer reactions and will use lactose as a glycosyl donor to transfer galactose onto other lactose molecules acting as acceptors (Prenosil *et al.*, 1987a, b; Záarate and López-Leiva, 1990). In this way, a complex series of higher oligosaccharides is built up.

4.2 Characteristics of GOS

The commercial GOS products are composed of oligosaccharides ranging from disaccharides to octasaccharides (Shin and Yang, 1998; Albayrak and Yang, 2002).

The commercial products are very impure and contain around 50 % oligosaccharides, 38 % glucose and 12 % lactose by weight. The major oligosaccharide components are disaccharides and trisaccharides (Table 4.1). A range of linkages occur in commercial GOS samples, principally β1-4 and β1-6 but also including β1-3 and β1-2 linkages (Matsumoto *et al.*, 1993; Rabiu *et al.*, 2001). It is important to note that the linkage composition and molecular weight distribution of GOS depend on the enzyme used in

Prebiotics: Development and Application Edited by G.R. Gibson and R.A. Rastall

Table 4.1 Composition of GOS

Component	Wt fraction of GOS
Disaccharide (other than lactose)	0.33
Trisaccharide	0.39
Tetrasaccharide	0.18
Pentasaccharide	0.07
Hexa-, hepta-, octasaccharides	0.03

their manufacture and on process conditions (Zárate and López-Leiva, 1990; Sako *et al.*, 1999; Rabiu *et al.*, 2001).

The nonglucose and lactose components in GOS are considered nondigestible oligosaccharides as judged by *in vitro* digestibility studies (Burvall *et al.*, 1980) and feeding studies in rats (Ohtsuka *et al.*, 1990). Good data in humans is scarce but there is evidence based on breath hydrogen tests that feeding GOS to volunteers results in colonic fermentation (Tanaka *et al.*, 1983; Alles *et al.*, 1999). This is, however, not a very reliable method for determining the degree of digestibility in humans and some studies have not shown an increase in breath hydrogen excretion (Bouhnik *et al.*, 1997; van Dokkum *et al.*, 1999).

4.3 Properties of GOS as a Food Ingredient

GOS are slightly sweet with a sweetness of 30–60 % relative to sucrose (Prapulla *et al.*, 2000). They are stable on long term storage and also have very good process stability remaining undegraded after 10 min at 160 °C and pH 7, after 10 min at 120 °C and pH 3 and after 10 min at 100 °C and pH 2 (Voragen, 1998; Sako *et al.*, 1999).

GOS have found application in Japan in a wide range of food products as sweeteners, bulking agents, and sugar substitutes. They have been used in a range of product types including bread, 'sports' drinks, jams and marmalades, fermented milk products, confectionary products and desserts (Playne and Crittenden, 1996; Prapulla *et al.*, 2000; Playne *et al.*, 2003). In Europe they are incorporated into infant formula foods (Savino *et al.*, 2003). Nutricia market a product called *Omneo* and Milupa market *Conformil*. These products contain 7.2 g GOS l^{-1} together with 0.8 g fructo-oligosaccharides (FOS) l^{-1}.

4.4 Functional Properties of GOS

GOS have been widely investigated for their prebiotic properties *in vitro* and *in vivo* in animal and human studies.

4.4.1 *In vitro* Studies

Early pure culture studies displayed the potential of GOS to stimulate the growth of bifidobacteria and lactobacilli. GOS at 1 % (w/v) stimulated growth of a range of bifidobacteria and lactobacilli over 24 h (Tanaka *et al.*, 1983). GOS also supported growth of several enterobacteriaceae and streptococci. Hopkins *et al.*, (1998) studied the growth of a range of isolates of *Bifidobacterium* and found significantly increased growth rates on GOS as compared with a range of other carbohydrates.

Using *in vitro* batch cultures with faecal inocula, Rycroft *et al.*, (2001) compared GOS with FOS, xylo-oligosaccharides (XOS), lactulose, inulin, soybean oligosaccharides (SOS) and isomalto-oligosaccharides (IMO). GOS resulted in significantly increased levels of bifidobacteria albeit to a lesser extent than XOS. GOS did, however, result in the largest decrease in clostridia and the highest production of short chain fatty acids (SCFA).

In a study of the fermentation of GOS in a three-stage continuous model of the human colon, McBain and Macfarlane (2001) evaluated the comparative performance of GOS and inulin. They found that GOS was weakly bifidogenic in vessel one, representing the ascending colon, but it did promote an increase in lactobacilli by an order of magnitude. Both prebiotics resulted in very minor changes in vessels two and three representing the transverse and descending colon, respectively. GOS did, however, strongly suppress the activities of β-glucosidase, β-glucuronidase and arylsulphatase, whereas inulin had very little effect on these enzymes which are considered to be undesirable metabolic activities in the colon.

Palframan *et al.*, (2002) used pH-controlled batch cultures to evaluate the effect of pH and substrate concentration on the selectivity of fermentation of a range of commercial prebiotics including XOS, GOS, lactulose, FOS and inulin. In this system, GOS demonstrated the greatest bifidogenic effect at pH 6 and 2 % (w/v), equivalent to a dose of approximately 8 g day^{-1} in adults.

4.4.2 Animal Studies

The majority of animal studies on GOS have been carried out in rats. Such studies have either used the standard rat model or have used gnotobiotic rats inoculated with a human faecal flora. Some such studies have established bifidogenic effects for GIS whilst others have not shown an effect. Two things must be borne in mind when evaluating these studies, however. Firstly, as the studies have used different experimental designs and have differed in the way they have reported the studies, it is often difficult to directly compare the doses received by the animals. In addition, most of the studies carried out on GOS to date have used culture-based methods to enumerate bacteria. The reliability of these methods is now in question (Greetham *et al.*, 2002) and there is a need for more studies using DNA-based methods.

Morishita and Konishi (1994) carried out a feeding study using 5 % (w/v) GOS in the diet over a 7-week period. They found a significant increase in bifidobacteria and significant decreases in staphylococci and streptococci and a significant increase in SCFA. In a study feeding 0.8 g kg^{-1} day^{-1} GOS, Kikuchi-Hayakawa *et al.* (1996) did not see any change in populations of lactobacilli or total bacteria, but they did report an increase in *Bacteroides* and a decrease in the enterobacteriaceae.

Rowland and Tanaka (1993) investigated the effect of feeding 2–3 g GOS kg^{-1} day^{-1} over 4 weeks to gnotobiotic rats associated with a human faecal flora. They found significant increases in bifidobacteria and lactobacilli and significant decreases in populations of the enterobacteriacae. There was no effect on the levels of *Bacteroides* spp., staphylococci, or clostridia. They also noted elevated β-glucosidase activity and decreased β-glucuronidase and nitrate reductase activities. Djouzi and Andrieux (1997) fed gnotobiotic rats with 2 g GOS kg^{-1} day^{-1} over a 4-week period. In this study, no effects were seen on total anaerobes or but bifidobacteria displayed a significant 2-log increase. They found no effect on the enzymes β-glucosidase, α-glucosidase or β-glucuronidase.

Another study in gnotobiotic rats, this time inoculated with bacteria representative of infant faeces fed GOS at 5.3 g kg^{-1} day^{-1} for 7 days (Suzuki *et al.*, 1999). This study did not show any effect on populations of *Staphylococcus epidermidis, Eubacteria aerofaciens, Bacteroides* spp., staphylococci, yeasts, or clostridia. There was, however a decrease in *Clostridium perfringens*. Similarly, in a gnotobiotic mouse model (Morishita *et al.*, 2002) fed with 7.5 g kg^{-1} day^{-1} GOS for 7 weeks, a decrease was seen in putrefactive *C. perfringens*.

4.4.3 Human Studies

There have been several human studies on the prebiotic properties of GOS with varying results. Some have shown clear prebiotic activities but other, well-designed studies have shown no effect.

One of the earliest studies is that of Tanaka *et al.*, (1983) who fed five adults with 43 mg kg^{-1} day^{-1} GOS for 7 days, followed by 143 mg kg^{-1} day^{-1} GOS for a further 7 days. Four of the five subjects had elevated bifidobacteria after the feeding period. They also examined synbiotic effects between GOS and *Bifidobacterium breve*. Bifidobacteria were elevated in the synbiotic group but not in the control group that only received *B. breve*. In addition, four out of the five subjects displayed a decrease in faecal ammonia on the synbiotic but not on the probiotic or GOS feeding alone.

Ito *et al.*, (1990) fed 12 adults with 36, 71 or 143 mg GOS kg^{-1} day^{-1} for 7 days in a crossover study. Faecal samples were analysed by cultivation-based methods and the levels of bifidobacteria increased as a proportion of the total bacterial count in a dose-dependent manner. The lowest dose tested, 36 mg GOS kg^{-1} day^{-1} (2.5 g GOS day^{-1}), resulted in a nonsignificant increase in bifidobacteria. In a subsequent paper, Ito *et al.* (1993) showed that the same dose could produce significant increases in subjects with low bifidobacterial numbers (log 9.46). They also found significant increases in lactobacilli with significant decreases in *Bacteroides* spp. and *Candida* spp. Faecal ammonia, indole and cresol were significantly reduced as were propionate, isobutyrate, valerate and isovalerate.

Bouhnik *et al.* (1997) carried out a limited feeding trial with GOS on eight healthy adults. They received 10 g GOS day^{-1} over 21 days and stool samples were collected every 7 days. Stool samples were analysed for bifidobacteria and enterobacteria by plate count. The GOS resulted in a significant increase in bifidobacteria of 1 log. No change was seen in the enterobacteria levels. The very limited microbiology carried out restricts the conclusions that can be drawn from this study, however. A similarly limited study was carried out by Teuri *et al.* (1998) who fed 15 g GOS day^{-1} to 12 healthy adults. No significant changes in bifidobacteria were seen in this study and the levels detected in the volunteers varied greatly at the start of the trial. GOS did lead to an increase in defaecation.

A much more substantial study was carried out by Alles *et al.* (1999) on feeding of GOS at 0, 7.5 and 15 g day^{-1} to 40 healthy adults divided between three treatment groups. All groups had a run-in diet with no prebiotics for 3 weeks followed by the dietary intervention for 3 weeks. The group receiving no GOS received a placebo of glucose and lactose in the amounts present in the GOS mixture. Faecal samples were analysed by plate count on selective agar for total bacteria (aerobes and anaerobes), bifidobacteria, clostridia, lactobacilli and *E. coli*. These authors also tested for bowel habit, stool frequency, concentration of short chain fatty acids, bile acids, ammonia, indole, skatole and for changes in faecal pH. No significant changes were seen between the three groups in any of the tested outcomes after consumption of the test diets. An absence of an effect was also seen by Van den Heuvel *et al.* (1998), who carried out a study to investigate the effects of GOS consumption on the absorption of minerals by healthy adults. They fed GOS, inulin, FOS or no supplementation to 12 volunteers at 15 g day^{-1} in a randomised crossover design. They then measured the absorption of iron and calcium by double stable isotope techniques. No changes were seen between any of the treatments used.

Other studies have failed to demonstrate a prebiotic effect in humans. Alander *et al.* (2001) fed 30 healthy adults diets containing either 116 mg GOS kg^{-1} day^{-1}, a probiotic *Bifidobacterium lactis* (Bb-12) or a combination of them both. Ten subjects were allocated to each group for 3 weeks. Faecal samples were analysed before, during and after feeding for lactic acid bacteria, *C. perfringens* and coliforms. No significant changes were seen in the GOS group but a significant increase in bifidobacteria was seen in the groups receiving the Bb-12 supplements. Similarly, Satokari *et al.* (2001) failed to demonstrate a bifidogenic effect from administration of 8 g day^{-1} (ca. 114 mg kg^{-1} day^{-1}) GOS for 2 weeks. This latter study is notable for its use of modern DNA-based techniques. These authors used polymerase chain reaction to amplify 16S rDNA sequences which were subsequently analysed by denaturing gradient gel electrophoresis (DGGE). Although this is essentially a qualitative technique, no changes in the bifdobacterial flora could be seen.

It is not clear at the present time why such inconsistent effects have been seen in GOS feeding trials.

4.4.4 Infant Studies

GOS have also been studied in the context of infant nutrition. Some studies, however, have used mixtures of GOS and FOS, making reliable conclusions about the efficacy of the GOS impossible.

Boehm *et al.* (2002) fed an infant formula feed containing a total of 10 g L^{-1} prebiotic oligosaccharides, of which 90 % was GOS and 10 % was FOS, to 30 pre-term infants. The subjects received either the prebiotic formula or a formula with maltodextrin as a control after their mother was no longer able to provide milk. A further 12 infants continued on mothers' milk by way of comparison.

The infants showed no significant differences in faecal bacteriology at the start of the study. After 4 weeks of feeding, however, the prebiotic formula group had significantly increased levels of bifidobacteria similar to those in the breast milk group. No difference was seen in populations of lactobacilli, *Escherichia coli*, bacteroides, *Candida* spp.,

Proteus spp., or clostridia. Bowel habit had also changed with the prebiotic formula group displaying increased defaecation. Stools were softer in the prebiotic and breast milk groups than in the maltodextrin control formula group.

A similar study was carried out by Moro *et al.* (2002, 2003). These authors also used a mixture of 90 % GOS and 10 % FOS, at either 4 or 8 g L^{-1}. Ninety infants were fed either the prebiotic formulae or a maltodextrin control formula. All infants were breast-fed prior to the study. The prebiotic formula-fed infants displayed significantly elevated levels of bifidobacteria after 4 weeks of feeding relative to the maltodextrin control group. The infants in the 8 g L^{-1} group had significantly higher (9.7 log) levels than did those in the 4 g L^{-1} group (9.3 log). Lactobacilli were also significantly increased, but not *E. coli*, bacteroides, *Candida* spp., *Proteus* spp., or clostridia. Stool frequency increased in infants receiving the highest dose of prebiotics.

There have also been studies on the feeding of GOS rather than GOS/FOS mixtures. Napoli *et al.* (2003) fed a formula containing 0.7 % GOS to 13 healthy full-term infants with a lactose containing formula as a control. Twenty-four infants receiving breast milk were included as a reference. The infants on the GOS formula had a 1-log increase in bifidobacteria to the same range as the breast-fed group, although no information is available on other bacterial groups enumerated.

Ben *et al.* (2004) fed full-term infants a commercial formula supplemented with 2.4 g GOS L^{-1} over 6 months. Sixty-nine infants received the GOS, 52 received the commercial formula with no supplement as a control and 26 received breast milk. The GOS formula-fed group had increased levels of bifidobacteria and lactobacteria at same level as the infants on the breast milk. There were no changes in *E. coli* populations. No other microbial groups were enumerated.

4.5 Future Developments

There is still much interest in developing novel GOS mixtures with enhanced functional effects. GOS are very complex mixtures and the exact oligosaccharide spectrum depends on the enzyme and conditions used in its manufacture. This leads to the idea that different GOS mixtures may have different properties. Rabiu *et al.* (2001) synthesised a range of GOS mixtures from lactose using crude β-galactosidase preparations from several probiotic bacteria. These had different glycosyl linkage compositions. The hypothesis was that, as GOS products are kinetically controlled, the oligosaccharide mixtures should be good substrates for the producing β-galactosidases. The GOS mixtures were then tested as substrates for the producing probiotic organisms and in most cases, the producing organism had a much higher growth rate on its own GOS mixture. This leads to the possibility of using probiotic enzymes to synthesise GOS that are very highly targeted at specific probiotics.

This idea has been taken further by Tzortzis *et al.* (2005a) who have synthesised a novel GOS mixture with whole cells of *Bifidobacterium bifidum* NCIMB 4117. This GOS is now being commercialised by Clasado. In a study using a three-stage model of the human colon (Tzortzis *et al.*, 2005b), the novel GOS resulted in significant increases in bifidobacteria in vessels one and two, representing the ascending and transverse colon, respectively. No changes were seen in the total bacterial population or the populations of lactobacilli, bacteroides or *Clostridium histolyticum* group. A further feeding trial in weaned pigs resulted in significant increases in bifidobacteria and lactobacilli, to a higher level than seen with inulin. Using an *in vitro* cell culture system the novel GOS also inhibited the adhesion of strains of *E. coli* and *Salmonella enterica* serotype Typhimurium.

4.6 Conclusions

GOS have repeatedly shown prebiotic properties *in vitro*. The bifidogenic effect is often greater than that seen with fructans. This is offset to some degree, however, by the rather mixed data obtained in human studies. There have been several, well-designed, human trials in which GOS has not shown a prebiotic effect. It is clear that there is a need for further study of the prebiotic effects of GOS in studies designed to explore the effects of such factors as diet and the starting populations and profiles of bifidobacteria in individuals. GOS are promising functional food ingredients and there are likely to be new, improved forms on the market in the future.

References

Alander, M., Mättö, J., Kneifel, W., Johansson, M., Kögler, B., Crittenden, R., Mattila-Sandholm, T., and Saarela, M. (2001) Effect of galacto-oligosaccharides supplementation on human faecal microflora and on survival and persistence on *Bifidobacterium lactis* Bb-12 in the gastrointestinal tract. *International Dairy Journal*, **11**, 817–825.

Albayrak, N., Yang, S.-T. (2002) Immobilization of β-galactosidase on fibrous matrix by polyethyleneimine for production of galacto-oligosaccharides from lactose. *Biotechnology Progress*, **18**, 240–251.

Alles, M. S., Hartemink, R., Meyboom, S., Harryvan, J. L., Van Laere, K. M. J., Nagengast, F. M., Hautvast, J. G. A. J. (1999) Effect of transgalactooligosaccharides on the composition of the human intestinal microflora and on putative risk markers for colon cancer. *American Journal of Clinical Nutrition*, **69**, 980–991.

Ben, X. M., Zhou, X. Y., Zhao, W. H., Yu, W. L., Pan, W., Zhang, W. L., Wu, S. M., Beusekom, C. R. V., Schaafsma, A. (2004) Supplementation of milk formula with galacto-oligosaccharides improves intestinal micro-flora and fermentation in term infants. *Chinese Medical Journal,* **117**, 927–931.

Boehm, G., Lidestri, M., Casetta, P., Jelinek, J., Negretti, F., Stahl, B., Marini, A. (2002) Supplementation of a bovine milk formula with an oligosaccharide mixture increases counts of faecal bifidobacteria in preterm infants. *Archives of Disease in Childhood: Fetal and Neonatal Edition,* **86**, F178–F181.

Bouhnik, Y., Flourie, B., D'Agay-Abensour, L., Pochart, P., Gramet, G., Durand, M., Rambaud, J. C. (1997) Administration of transgalacto-oligosaccharides increases fecal bifidobacteria and modifies colonic fermentation metabolism in healthy humans. *Journal of Nutrition*, **127**, 444–448.

Burvall, A., Asp, N.-G., Dahlqvist, A. (1980) Oligosaccharide formation during hydrolysis of lactose with *S. lactis* lactase (Maxilact): Part III: Digestibility by human intestinal enzymes *in vitro*. *Food Chemistry*, **5**, 189–194.

Djouzi, Z., Andrieux, C. 1997. Compared effects of three oligosaccharides on metabolism of intestinal microflora in rats inoculated with a human faecal flora. *British Journal of Nutrition*, **78**, 313–324.

Greetham, H. L., Giffard, C., Hutson, R. A., Collins, M. D., Gibson, G. R. (2002) Bacteriology of the Labrador dog gut: a cultural and genotypic approach. *Journal of Applied Microbiology*, **93**, 640–646.

Hopkins, M. J., Cummings, J. H., Macfarlane, G. T. (1998) Inter-species differences in maximum specific growth rates and cell yields of bifidobacteria cultured on oligosaccharides and other simple carbohydrate sources. *Journal of Applied Microbiology,* **85**, 381–386.

Ito, M., Deguchi, Y., Miyamori, A., Matsumoto, K., Kikuchi, H., Kobayashi, Y., Yajima, T., Kan, T. (1990) Effects of administration of galactooligosaccharides on the human fecal microflora, stool weight and abdominal sensation. *Microbial Ecology in Health and Disease*, **3**, 285–292.

Ito, M., Kimura, M., Deguchi, Y., Miyamori-Watabe, A., Yajima, T., Kan, T. (1993) Effects of transgalactosylated disaccharides on the human intestinal microflora and their metabolism. *Journal of Nutritional Science and Vitaminology*, **39**, 279–288.

Kikuchi, H., Andrieux, C., Riottot, M., Bensaada, M., Popot, F., Beaumatin, P., Szylit, O. (1996) Effect of two levels of transgalactosylated oligosaccharide intake in rats associated with human faecal microflora on bacterial glycolytic activity, end-products of fermentation and bacterial steroid transformation. *Journal of Applied Bacteriology,* **80**, 439–446.

Matsumoto, K., Kobayashi, Y., Ueyama, S., Watanabe, T., Tanaka, R., Kan, T., Kuroda, A., Sumihara, Y. (1993) Oligosaccharides: production, properties and application. In *Japanese Technology Reviews Section E: Biotechnology Volume 3 Number 2*, T. Nakakuki (ed.). Gordon and Breach Science, Yverdon, pp. 90–106.

McBain, A. J., Macfarlane, G. T. (2001) Modulation of genotoxic enzyme activities by non-digestible oligosaccharide metabolism in in-vitro human gut bacterial ecosystems. *Journal of Medical Microbiology,* **50**, 833–842.

Morishita, Y., Konishi, Y. (1994) Effects of high dietary cellulose on the large intestinal microflora and short-chain fatty acids in rats. *Letters in Applied Microbiology*, **19**, 433–435.

Morishita, Y., Oowada, T., Ozaki, A., Mizutani, T. (2002) Galactooligosaccharide in combination with *Bifidobacterium* and *Bacteroides* affects the population of *Clostridium perfringens* in the intestine of gnotobiotic mice. *Nutrition Research*, **22**, 1333–1341.

Moro, G., Minoli, I., Mosca, M., Fanaro, S., Jelinek, J., Stahl B., Boehm, G. (2002) Dosage-related bifidogenic effects of galacto- and fructooligosaccharides in formula-fed term infants. *Journal of Pediatric Gastroenterology and Nutrition,* **34**, 291–295.

Moro, G. E., Mosca, F., Miniello, V., Fanaro, S., Jelinek, J., Stahl B., Boehm, G. (2003) Effects of a new mixture of prebiotics on faecal flora and stools in term infants. *Acta Paediatrica*, Supplement **91**(441), 77–79.

Napoli, J. E., Brand-Miller, J. C., Conway, P. (2003) Bifidogenic effects of feeding infant formula containing galacto-oligosaccharides in healthy formula-fed infants. *Asia Pacific Journal of Clinical Nutrition*, **12** (Suppl.), S60, Abstract.

Ohtsuka, K., Tsuji, K., Nakagawa, Y., Ueda, H., Ozawa, O., Uchida, T., Ichikawa, T. (1990) Availability of 4′galactosyllactose (O-β-D-galactopyranosyl-(1$\rightarrow$4)-O-β-D-galactopyranosyl-(1$\rightarrow$4)-D-glucopyranose) in rat. *Journal of Nutritional Science and Vitaminology*, **36**, 265–276.

Palframan, R. J., Gibson, G. R., Rastall, R. A. (2002) Effect of pH and dose on the growth of gut bacteria on prebiotic carbohydrates *in vitro*. *Anaerobe*, **8**(5), 287–292.

Playne, M. J., Crittenden, R. (1996) Commercially available oligosaccharides. *Bulletin of the IDF*, **313**, 10–22.

Playne, M. J., Bennett, L. E., Smithers, G. W. (2003) Functional dairy foods and ingredients. *The Australian Journal of Dairy Technology*, **58**(3), 242–264.

Prapulla, S. G., Subhaprada, V., Karanth, N. G. (2000) Microbial production of oligosaccharides: a review. *Advances in Applied Microbiology*, **47**, 299–434.

Prenosil, J. E., Stuker, E., Bourne, J. R. (1987a) Formation of oligosaccharides during enzymatic lactose: Part I: State of art. *Biotechnology and Bioengineering*, **30**, 1019–1025.

Prenosil, J. E., Stuker, E., Bourne, J. R. (1987b) Formation of oligosaccharides during enzymatic lactose hydrolysis and their importance in whey hydrolysis process: Part II: Experimental. *Biotechnology and Bioengineering*, **30**, 1026–1031.

Rabiu, B. A., Jay, A. C., Gibson, G. R., Rastall, R. A. (2001) Synthesis and fermentation properties of novel galactooligosaccharides by β-galactosidases from *Bifidobacterium* spp. *Applied and Environmental Microbiology*, **67**, 2526–2530.

Rowland, I. R., Tanaka, R. (1993) The effect of transgalactosylated oligosaccharides on the gut flora metabolism in rats associated with human faecal microflora. *Journal of Applied Bacteriology*, **74**, 667–674.

Rycroft, C. E., Jones, M. R., Gibson, G. R., Rastall, R. A. (2001) A comparative *in vitro* evaluation of the fermentation properties of prebiotic oligosaccharides. *Journal of Applied Microbiology*, **91**, 878–887.

Sako, T., Matsumoto, K., Tanaka, R. (1999) Recent progress on research and applications of non-digestible galacto-oligosaccharides. *International Dairy Journal*, **9**, 69–80.

Satokari, R. M., Vaughan, E. E., Akkermans, A. D. L., Saarela, M., de Vos, W. M. (2001) Polymerase chain reaction and denaturing gradient gel electrophoresis monitoring of fecal *Bifidobacterium* populations in a prebiotic and probiotic feeding trial. *Systemic and Applied Microbiology*, **24**, 227–231.

Savino, F., Cresi, F., Maccario, S., Cavallo, F., Dalmasso, P., Fanaro, S., Oggero, R. Vigi, V., Silvestro, L. (2003) 'Minor' feeding problems during the first months of life: effect of a partially hydrolysed milk formula containing fructo- and galactooligosaccharides. *Acta Paediatrica*, Supplement **91**(441), 86–90.

Shin, H. J., Yang, J. W. (1998) Enzymatic production of galactooligosaccharide by *Bullera singularis* β-galactosidase. *Journal of Microbiology and Biotechnology*, **8**, 484–489.

Suzuki, M., Sato, S., Oowada, T., Ozaki, A., Mizutani, T., Morishita, Y. (1999) The effect of dietary galactooligosaccharide on the formation of the infant intestinal microflora in gnotobiotic mouse models. *Journal of Intestinal Microbiology*, **12**, 81–88.

Tanaka, R., Takayama, H., Morotomi, M., Kuroshima, T., Ueyama, S., Matsumoto, K., Kuroda, A., Mutai, M. (1983) Effects of administration of TOS and *Bifidobacterium breve* 4006 on the human fecal flora. *Bifidobacteria Microflora*, **2**, 17–24.

Teuri, U., Korpela, R., Saxelin, M., Montonen, L., Salminen, S. (1998) Increased fecal frequency and gastrointestinal symptoms following ingestion of galactooligosaccharide-containing yoghurt. *Journal of Nutritional Science and Vitaminology*, **44**, 465–471.

Tzortzis, G., Goulas, A. K., Gibson, G. R. (2005a) Synthesis of prebiotic galactooligosaccharides using whole cells of a novel strain *Bifidobacterium bifidum* NCIMB 41171. *Applied Microbiology and Biotechnology*, **68**, 412–416.

Tzortzis, G., Goulas, A. K., Gee, J. M., Gibson, G. R. (2005b) A novel galactooligosaccharide mixture increases the bifidobacterial population numbers in a continuous *in vitro* fermentation system and in the proximal colonic contents of pigs *in vivo*. *Journal of Nutrition*, **135**, 1726–1731.

Van den Heuvel, E. G. H. M., Schaafsma, G., Murs, T., van Dokkum, W. (1998) Nondigestible oligosaccharides do not interfere with calcium and nonheme-iron absorption in young healthy men. *American Journal of Clinical Nutrition*, **67**, 445–451.

van Dokkum, W., Wezendonk, B., Srikumar, T. S., Van den Heuvel, E. G. H. M. (1999) Effect of non digestible oligosaccharides on large-bowel functions, blood lipid concentrations, and glucose absorption in young healthy male subjects. *European Journal of Clinical Nutrition*, **53**, 1–7.

Voragen, A. G. J. (1998) Technological aspects of functional food related carbohydrates. *Trends in Food Science and Technology*, **9**, 328–335.

Zárate, S., López-Leiva, M. H. (1990) Oligosaccharide formation during enzymatic lactose hydrolysis: a literature review. *Journal of Food Protection*, **53**(3), 262–268.

5

Emerging Prebiotic Carbohydrates

Ross Crittenden

5.1 Introduction

While studies exploring the benefits of dietary fructo-oligosaccharides (FOS) and galacto-oligosaccharides (GOS) dominate the scientific literature in the field of prebiotics, a variety of other carbohydrates also show promise as prebiotics. These span a wide diversity of carbohydrate structures from small sugars such as lactulose and various oligosaccharides, through to polysaccharides including arabinoxylans and resistant starches. The common factors with all of these carbohydrates is that they survive digestion during passage through the small intestine to reach the colon where they are fermentable by populations of bacteria purportedly advantageous to the health of the host. This chapter summarises the scientific evidence supporting the benefit of these emerging prebiotics as functional ingredients in the human diet and anticipates some of the future developments in prebiotic technologies.

5.2 The Prebiotic Concept

The two genera most often proposed as beneficial bacteria with which to augment the intestinal microbiota are the lactobacilli and bifidobacteria, both of which are numerically common members of the human intestinal microbiota. These bacteria are non-pathogenic, nonputrefactive, nontoxigenic, saccharolytic organisms that appear from available knowledge to provide little opportunity for deleterious activity in the intestinal tract. Indeed, there is a growing body of evidence that directly consuming these bacteria as probiotics can exert positive impacts on human health (Crittenden, 2004; Salminen

Prebiotics: Development and Application Edited by G.R. Gibson and R.A. Rastall

et al., 2004). As such, they are reasonable candidates to target in terms of restoring a favourable balance of intestinal species.

While the probiotic strategy aims to improve the intestinal microbiota through consumption of live, health-promoting bacteria, prebiotics are dietary supplements that selectively stimulate the proliferation and/or activity of beneficial populations *already resident* in the intestine (Gibson and Roberfroid, 1995). Successful prebiotics should remain largely undigested during passage through the upper gastrointestinal tract and be fermentable by the intestinal microbiota. Importantly, they should only stimulate beneficial populations and not the growth, pathogenicity, or putrefactive activity of potentially deleterious micro-organisms.

5.3 The Bifidogenic Effect of Prebiotics

The prebiotics identified to date all promote the proliferation of bifidobacteria in particular. Hence, they are often referred to as bifidogenic or bifidus factors. The mechanism(s) by which prebiotics promote the selective proliferation of bifidobacteria amongst the multitude of genera within the intestinal microbiota remain speculative. It is probably due to the ability of bifidobacteria to utilise these growth substrates relatively efficiently compared with other members of the microbiota, and their tolerance to the short-chain fatty acids (SCFA) and acidification of the microenvironment resulting from fermentation. Bifidobacteria are not fastidious with respect to carbohydrate carbon sources. Analysis of the *Bifidobacterium longum* genome revealed sequences for a large number of proteins specialised for the catabolism of a diverse range of carbohydrates (Schell *et al.*, 2002), no doubt contributing to its competitiveness in the colonic environment.

Prebiotics typically induce 10- to 100-fold increases in the size of the intestinal *Bifidobacterium* population (Crittenden, 1999). However, a range of factors may influence the magnitude of any increase in *Bifidobacterium* numbers, the most important being the initial size of the population within the intestinal tract. In comparing different trials conducted using FOS, Rao (1999) concluded that the scale of the bifidogenic response was inversely proportional to the size of the initial *Bifidobacterium* population rather than showing a strong dose response. In individuals colonised with an already large population of bifidobacteria (in the order of 10^8 CFU g^{-1} in faeces), prebiotic consumption appears not to increase *Bifidobacterium* numbers further.

5.4 Health Effects of Prebiotics

A number of largely prophylactic health targets have been proposed for prebiotics that, as might be expected, overlap considerably with the targets of probiotic interventions. The mechanisms of action remain largely theoretical, but rational hypotheses can be applied involving modifications to microbial population dynamics (Figure 5.1) and/or modifications to the metabolic activity of the microbiota (Figure 5.2). Proposed benefits in the gut include improved colonisation resistance, increased mineral absorption, immunomodulation, trophic and anti-neoplastic effects of SCFA, faecal bulking and reduced toxigenic

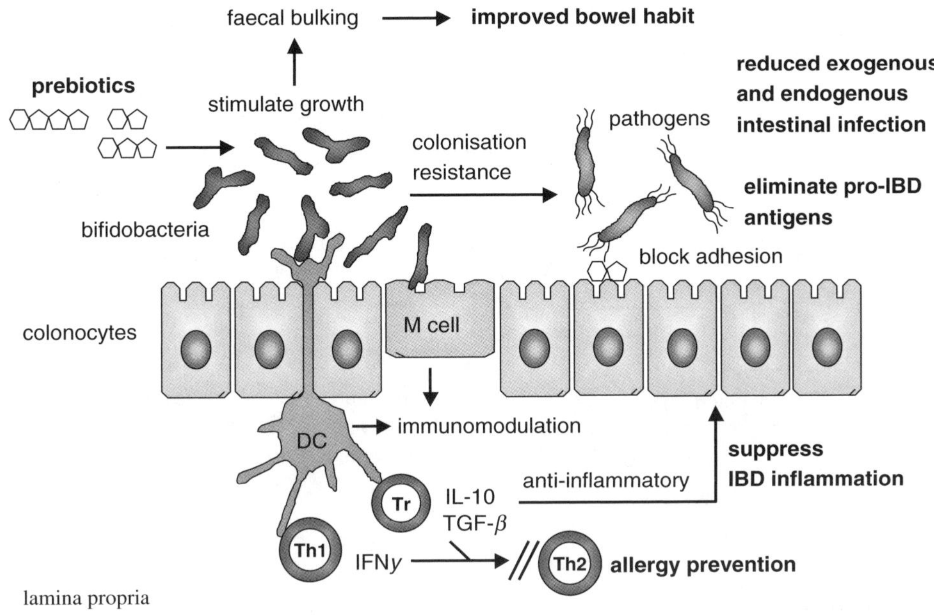

Figure 5.1 *Postulated mechanisms of health benefits induced by prebiotics via the selective proliferation of beneficial populations of intestinal bacteria. DC, dendritic cell; Th1, T helper cell type 1; Th2, T helper cell type 2; Tr, regulatory T cell*

microbial metabolism. The following sections outline the experimental evidence for impacts on intestinal microbiota and health benefits elicited by a number of emerging prebiotics.

5.5 Emerging Prebiotics

5.5.1 Lactulose

5.5.1.1 Synthesis. Alkaline isomerisation of lactose converts the glucose moiety to fructose yielding the indigestible disaccharide lactulose (4-*O*-β-galactopyranosyl-D-fructose) (Harju, 1986) (Figure 5.3). Lactulose has a long history of pharmaceutical use as a laxative and as a treatment for hepatic encaphalothopy and is still manufactured predominantly for these applications (Schumann, 2002). Through escaping hydrolysis by digestive enzymes, lactulose is readily fermented by the intestinal microbiota including bifidobacteria.

5.5.1.2 Impact on Intestinal Microecology. Lactulose was the first recognised bifidogenic carbohydrate, being reported to show this effect as early as the 1950s (Petuely, 1957). Besides the fructans (FOS and inulin) and GOS, it has the best supporting

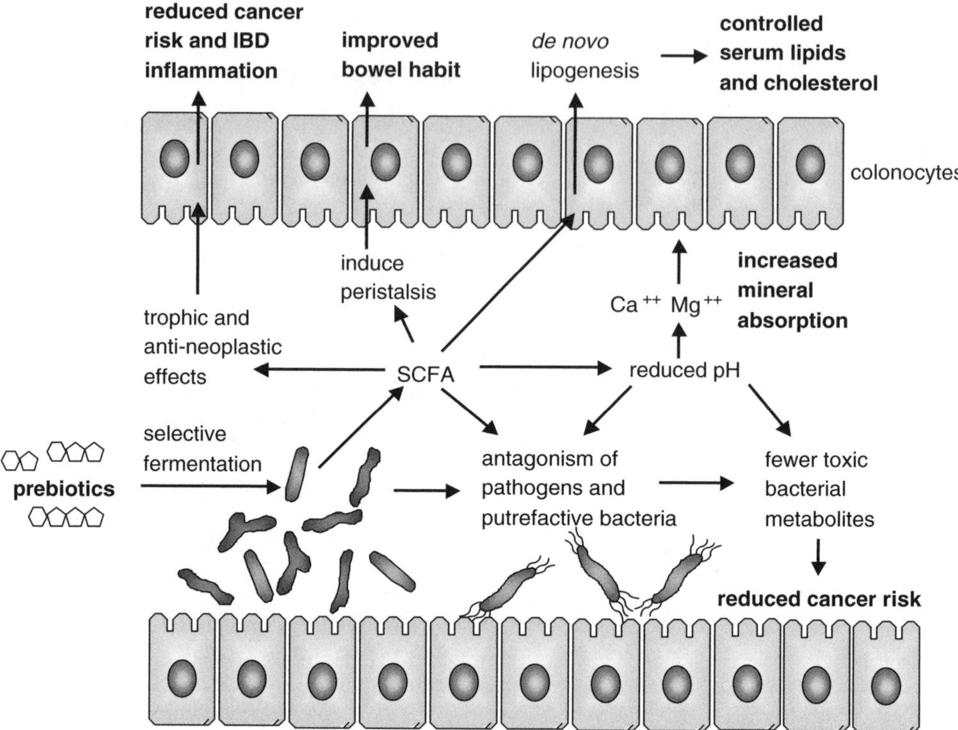

Figure 5.2 Postulated mechanisms of health benefits induced by prebiotics via the stimulation of beneficial microbial activities within selected populations of intestinal bacteria. IBD, inflammatory bowel diseases; SCFA, short-chain fatty acids

evidence of prebiotic effects in the human gastrointestinal tract. Four human studies using much lower doses of lactulose than used in pharmaceutical applications have demonstrated prebiotic action with minimal adverse intestinal effects such as bloating or excessive flatulence. Terada *et al.* (1992), showed that a dose as low as 3 g day^{-1} consumed for 14 days by healthy adults significantly raised the faecal numbers of bifidobacteria while lowering numbers of bacteriodes, streptococci, *Clostridium perfringens* and *Enterobacteriaceae*.

Three later double-blind, placebo-controlled, studies involving a further 28 individuals receiving lactulose treatments at between 10 and 20 g day^{-1} (Ballongue *et al.*, 1997; Tuohy *et al.*, 2002; Bouhnik *et al.*, 2004) reported similar bifidogenic responses with concomitant reductions in the numbers of potentially deleterious intestinal bacteria. Each study compared changes in the microbiota of a lactulose-fed group of volunteers to changes in a parallel group fed placebo (a readily digestible sugar), a superior experimental design than many of the early prebiotic studies that were not placebo controlled or blinded. Tuohy *et al.* (2002) also included analysis using fluorescent *in-situ* hybridisation (FISH) to complement traditional culturing methods in the microbiological analysis. In this study, lactulose was fed to 10 adult volunteers for 26–33 days at a dose of

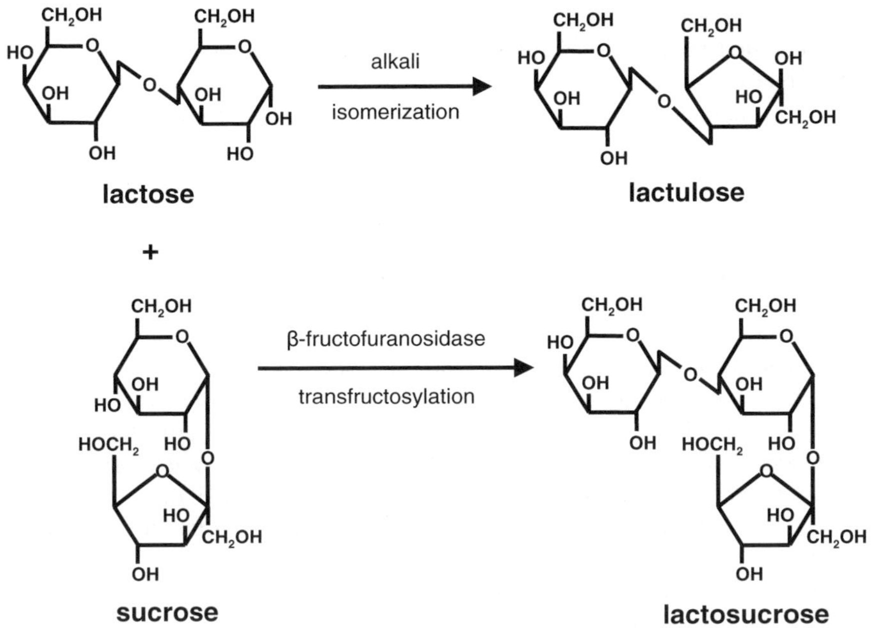

Figure 5.3 *Lactose is used as a substrate in the manufacture of the emerging prebiotics lactulose and lactosucrose*

10 g day^{-1} while a placebo (glucose/lactose) was fed to a parallel group. Lactulose consumption stimulated a significant increase in the numbers of faecal bifidobacteria (observed using both the culture and FISH techniques). A significant reduction in the numbers of clostridia was evident from the FISH analysis. Taken together, these four human studies provide good evidence to support the status of lactulose as a prebiotic carbohydrate. Lactulose is currently used in Japan as a prebiotic in a number of food and beverage applications with FOSHU (Foods for Specified Health Use) status, Japan's regulatory system governing functional food claims.

5.5.1.3 Health Effects. The four above-mentioned human studies have also investigated the impact of lactulose on putative risk factors for colon cancer stemming from toxigenic metabolic activities of the microbiota. The researchers used differing approaches to measure the impact of lactulose on the genotoxicity of the faecal environment, with varying results. Using the Comet assay to investigate DNA damage to colonocytes, Tuohy *et al.* (2002) showed no protective effect from lactulose consumption. Likewise, the study by Bouhnik *et al.* (2004), in which 2×5 g doses of lactulose were consumed per day for 6 weeks, also failed to show any modification to the levels of faecal bile acids nor neutral sterols. In contrast, the studies by Terada *et al.* (1992) and Ballongue *et al.* (1997) reported improvements in the metabolic activity of the intestinal microbiota through reductions in a number of mutagenic microbial metabolites and enzyme activities. Supporting these findings, De Preter *et al.* (2004) recently presented an approach using stable isotope-labelled biomarkers ($n = 8$) to demonstrate that feeding lactulose (2 × 10 g doses per day) significantly reduced the production of toxic fermentation metabolites in the human intestine. While reductions in the levels of toxic microbial metabolites is indicative of an improved colonic environment, direct links between lowering their synthesis through prebiotics to reductions in colon carcinogenesis remain to be substantiated.

Rodent models of colon cancer have provided some indications that feeding lactulose could provide benefit. Rowland *et al.* (1996) reported that lactulose consumption reduced DNA damage in the intestinal mucosa of human microbiota-associated rats dosed with carcinogenic 1,2-dimethylhydrazine dihydrochloride, while Challa *et al.* (1997) showed that feeding lactulose protected against the formation of aberrant crypt foci in rats challenged with the carcinogen azoxymethane. Lactulose has also been reported to reduce colonic adenoma recurrence in humans (Ponz de Leon and Roncucci, 1997). Lactulose was administered to 61 patients who had colonic adenomas removed, and recurrence was investigated over approximately 1 year. A significant reduction in recurrence was observed in the lactulose treated group (14.7 %) compared with a control group of 78 patients (35.9 %). However, as for all prebiotics, a great deal more research is required to discern if lactulose consumption has any protective function against colorectal cancer.

The ability of lactulose to improve colonisation resistance and prevent bacterial infections from the gut has been only scantly explored, but results to date provide indications of a potential role for lactulose in this capacity. Özaslan *et al.* (1997) observed lower caecal overgrowth and translocation of *Escherichia coli* in rats with obstructive jaundice when they were fed lactulose, while Bovee-Oudenhoven *et al.* (1997) reported that consumption of lactulose increased colonisation resistance against the invasive pathogen *Salmonella enteritidis* in a rat model. Indeed, lactulose consumption at high does (up to 60 g day^{-1}) is effective in eliminating salmonella from the intestinal tract of chronic human carriers and is used as a pharmaceutical for this purpose in some countries (Schumann, 2002). The mode of action is speculated to be acidification of the gut that prevents growth of the acid-sensitive pathogen. However, a potential down side to rapidly fermented, acidogenic sugars in the colon was reported in a second study by Bovee-Oudenhoven *et al.* (2003) while investigating the effect of lactulose on the translocation of *S. enteritidis* in rats. Feeding lactulose (and FOS) inhibited intestinal colonisation by *S. enteritidis,* but also left the animals more susceptible to pathogen translocation from the gut. Intestinal acidification was observed due to the rapid fermentation of lactulose, and while this inhibited the acid-sensitive pathogen in the intestinal lumen, it possibly also damaged the mucosa leading to an impaired barrier effect. Further research is needed to investigate possible negative impacts of rapidly fermented sugars on intestinal mucosa.

Elimination of specific bacterial antigens, immunomodulation, and trophic effects of SCFA on the intestinal epithelium have all been proposed as mechanisms by which prebiotics could provide benefit to sufferers of inflammatory bowel diseases (IBD) (Figures 5.1 and 5.2). Lactulose has recently been trialled in a rodent model of ulcerative colitis where consumption of this prebiotic successfully ameliorated colitis induced by sodium dextran sulphate (Rumi *et al.*, 2004). A dose-dependent improvement in a range of disease parameters was observed. Further research into the efficacy of lactulose in the treatment of IBD is warranted and should incorporate examinations of its impact upon the composition and activity of the microbiota.

5.5.2 Nondigestible Oligosaccharides

Many types of nondigestible oligosaccharides (NDOs) are manufactured for food applications (Crittenden and Playne, 1996). These ingredients have many physicochemical properties that make them attractive replacements for simple sugars, such as reduced sweetness, increased viscosity, lower freezing points and reduced Maillard reactivity. Additionally, they have numerous physiological benefits including low calorific value, low cariogenicity and low glycaemic index. They are used in a wide range of foods and beverages such as juices, yoghurt, confectionery, ice cream, and breads. Commercially available oligosaccharide products are not homogenous, but usually contain a number of oligosaccharides of differing molecular sizes (number of sugar moieties) or heterogeneous glycosidic linkages (Playne and Crittenden, 1996).

FOS and GOS in particular have been studied for their prebiotic effects. However, a number of other NDOs, to which less rigorous study has been so far applied, have at least indications of prebiotic potential. Those with the most accumulated evidence to date are isomalto-oligosaccharides (IMO), soybean oligosaccharides (SBO), xylo-oligosaccharides (XOS) and lactosucrose. Together with FOS, GOS, and lactulose, all of these oligosaccharides are recognised in Japan's FOSHU functional food regulation system as ingredients with beneficial health effects. This section outlines the evidence for prebiotic actions by these NDOs.

5.5.2.1 Isomalto-oligosaccharides

Synthesis. Of the emerging prebiotic oligosaccharides, IMO are used in the largest quantities for food applications. In Japan, the volume of IMOs manufactured is estimated to be three times greater than for either FOS or GOS (Nakakuki, 2002). They are produced from starch using a combination of immobilised enzymes in a two-stage process. In the first stage, starch is liquefied using α-amylase (EC 3.2.1.1). The liquefied starch is then processed in a second stage involving reactions catalysed by both β-amylase (EC 3.2.1.2) and α-glucosidase (EC 3.2.1.20). The β-amylase hydrolyses the liquefied starch to maltose and the transglucosidase activity of α-glucosidase then produces a mixture of IMO. The IMO mixtures produced contain oligosaccharides with predominantly α(1-6)-linked glucose residues with a degree of polymerisation (DP) ranging from 2–6, and also oligosaccharides with a mixture of α(1-6) and α(1-4) bonds such as panose (Playne and Crittenden, 2004) (Figure 5.4).

Impact on intestinal microecology. Unlike other prebiotic oligosaccharides, considerable digestion of IMO occurs during intestinal transit (Oku and Nakamura, 2003). However, at least some IMO is able to survive gastric transit to be fermented by the intestinal microbiota (Kohmoto *et al.*, 1992). *In vitro* fermentation studies using FISH analysis of changes to bacterial population sizes have shown that IMO promote the proliferation of bifidobacteria within the faecal microbiota (Rycroft *et al.*, 2001; Palframan *et al.*, 2002). Four small, noncontrolled human feeding trials, with a total of 48 healthy adults, have reported investigations of the effects of IMO consumption on gut microecology (Kohmoto *et al.*, 1988, 1992; Kaneko *et al.*, 1993, 1994). All showed a significant bifidogenic effect at doses of 10–20 g day^{-1}. Kohmoto *et al.* (1988) examined the impact on a range of bacterial populations using culture techniques and noted a selective effect on bifidobacteria. However, further controlled human feeding studies

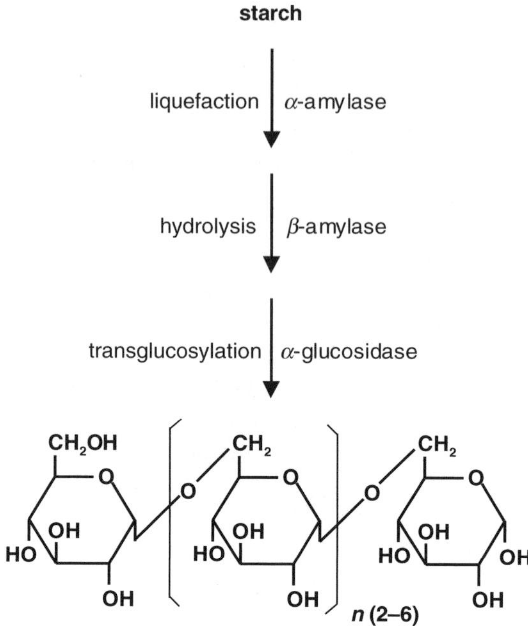

Figure 5.4 *Isomalto-oligosaccharides are manufactured from starch. In addition to the structures shown, commercial isomalto-oligosaccharide mixtures usually also contain some mixed α-(1-4) and α-(1-6) linked gluco-oligosaccharides*

employing culture and molecular techniques are required to determine the impact of IMO on the intestinal microbiota.

Health effects. Beneficial effects of IMO consumption have been reported in two human feeding studies investigating health parameters in specific populations. Chronic constipation and hyperlipidemia are common complications in hemodialysis patents. Wang *et al.* (2001) showed that feeding IMO to 20 hemodialysis patients for 4 weeks relieved constipation and significantly lowered total cholesterol and triglycerides in these patients, while elevating high density lipoprotein-cholesterol. Chen *et al.* (2001) also showed, in a small trial with seven elderly subjects, that consumption of IMO improved faecal frequency and stool bulk via increases in microbial biomass.

5.5.2.2 Soybean Oligosaccharides

Synthesis. SBO consist mainly of the trisaccharide raffinose [α-D-Gal-(1-6)-α-D-Glu-(1-2)-β-D-Fru] and the tetrasaccharide stachyose [α-D-Gal-(1-6)-α-D-Gal-(1-6)-α-D-Glu-(1-2)-β-D-Fru]. They are extracted directly from soybean whey rather than being commercially synthesised using enzymatic processes (Figure 5.5).

Impact on intestinal microecology. Both raffinose and stachyose are resistant to digestion and are readily fermented by bifidobacteria *in vitro* (Hayakawa *et al.*, 1990). The bifidogenic effects of SBO have been observed in a number of small, noncontrolled

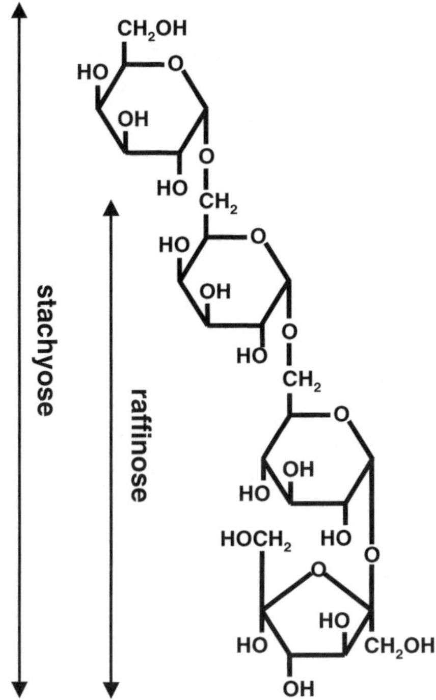

Figure 5.5 *Soybean oligosaccharides are extracted directly from soybean whey. The trisaccharide raffinose and the tetrasaccharide stachyose are the major oligosaccharide structures*

human feeding studies. Hara *et al.* (1997), showed that consumption of levels as low as 1–2 g day^{-1} could induce a substantial increase in the size of the faecal *Bifidobacterium* population in individuals with a low number of these bacteria in their intestines. Hayakawa *et al.* (1990) demonstrated the bifidogenic impact of SBO in a group of six volunteers fed 10 g day^{-1} for 3 weeks and also noted significant reductions in the sizes of the clostridia and peptostreptococci populations. Similar results were observed in studies by Wada *et al.* (1992), who fed seven volunteers 0.6 g day^{-1} SBO for 3 weeks, and by Benno *et al.* (1987) in which seven healthy adult volunteers were fed 15 g day^{-1} raffinose for 4 weeks. Taken together, these studies provide an indication that SBO have prebiotic activity which might be borne out in future controlled feeding studies.

Health effects. In addition to demonstrating a bifidogenic effect, Wada *et al.* (1992) also reported significant falls in the concentrations of toxigenic bacterial enzyme activities and toxic metabolites in the volunteers fed SBO. However, perhaps the most interesting finding for a potential benefit from SBO consumption has been demonstrated in a murine model of allergy. Allergic reactions mediated by the production of the antibody IgE are thought to be caused by a skewing of the specific immune response to the allergen at the T helper (Th) cell level towards an aberrant dominance of allergenic T helper cell type 2 (Th2) cell activity. Gram-positive bacteria, including probiotic lactic acid bacteria, have been demonstrated in animal models to redress this allergenic Th2 skew by inducing a counterbalancing, tolerogenic T helper cell type 1 (Th1) response (Shida *et al.*, 2002). In humans, probiotic intervention in infancy has been demonstrated to reduce the development of atopic eczema (Kalliomäki *et al.*, 2001, 2003), while breast-feeding (which induces a *Bifidobacterium* dominant intestinal microbiota) is known to be protective against the development of food allergies (Halken, 2004). Therefore, a number of researchers have speculated that prebiotic intervention might be effective in preventing the development of allergies.

Nagura *et al.* (2002) tested the ability of raffinose consumption to re-balance a Th2-biased immune response in a controlled study using an engineered murine model of IgE-mediated allergy to ovalbumin. Feeding a relatively high dose of raffinose stimulated a counterbalancing Th1-type immune response, reduced Th2 cell activity (measured via IL-4) and suppressed the synthesis of serum IgE to ovalbumin in response to long-term allergen challenge. Using a similar model, Yoshida *et al.* (2004) recently reported similar positive results for bifidogenic alginate-oligosaccharides, indicating that prebiotics could replicate the benefits seen for probiotics in allergy prevention.

5.5.2.3 Xylo-oligosaccharides

Synthesis. At present, XOS represent only a small proportion of the total oligosaccharide market (Nakakuki, 2002). The raw material for XOS synthesis is the polysaccharide xylan, extracted mainly from corncobs. The xylan is hydrolysed to XOS by controlled activity of the enzyme endo-1,4-β-xylanase (3.2.1.8). To produce higher purity oligosaccharide products, the monosaccharide xylose and high molecular mass carbohydrates are removed from the oligosaccharides using ultrafiltration and reverse osmosis processes (Crittenden and Playne, 1996). The process yields predominantly linear β(1-4)-linked XOS (mainly xylobiose, xylotriose and xylotetraose) as well as some oligosaccharides with branched arabinose residues (Figure 5.6).

Impact on intestinal microecology. A number of *in vitro* studies have shown that XOS are efficiently and relatively selectively fermented by bifidobacteria with the intestinal microbiota (Okazaki *et al.*, 1990; Yamada *et al.*, 1993; Jaskari *et al.*, 1998; Van Laere *et al.*, 2000; Rycroft *et al.*, 2001; Crittenden *et al.*, 2002). In a noncontrolled human study, Okazaki *et al.* (1990) fed 5 g day^{-1} of XOS to nine adults for 3 weeks and examined the impact on the population size of a range of intestinal bacteria including potential pathogens and putrefactive bacteria. They observed a significant change only for bifidobacteria and megasphera (both increased), indicating a selective proliferative effect. The bifidogenic effect of XOS was also observed in a later rat study (Campbell *et al.*, 1997), which also showed the XOS induced a larger increase in *Bifidobacterium* numbers compared with an equivalent dose of FOS. However, to date there has been a dearth of well-controlled animal and human feeding studies to confirm the prebiotic activity of XOS. While they show promise, more research is required before XOS can conclusively be claimed as prebiotics.

5.5.2.4 Lactosucrose

Synthesis. Lactosucrose is a nondigestible trisaccharide [β-D-galactopyranosyl-(1$\rightarrow$4)-α-D-glucopyranosyl β-D-frucrofuranoside] manufactured from a mixture of lactose and

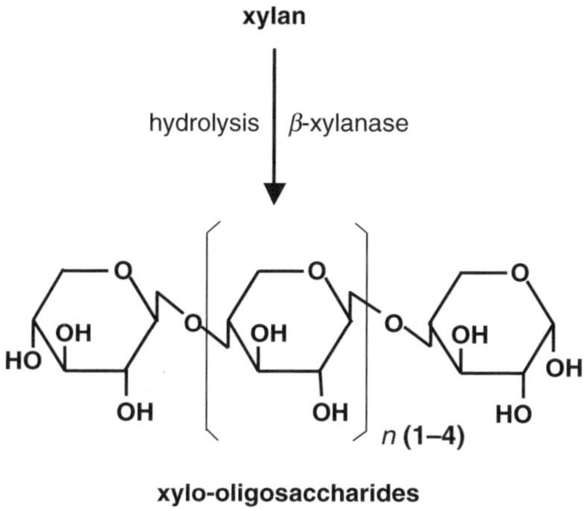

xylan

hydrolysis | β-xylanase

xylo-oligosaccharides

Figure 5.6 *Xylo-oligosaccharides are manufactured by the controlled enzymatic hydrolysis of the polysaccharide xylan*

sucrose using the transfructosylation activity of the enzyme β-fructofuranosidase (EC 3.2.1.26) (Figure 5.3). It is used mainly in Japan in a range of drinks, confectioneries, pet foods and as a table sugar (Playne and Crittenden, 2004).

Impact on the intestinal microbiota. Bifidobacteria ferment lactosucrose well *in vitro* relative to most enterobacteriaceae (Minami *et al.*, 1983) and its consumption has been observed to increase the size of the faecal *Bifidobacterium* population in several small, noncontrolled trials on healthy adults (Fujita *et al.*, 1991; Yoneyama *et al.*, 1992; Hara *et al.*, 1994; Ohkusa *et al.*, 1995) and in patients with inflammatory bowel disease (Teramoto *et al.*, 1996). Hara *et al.* (1994) reported that lactosucrose consumption induced a drop in the numbers of faecal clostridia and bacteroides and a concomitant reduction in toxigenic enzyme activities and metabolites. As with the other oligosaccharides described in this section, lactosucrose shows potential to act as a prebiotic but requires further confirmatory studies.

5.5.2.5 Other Potentially Prebiotic Saccharides. A wide range of other saccharides have been reported to show some potential as prebiotics. These include *inter alia* lactitol, which produced similar effects to lactulose in a double-blind, placebo-controlled human feeding study, but was slightly less potent (Ballongue *et al.*, 1997), and glucono-δ-lactone, which induced a selective increase in faecal *Bifidobacterium* numbers in humans when dosed at 3 or 9 g day^{-1} in a small, noncontrolled study (Asano *et al.*, 1994). Depolymerised pyrodextrin (Satouchi *et al.*, 1996); partially hydrolysed guar gum (Okubo *et al.*, 1994); palatinose polycondensates (Kashimura *et al.*, 1989); α-gluco-oligosaccharides (Djouz *et al.*, 1995) and chitosan-oligosaccharides (Lee *et al.*, 2002) have all been proposed as potential bifidogenic factors. All have only preliminary evidence and require further investigation with properly controlled human studies to assess their prebiotic potential.

5.6 Polysaccharides

Large, slowly fermented polysaccharides with prebiotic effects may have advantages over small, rapidly fermented sugars such as lactulose and NDOs. These include the ability to be tolerated at higher doses by consumers with reduced risk of side effects such as intestinal discomfort and flatulence caused by excessive gas formation; mucosal damage from rapid acidification; and the laxative effect of too high concentrations of small sugars in the colon. Perhaps more importantly, large polysaccharides supply a persistent source of fermentable carbohydrate throughout the length of the colon rather than being completely fermented proximally. This is particularly pertinent to the prevention of colon cancer since the distal colon and rectum are the major sites of disease in humans.

5.6.1 Resistant Starches

5.6.1.1 Classification of Resistant Starch. Not all starch that is consumed in the diet is hydrolysed and absorbed in the small intestine. Some starches are relatively resistant to enzymatic hydrolysis and pass undigested to the colon. This 'resistant starch' (RS) provides a major source of fermentable carbohydrate to the bacteria in the colon (Cummings and Macfarlane, 1997). Resistant starch cannot be regarded as a homogeneous material, and the multitudes of sources and different structures are generally classified into four categories (Brown, 1996). Type I are physically inaccessible starch granules such as whole or partially milled grains. Type II are native starch granules found in, for example, potato, banana and high-amylose maize. Retrograded starch formed during starch processing make up Type III, while Type IV are chemically modified starches altered by cross-linking, esterification, or etherification. The degree to which a RS variety remains unhydrolysed in the small intestine and its accessibility for bacterial fermentation in the colon is dependent on its source and structure.

5.6.1.2 Impact on Intestinal Microecology. Although the focus of much of the earlier research on RS related to its butyrogenic capacity in the colon, there is now growing interest in the potential of RS to act as a prebiotic. Evidence for prebiotic effects is accumulating, but to date remains limited to animal studies. A number of rodent and pig studies have demonstrated that various RSs stimulate the proliferation of bifidobacteria and/or lactobacilli in the intestinal tract (Kleessen *et al.*, 1997; Brown *et al.*, 1998; Silvi *et al.*, 1999; Wang *et al.*, 1999; Bielecka *et al.*, 2002; Wang *et al.*, 2002; Le Blay *et al.*, 2003). However, not all RSs are similarly potent (Wang *et al.*, 2002) and some varieties do not appear to stimulate the proliferation of these bacteria at all (Bird *et al.*, 2004; Crittenden *et al.*, 2005).

The effectiveness of RS in a synbiotic combination has been examined in a pig study involving concurrent feeding of high amylose maize starch and a probiotic *Bifidobacterium.* Inclusion of RS in the formulation significantly increased *Bifidobacterium* numbers in the colon compared with feeding the probiotic alone. Continued feeding of the RS following cessation of probiotic administration lengthened persistence of the *Bifidobacterium* in the intestinal tract (Figure 5.7) (Crittenden *et al.*, 2005), indicating that the probiotic was utilising the RS as a growth substrate in the intestinal tract.

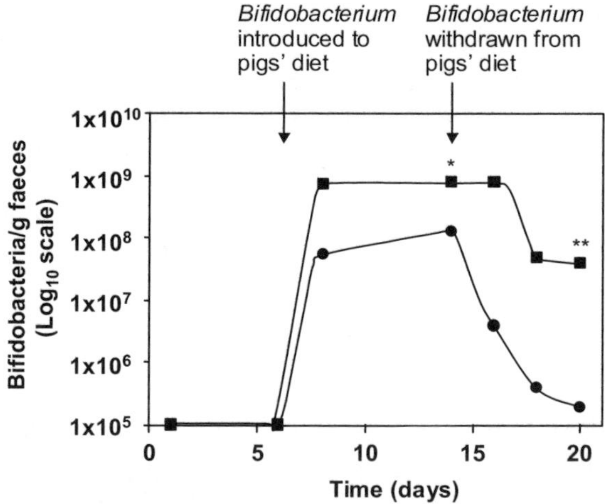

Figure 5.7 *The effect of resistant starch on the intestinal colonisation and persistence of a co-administered probiotic Bifidobacterium. In a synbiotic feeding trial (described by Brown et al., 1997), a probiotic Bifidobacterium was fed to pigs in combination with either high amylose maize resistant starch (RS) (■) or a readily digestible starch (DS) (•). After 1 week, the pigs fed the RS/probiotic combination had significantly higher faecal Bifidobacterium numbers (*$P < 0.05$) than pigs fed DS/probiotic. At day 14, the probiotic was removed from the diet, but feeding of the starch sources continued for a further 6 days. Pigs fed RS retained significantly higher numbers of bifidobacteria in their faeces compared with pigs fed DS (**$P < 0.01$)*

Despite data in animal models showing bifidogenic effects of feeding resistant starch, the promotion of RS as a potential prebiotic remains controversial largely because a wide range of intestinal bacteria are amylolytic, drawing into question the selectivity of RS fermentation. Thus far, controlled human feeding trials to assess the impact of RS consumption on population sizes within the human intestinal microbiota remain to be reported. Since many of the potential health benefits arising from dietary RS appear likely to stem from production of butyrate in the colon, future studies should focus on both the numbers and activities of butyrate producing bacteria in addition to lactobacilli and bifidobacteria.

5.6.1.3 Health Effects. While FOS is rapidly fermented proximately in the intestinal tract, RS is fermented more slowly and distally (Le Blay *et al.*, 2003), thereby supplying SCFA to colonocytes in the colorectum. RS is known to promote microbial production of butyrate in the colon, which has beneficial trophic effects on the colonic epithelium (Topping and Clifton, 2001). In experimental animal models of colon cancer, RS has been shown to promote apoptosis (Le Leu *et al.*, 2002) and reduce the formation of aberrant crypt foci (Perrin *et al.*, 2001; Nakanishi *et al.*, 2003). Consumption of RS has also been reported to decrease putative markers of colon cancer risk stemming from microbial activity in the colon in both animal and human studies (Phillips *et al.*, 1995; Hylla *et al.*, 1998; Silvi *et al.*, 1999; Grubben *et al.*, 2001). However, as for all emerging prebiotics, the capacity of dietary RS to significantly contribute to a reduced incidence of colorectal cancer remains unproven.

The ability of dietary RS to stimulate butyrate synthesis by the microbiota and to alter intestinal microecology has prompted studies into its efficacy in controlling IBD. Using different rodent models of ulcerative colitis, both Jacobasch *et al.* (1999) and Moreau *et al.* (2003) showed amelioration of disease in RS-fed animals compared with controls. In the latter study, the RS diet (retrograded high amylose maize starch) significantly outperformed a diet with an equivalent dose of FOS (Actilight) in terms of reducing inflammation.

5.6.2 Dietary Fibres

5.6.2.1 Cereal Fibres. Since they escape digestion and supply fermentable carbohydrate in the colon, there is increasing interest in the impact of dietary fibres on intestinal microecology. To date, there have only been a few studies exploring the prebiotic potential of dietary fibres, with fibres from cereal crops perhaps the most studied. In cereals, β-glucans and arabinoxylans are the major dietary fibre constituents that are fermentable by bacteria in the human gastrointestinal tract (Fleming *et al.*, 1983; Fincher and Stone, 1986; Henry, 1987; Stevens and Selvendran, 1988). In general, cereal β-glucans are unsubstituted glucose polymers with a mixture of β(1-3) and β(1-4) glycosidic linkages, while cereal arabinoxylans consist of a backbone of (1-4)-linked β-D-xylose residues that are either non, singularly or doubly substituted with predominantly terminal α-L-arabinose moieties.

5.6.2.2 Impact on Intestinal Microecology. Bifidobacteria and lactobacilli cannot ferment cereal β-glucans well *in vitro* (Crittenden *et al.*, 2002), but can utilise oligosaccharides resulting from its partial hydrolysis (Jaskari *et al.*, 1998; Kontula *et al.*, 1998). In contrast, *in vitro* studies have shown that *Bifidobacterium longum* and *Bifidobacterium adolescentis* are able to ferment arabinoxylan from a variety of cereal sources in addition to arabinoxylan oligosaccharides (Van Leare *et al.*, 2000; Crittenden *et al.*, 2002). Importantly, potentially deleterious intestinal bacteria such as *Escherichia coli*, *Clostridium perfringens* or *Clostridium difficile*, do not directly ferment these substrates. When fed to mice, rye bran rich in arabinoxylan produced a strong bifidogenic effect (Oikarinen *et al.*, 2003). Arabinoxylan is slowly fermented by the intestinal microbiota compared with inulin (Karppinen *et al.*, 2000), and behaves like a soluble, fermentable fibre (Lu *et al.*, 2000). Further research into the prebiotic potential of arabinoylan is warranted.

5.6.2.3 Health Effects. An arabinoxylan-rich germinated barley product has been reported by Kanauchi *et al.* (1999) to induce the proliferation of bifidobacteria in the human intestine. In rodent models of IBD, and in two, small, human studies of subjects with ulcerative colitis, consumption of the germinated barley product ameliorated inflammation (Bamba *et al.*, 2002; Fukuda *et al.*, 2002; Kanauchi *et al.*, 2002, 2003). However, as for all known and emerging prebiotics, convincing evidence of a consistent clinical benefit in the treatment of IBD remains to be demonstrated in large, randomised, double-blind, placebo-controlled studies.

5.7 On the Horizon

5.7.1 Synthesising Human Milk Oligosaccharides

Human milk oligosaccharides (HMO) are the original prebiotics. They stimulate the development of an intestinal microbiota dominated by bifidobacteria in breast-fed infants (Harmsen *et al.*, 2000). While the bifidogenic effect of HMO can be emulated using FOS and GOS (Bohem *et al.*, 2002; Moro *et al.*, 2002) there is increasing evidence for roles of HMO outside their bifidogenic impact in the gut. These include blocking adhesion of pathogens to the intestinal mucosa (Newburg, 1999; Martin-Sosa *et al.*, 2002; Morrow *et al.*, 2004) and roles in developing cognition (Wang and Brand-Miller, 2003). HMO are complex with more than 130 identified structures (Brand-Miller and McVeagh, 1999). Each individual oligosaccharide is based on a variable combination of glucose, galactose, sialic acid, fucose and/or *N*-acetylglucosamine, with varied sizes and linkages accounting for the considerable variety (Kunz *et al.*, 2000).

Attempts to synthesise the full range of HMO for commercial production have been thwarted by the complexity of the oligosaccharide structures, although endeavours to develop suitable manufacturing processes continue. An Italian research team have reported chemico-enzymatic methods to produce a number of smaller fucosylated and sialylated HMO structures (La Ferla *et al.*, 2002; Rencurosi *et al.*, 2002), while Dumon *et al.* (2004) describe an *in vivo* approach using *E. coli* cells engineered with fucosyltransferase genes from *Helicobacter pylori* to produce fucosylated HMO. Bacterial exopolysaccharides have remained largely unexplored as a source of prebiotic substrates, but a polysaccharide produced by *Streptococcus macedonicus* has been noted to contain a repeated oligosaccharide unit that forms the backbone to many HMOs (Vincent *et al.*, 2001). There is a ready market in infant milk formulas for oligosaccharides that more closely replicate all of the properties of HMOs and research to synthesise them will no doubt continue.

5.7.2 Patent Activity

Much of the patent literature in the mid-1990s focused on discovering new prebiotic forms and developing novel manufacturing processes to produce them (Playne and Crittenden, 1996). A decade later, new prebiotics with novel properties continue to be registered although the focus of innovation has now slanted towards novel nutraceutical and food formulations that include known prebiotics and complementary synbiotic combinations. On the theme of replicating activities of HMO, a patent from Wyeth reports on the combination of FOS and sialylactose having a synergistic prebiotic effect (Wilson, 2003), while novel acetylated oligosaccharide forms with the ability to selectively inhibit pathogen and bacterial toxin binding have also been described (Natunen *et al.*, 2003). New bioprocesses include methods to produce longer chain IMO and new gluco-oligosaccharides (van Geel-Schutten, 2003; Vercauteren and Nguyen, 2004), while emerging applications include antioxidant activity (Gueuxe *et al.*, 2004), treatment of chronic hepatitis C infection (Morinaga Milk Industry Co., 2002), and as ingredients to prevent the development of allergies (Rautonen *et al.*, 2004).

5.7.3 Prebiotic Effects in Bacterial Groups Beyond Bifidobacteria

Bifidobacteria and lactobacilli have been the dominant intestinal genera targeted with both probiotic and prebiotics. Evidence to date supports their beneficial role in the intestinal tract (Crittenden, 2004; Salminen *et al.*, 2004). However, they are but two of a vast multitude of bacteria within the intestinal microbiota that may potentially confer benefits on the host. The development of culture-independent molecular tools to examine the diversity of organisms within the intestinal microbiota has unveiled a remarkable complexity. Whole new bacterial taxa continue to be identified. As we gradually shed light on the activities of these organisms and their interactions with the host in health and disease new beneficial organisms and activities will undoubtedly be identified. The challenge will be to find or design selective prebiotics to stimulate these particular organisms.

In addition to modulating bacterial numbers, it is the effects of prebiotics on the metabolic activities of the microbiota that are anticipated to underpin many of the health benefits to the host. Of the SCFA produced during fermentations of carbohydrates in the colon, butyrate has attracted particular interest due to its anti-neoplastic and tropic effects on the colonic mucosa (Topping and Clifton, 2001). Butyrate produced by the intestinal microbiota supplies approximately 70 % of the metabolic energy requirements of colonocytes. Lactobacilli and bifidobacteria are not directly butyrogenic and hence other bacteria are primarily responsible for producing this SCFA in the colon. Stimulating the specific proliferation, and more importantly, the activity, of butyrogenic bacteria within the colon may well be a valid target for future prebiotics. This will require new tools to examine the impact of prebiotics on the *in situ* metabolic activities of specific bacterial communities.

5.7.4 Synergies between Prebiotics with Different Fermentation Kinetics

A number of researchers have hypothesised that synergies might exist between NDOs that stimulate a bifidogenic response and SCFA production in the proximal colon and larger polysaccharides that sustain a source of fermentable carbohydrate through to the distal colon. ORAFTI (Belgium) market a prebiotic (Synergy 1) that includes both short chain FOS and the longer chain fructan inulin and have shown synergistic effects in this combination for a range of physiological effects (Van Loo, 2004). Similarly, complementary effects have been noted for FOS/inulin and resistant starches (Younes *et al.*, 2001; Le Blay *et al.*, 2003). Development of synergistic prebiotic combinations to optimise the composition and activity of the microbiota throughout the length of the intestinal tract, or to target specific intestinal regions (e.g. for treatment of IBD) is set to provide new avenues for future research.

5.8 Conclusions

Though the greatest volume of research and evidence for prebiotic effects has been accrued for FOS and inulin, there is accumulating evidence of prebiotic actions by a number of nondigestible carbohydrates. Lactulose has strong claims to be a prebiotic, while there is promising evidence for prebiotic activity by IMO, XOS and SBO.

Lactosucrose is a probable candidate, but with currently less published evidence than for the other oligosaccharides.

Indications of prebiotic action are not limited to small, rapidly fermented sugars, but also include large size polysaccharides. Studies of prebiotic action by resistant starches have been limited to animal studies, but evidence collected to date provides a persuasive case for continued exploration of their prebiotic potential. Likewise, there is growing interest in the impact of dietary fibres on the composition and activity of the intestinal microbiota, and arabinoxylans in particular warrant further study.

Bifidobacteria are the dominant group of bacteria stimulated by all prebiotics developed so far. That such a range of diverse carbohydrate structures can promote the selective proliferation of bifidobacteria is testament to the remarkable metabolic agility of these organisms. However, as we learn more about the intestinal microbiota and the roles of various bacteria in health and disease, new beneficial bacterial groups are likely to be identified. Stimulating specific populations of butyrogenic bacteria in the colon may well be the next target for prebiotics.

Although traditional microbiology culture methods have enabled some assessment of the selectivity of prebiotics, new molecular techniques that enable analysis of noncultivable bacteria should be applied to future studies investigating the impact of emerging prebiotics on the colonic microbiota. It should be emphasised that altering the microbial population dynamic is only one aspect of prebiotic action. While stimulating the proliferation of particular groups of bacteria might be important for some health effects (e.g. immunomodulation), this may be secondary to specifically altering the metabolic activity of groups within the microbiota for other effects (e.g. anti-cancer). *In situ* measurement of bacterial activities remains problematic, but functional genomics may provide a new avenue to explore the interactions between prebiotics, the intestinal microbiota and the host in health and disease.

Preliminary evidence for specific health benefits has been gathered for many of the emerging prebiotics discussed in this chapter, but has been largely confined to animal studies so far. As for all prebiotics, the challenge remains to demonstrate clinically relevant benefits to health in well designed and controlled human trials.

References

Asano, T., Yuasa, K., Kunugita, K., Teraja, T. and Mitsuoka, T. (1994) Effects of gluconic acid on human faecal bacteria. *Microbial Ecology in Health and Disease* **7**, 247–256.

Ballongue, J., Schumann, C. and Quignon, P. (1997) Effects of lactulose and lactitol on colonic microflora and enzymatic activity. *Scandinavian Journal of Gastroenterology* **32** (Suppl. 222) 41–44.

Bamba, T., Kanauchi, O., Andoh, A. and Fujiyama, Y. (2002) A new prebiotic from germinated barley for nutraceutical treatment of ulcerative colitis. *Journal of Gastroenterology and Hepatology* **17**, 818–824.

Benno, Y., Endo, K., Shiragami, N., Sayama, K. and Mitsuoka, T. (1987) Effects of raffinose intake on human fecal microflora. *Bifidobacteria Microflora* **6**, 59–63.

Bielecka, M., Biedrzycka, E., Majkowska, A., Juskiewicz, J. and Wroblewska, M. (2002) Effect of non-digestible oligosaccharides on gut microecosystem in rats. *Food Research International* **35**, 139–144.

Bird, A. R., Jacksom, M., King, R. A., Davies, D. A., Usher, S. and Topping, D. L. (2004) A novel high-amylose barley cultivar (*Hordeum vulgare* var. Himalaya 292) lowers plasma cholesterol and alters indices of large-bowel fermentation in pigs. *British Journal of Nutrition* **92**, 607–615.

Bohem, G., Lidestri, M., Casetta, P., Jelinek, J., Negretti, F., Stahl, B. and Marini, A. (2002) Supplementation of a bovine milk formula with an oligosaccharide mixture, increases counts of faecal bifidobacteria in preterm infants. *Archives of Disease in Childhood Fetal and Neonatal Edition* **86**, F178–F181.

Bouhnik, Y., Attar, A., Joly, F. A., Riottot, M., Dyard, F. and Flourie, B. (2004) Lactulose ingestion increases faecal bifidobacterial counts: a randomised double-blind study in healthy humans. *European Journal of Clinical Nutrition* **58**, 462–466.

Bovee-Oudenhoven, I. M. J., Termont, D. M. S. L., Heidt, P. J. and Van der Meer, R. (1997) Increasing the intestinal resistance of rats to the invasive pathogen *Salmonella enteritidis*: additive effects of dietary lactulose and calcium. *Gut* **40**, 497–504.

Bovee-Oudenhoven, I. M. J., ten Bruggencate, S. J. M., Lettink-Wissink, M. L. G. and van der Meer, R. (2003) Dietary fructo-oligosaccharides and lactulose inhibit intestinal colonisation but stimulate translocation of salmonella in rats. *Gut* **52**, 1572–1578.

Brown, I. (1996) Complex carbohydrates and resistant starch. *Nutrition Reviews* **54**, S115–S119.

Brand-Miller, J. and McVeagh, P. (1999) Human milk oligosaccharides: 130 reasons to breast-feed. *British Journal of Nutrition* **82**, 333–335.

Brown, I., Warhurst, M., Arcot, J., Playne, M., Illman, R. J. and Topping, D. L. (1997) Fecal numbers of bifidobacteria are higher in pigs fed *Bifidobacterium longum* with a high amylose cornstarch than with a low amylose cornstarch. *Journal of Nutrition* **127**, 1822–1827.

Brown, I. L., Wang, X., Topping, D. L., Playne, M. J. and Conway, P. L. (1998) High amylose maize starch as a versatile prebiotic for use with probiotic bacteria. *Food Australia* **50**, 603–610.

Campbell, J. M., Fahey Jr., G. C. and Wolf, B. W. (1997) Selected indigestible oligosaccharides affect large bowel mass, cecal and fecal short-chain fatty acids, pH and microflora in rats. *Journal of Nutrition* **127**, 130–136.

Challa, A., Rao, D. R., Chawan, C. B. and Shackelford, L. (1997) *Bifidobacterium longum* and lactulose suppress azoxymethane-induced colonic aberrant crypt foci in rats. *Carcinogenesis* **18**, 517–521.

Chen, H. L., Lu, Y. H., Lin, J. J. and Ko, L. Y. (2001) Effects of isomalto-oligosaccharides on bowel functions and indicators of nutritional status in constipated elderly men. *Journal of the American College of Nutrition* **20**, 44–49.

Crittenden, R. G. (1999) Prebiotics. In: Tannock, G. W. (ed.), *Probiotics: a Critical Review*. Horizon Scientific Press, Wymondham, pp. 141–156.

Crittenden, R. (2004) An update on probiotic bifidobacteria. In: Salminen, S., von Wright A. and Ouwerhand A. (eds), *Lactic Acid Bacteria: Microbiological and Functional Aspects*. Marcel Dekker, New York, pp. 125–157.

Crittenden, R., Bird, A. R., Gopal, P., Henriksson, A., Lee, Y. K. and Playne, M. J. (2005) Probiotic research in Australia, New Zealand and the Asia-Pacific region. *Current Pharmaceutical Design* **11**, 37–53.

Crittenden, R., Karppinen, S., Ojanen, S., Tenkanen, M., Fagerström, R., Mättö, J., Saarela, M., Mattila-Sandholm, T. and Poutanen, K. (2002) *In vitro* fermentation of cereal dietary fibre carbohydrates by probiotic and intestinal bacteria. *Journal of the Science of Food and Agriculture* **82**, 1–9.

Crittenden, R. G. and Playne, M. J. (1996) Production, properties, and applications of food-grade oligosaccharides. *Trends in Food Science and Technology* **7**, 353–361.

Cummings, J. H. and Macfarlane, G. T. (1997) Role of intestinal bacteria in nutrient metabolism. *Journal of Clinical Nutrition* **16**, 3–11.

De Preter, V., Geboes, K., Verbrugghe, K., De Vuyst, L., Vanhoutte, T., Huys, G., Swings, J., Pot, B. and Verbeke, K. (2004) The *in vivo* use of the stable isotope-labelled biomarkers lactose-[N-15]ureide and [H-2(4)]tyrosine to assess the effects of pro- and prebiotics on the intestinal flora of healthy human volunteers. *British Journal of Nutrition* **92**, 439–446.

Djouzi, Z., Andrieux, C., Pelenc, V., Somarriba, S., Popot, F., Paul, F., Monsan, P. and Szylit, O. (1995) Degradation and fermentation of alpha-gluco-oligosaccharides by bacterial strains from human colon: *in vitro* and *in vivo* studies in gnotobiotic rats. *Journal of Applied Bacteriology* **79**, 117–127.

Dumon, C., Samain, E. and Priem, B. (2004) Assessment of the two *Helicobacter pylori* alpha-1,3-fucosyltransferase ortholog genes for the large-scale synthesis of LewisX human milk oligosaccharides by metabolically engineered *Escherichia coli*. *Biotechnology Progress* **20**, 412–419.

Fincher, G. B. and Stone, B. A. (1986) Cell walls and their components in cereal grain technology. *Advances in Cereal Science and Technology* **6**, 207–295.

Fleming, S. E., Marthinsen, D. and Kuhnlein, H. (1983) Colonic function and fermentation in men consuming high fiber diets. *Journal of Nutrition* **113**, 2535–2544.

Fujita, K., Hara, K., Sakai, S., Miyake, T., Yamashita, M., Tsunstomi, Y. and Mitsuoka, Y. (1991) Effects of 4-β-D-galactosylsucrose (lactosucrose) on intestinal flora and its digestibility in humans. *Journal of the Japanese Society for Starch Science* **38**, 249–255.

Fukuda, M., Kanauchi, O., Araki, Y., Andoh, A., Mitsuyama, K., Takagi, K., Toyonaga, A., Sata, M., Fujiyama, Y., Fukuoka, M., Matsumoto, Y. and Bamba, T. (2002) Prebiotic treatment of experimental colitis with germinated barley foodstuff: a comparison with probiotic or antibiotic treatment. *International Journal of Molecular Medicine* **9**, 65–70.

Gibson, G. R. and Roberfroid, M. B. (1995) Dietary modulation of the human colonic microbiota — introducing the concept of prebiotics. *Journal of Nutrition* **125**, 1401–1412.

Grubben, M. J., van den Braak, C. C., Essenberg, M., Olthof, M., Tangerman, A., Katan, M. B. and Nagengast, F. M. (2001) Effect of resistant starch on potential biomarkers for colonic cancer risk in patients with colonic adenomas: a controlled trial. *Digestive Diseases and Sciences* **46**, 750–756.

Gueuxe, E., Rayssiguier, Y., Bussoerolles, J. and Mazur, A. (2004) Use of prebiotics to prepare food, neutraceutical or pharmaceutical products for preventing or treating oxidative stress, especially caused by high fructose consumption. *Patent WO2004056210-A1*.

Halken, S. (2004) Prevention of allergic disease in childhood: clinical and epidemiological aspects of primary and secondary allergy prevention. *Pediatric Allergy and Immunology* **15** (Suppl) 16., 4–5, 9–32.

Hara, H., Li, S.-T., Sasaki, M., Maruyama, T., Terada, A., Ogata, Y., Fujita, K., Ishigami, H., Hara, K., Fujimori, I. and Mitsuoka, T. (1994) Effective dose of lactosucrose on fecal flora and fecal metabolites of humans. *Bifidobacteria Microflora* **13**, 51–63.

Hara, T., Ikeda, N., Hatsumi, K., Watabe, J., Iino, H. and Mitsuoka, T. (1997) Effects of small amount ingestion of soybean oligosaccharides on bowel habits and fecal flora of volunteers. *Japanese Journal of Nutrition* **55**, 79–84.

Harmsen, H. J. M., Wildeboer-Veloo, A. C. M., Raangs, G. C., Wagendorp, A. A., Klijn, N., Bindels, J. G. and Welling, G. W. (2000) Analysis of intestinal flora development in breast-fed and formula-fed infants by using molecular identification and detection methods. *Journal of Pediatric Gastroenterology and Nutrition* **30**, 61–67.

Harju, M. (1986) Lactulose as a substrate for β-galactosidases I. *Milchwissenschaft* **41**, 281–282.

Hayakawa, K., Mizutani, J., Wada, K., Masai, T., Yoshihara, I. and Mitsuoka, T. (1990) Effects of soybean oligosaccharides on human faecal microflora. *Microbial Ecology in Health and Disease* **3**, 293–303.

Henry, R. J. (1987) Pentosan and $(1\rightarrow3)$, $(1\rightarrow4)$-β-glucan concentrations in endosperm and whole grain of wheat, barley, oats and rye. *Journal of Cereal Chemistry* **6**, 253–258.

Hylla, S., Gostner, A., Dusel, G., Anger, H., Bartram, H. P., Christl, S. U., Kasper, H. and Scheppach, W. (1998) Effects of resistant starch on the colon in healthy volunteers: possible implications for cancer prevention. *American Journal of Clinical Nutrition* **67**, 136–142.

Jacobasch, G., Schmiedl, D., Kruschewski, M. and Schmehl, K. (1999) Dietary resistant starch and chronic inflammatory bowel diseases. *International Journal of Colorectal Diseases* **14**, 201–211.

Jaskari, J., Kontula, P., Siitonen, A., Jousimies-Somer, H., Mattila-Sandholm, T. and Poutanen, K. (1998) Oat β-glucan and xylan hydrolysates as selective substrates for *Bifidobacterium* and *Lactobacillus* strains. *Applied Microbiology and Biotechnology* **49**, 175–181.

Kalliomäki, M., Salminen, S., Arvilommi, H., Kero, P., Koskinen, P. and Isolauri, E. (2001) Probiotics in primary prevention of atopic disease: a randomised placebo-controlled trial. *Lancet* **357**, 1076–1079.

Kalliomäki, M., Salminen, S., Poussa, T., Arvilommi, H. and Isolauri, E. (2003) Probiotics and prevention of atopic disease: 4-year follow-up of a randomised placebo-controlled trial. *Lancet* **361**, 1869–1871.

Kanauchi, O., Fujiyama, Y., Mitsuyama, K., Araki, Y., Ishii, T., Nakamura, T., Hitomi, Y., Agata, K., Saiki, T., Andoh, A., Toyonaga, A. and Bamba, T. (1999) Increased growth of *Bifidobacterium* and *Eubacterium* by germinated barley foodstuff, accompanied by enhanced butyrate production in healthy volunteers. *International Journal of Molecular Medicine* **3**, 175–179.

Kanauchi, O., Mitsuyama, K., Araki, Y. and Andoh, A. (2003) Modification of intestinal flora in the treatment of inflammatory bowel disease. *Current Pharmaceutical Design* **9**, 333–346.

Kanauchi, O., Suga, T., Tochihara, M., Hibi, T., Naganuma, M., Homma, T., Asakura, H., Nakano, H., Takahama, K., Fujiyama, Y., Andoh, A., Shimoyama, T., Hida, N., Haruma, K., Koga, H., Mitsuyama, K., Sata, M., Fukuda, M., Kojima, A. and Bamba, T. (2002) Treatment of ulcerative colitis by feeding with germinated barley foodstuff: first report of a multicenter open control trial. *Journal of Gastroenterology* **37** (Suppl) 14., 67–72.

Kaneko, T., Kohmoto, T., Kikuchi, H., Shiota, M., Iino, H. and Mitsuoka, T. (1994) Effects of isomaltooligosaccharides with different degrees of polymerisation on human fecal bifidobacteria. *Bioscience, Biotechnology and Biochemistry* **58**, 2288–2290.

Kaneko, T., Kohmoto, T., Kikuchi, H., Shiota, M., Yatake, T., Iino, H. and Tsuji, K. (1993) Effects of isomaltooligosaccharides intake on defecation and intestinal environment in healthy volunteers. *Journal of Home Economics Japan* **44**, 245–254 (in Japanese).

Karppinen, S., Liukkonen, K., Aura, A. M., Forssell, P. and Poutanen, K. (2000) *In vitro* fermentation of polysaccharides of rye, wheat and oat brans and inulin by human faecal bacteria. *Journal of the Science of Food and Agriculture* **80**, 1469–1476.

Kashimura, J., Nakajima, Y., Benno, Y., Endo, K. and Mitsuoka, T. (1989) Effects of palatinose and its condensate intake on human fecal microflora. *Bifidobacteria Microflora* **8**, 45–50.

Kleessen, B., Stoof, G., Proll, J., Schmiedl, D., Noack, J. and Blaut, M. (1997) Feeding resistant starch affects fecal and cecal microflora and short chain fatty acids in rats. *Journal of Animal Sciences* **75**, 2453–2462.

Kohmoto, T., Fukui, F., Takaku, H., Machida, Y., Arai, M. and Mitsuoka, T. (1988) Effect of isomalto-oligosaccharides on human facal flora. *Bifidobacteria Microflora* **7**, 61–69.

Kohmoto, T., Tsuji, K., Kaneko, T., Shiota, M., Fukui, F., Takaku, H., Nakagawa, Y., Ichikawa, T. and Kobayashi, S. (1992) Metabolism of ^{13}C-isomaltooligosaccharides in healthy men. *Biosciences, Biotechnology and Biochemistry* **56**, 937–940.

Kontula, P., von Wright, A. and Mattila-Sandholm, T. (1998) Oat bran beta-gluco- and xylo-oligosaccharides as fermentative substrates for lactic acid bacteria. *International Journal of Food Microbiology* **45**, 163–169.

Kunz, C., Rudloff, S., Baier, W., Klein, N. and Strobel, S. (2000) Oligosaccharides in human milk: structural, functional, and metabolic aspects. *Annual Review of Nutrition* **20**, 699–722.

La Ferla, B., Prosperi, D., Lay, L., Russo, G. and Panza, L. (2002) Synthesis of building blocks of human milk oligosaccharides. Fucosylated derivatives of the lacto- and neolacto-series. *Carbohydrate Research* **337**, 1333–1342.

Le Blay, G., Michel, C., Blottiere, H. and Cherbut, C. (2003) Raw potato starch and short-chain fructo-oligosaccharides affect the composition and metabolic activity of rat intestinal microbiota differently depending on the caecocolonic segment involved. *Journal of Applied Microbiology* **94**, 312–320.

Le Leu, R. K., Hu, Y. and Young, G. P. (2002) Effects of resistant starch and nonstarch polysaccharides on colonic luminal environment and genotoxin-induced apoptosis in the rat. *Carcinogenesis* **23**, 713–719.

Lee, H. W., Park, Y. S., Jung, J. S. and Shin, W. S. (2002) Chitosan oligosaccharides, dp 2–8, have prebiotic effect on the *Bifidobacterium bifidium* and *Lactobacillus* sp. *Anaerobe* **8**, 319–324.

Lu, Z. X., Walker, K. Z., Muir, J. G., Mascara, T. and O'Dea, K. (2000) Arabinoxylan fiber, a byproduct of wheat flour processing, reduces the postprandial glucose response in normoglycemic subjects. *American Journal of Clinical Nutrition* **71**, 1123–1128.

Martin-Sosa, S., Martin, M. J. and Hueso, P. (2002) The sialylated fraction of milk oligosaccharides is partially responsible for binding to enterotoxigenic and uropathogenic *Escherichia coli* human strains. *Journal of Nutrition* **132**, 3067–3072.

Minami, Y., Yazawa, K., Tamura, Z., Tanaka, T. and Yamamoto, T. (1983) Selectivity of utilization of galactosyl-oligosaccharides by bifidobacteria. *Chemical and Pharmaceutical Bulletin* **31**, 1688–1691.

Moreau, N. M., Martin, L. J., Toquet, C. S., Laboisse, C. L., Nguyen, P. G., Siliart, B. S., Dumon, H. J. and Champ, M. M. J. (2003) Restoration of the integrity of rat caeco-colonic mucosa by resistant starch, but not by fructo-oligosaccharides, in dextran sulphate sodium-induced experimental colitis. *British Journal of Nutrition* **90**, 75–85.

Morinaga Milk Industry Co. (2002) Composition for treating chronic hepatitis C, contains a mixture of lactoferrin, probiotics and/or prebiotics as active ingredient. *Patent JP2002332242-A*.

Moro, G., Minoli, I., Mosca, M., Jelinek, J., Stahl, B. and Boehm, G. (2002) Dosage-related bifidogenic effects of galacto- and fructo-oligosaccharides in formula-fed term infants. *Journal of Pediatric Gastroenterology and Nutrition* **34**, 291–295.

Morrow, A. L., Ruiz-Palacios, G. M., Altaye, M., Jiang, X., Guerrero, M. L., Meinzen-Derr, J. K., Farkas, T., Chaturvedi, P., Pickering, L. K. and Newburg, D. S. (2004) Human milk oligosaccharides are associated with protection against diarrhea in breast-fed infants. *Journal of Pediatrics* **145**, 297–303.

Nagura, T., Hachimura, S., Hashiguchi, M., Ueda, Y., Kanno, T., Kikuchi, H., Sayama, K. and Kaminogawa, S. (2002) Suppressive effect of dietary raffinose on T-helper 2 cell-mediated immunity. *British Journal of Nutrition* **88**, 421–427.

Nakakuki, T. (2002) Present status and future of functional oligosaccharide development in Japan. *Pure and Applied Chemistry* **74**, 1245–1251.

Nakanishi, S., Kataoka, K., Kuwahara, T. and Ohnishi, Y. (2003) Effects of high amylose maize starch and *Clostridium butyricum* on metabolism in colonic microbiota and formation of azoxymethane-induced aberrant crypt foci in the rat colon. *Microbiology and Immunology* **47**, 951–958.

Natunen, J., Miller-Podraza, H., Teneberg, S., Angstroem, J., Karlsson, K., Ångström, J. M. and Angstrom, J. (2003) A glycoinhibitor substance comprising a pyranose formed ring structure useful for treating infectious diseases e.g. bacterial infection. *Patent WO2003002127-A*.

Newburg, D. S. (1999) Human milk glycoconjugates that inhibit pathogens. *Current Medicinal Chemistry* **6**, 117–127.

Oku, T. and Nakamura, S. (2003) Comparison of digestibility and breath hydrogen gas excretion of fructo-oligosaccharide, galactosyl-sucrose, and isomalto-oligosaccharide in healthy human subjects. *European Journal of Clinical Nutrition* **57**, 1150–1156.

Okubo, T., Ishihara, N., Takahashi, H., Fujisawa, T., Kim, M., Yamamoto, T. and Mitsuoka, T. (1994) Effects of partially hydrolyzed guar gum intake on human intestinal microflora and its metabolism. *Biosciences, Biotechnology and Biochemistry* **58**, 1364–1369.

Ohkusa, T., Ozaki, Y., Sato, C., Mikuni, K. and Ikeda, H. (1995) Long-term ingestion of lactosucrose increases *Bifidobacterium* sp. in human fecal flora. *Digestion* **56**, 415–420.

Oikarinen, S., Heinonen, S., Karppinen, S., Mättö, J., Adlecreutz, H., Poutanen, K. and Mutanen, M. (2003) Plasma enterolactone or intestinal *Bifidobacterium* levels do not explain adenoma formation in multiple intestinal neoplasia (Min) mice fed with two different types of rye-bran fractions. *British Journal of Nutrition* **90**, 119–125.

Okazaki, M., Fujikawa, S. and Matsumomo, N. (1990) Effect of xylooligosaccharide on the growth of bifidobacteria. *Bifidobacteria Microflora* **9**, 77–86.

Özaslan, C., Türkçapar, A. G., Kesenci, M., Karayalçin, K., Yerdel, M. A., Bengisun, S. and Törüner, A. (1997) Effect of lactulose on bacterial translocation. *European Journal of Surgery* **163**, 463–467.

Palframan, R. J., Gibson, G. R. and Rastall, R. A. (2002) Effect of pH and dose on the growth of gut bacteria on prebiotic carbohydrates *in vitro*. *Anaerobe* **8**, 287–292.

Perrin, P., Pierre, F., Patry, Y., Champ, M., Berreur, M., Pradal, G., Bornet, F., Meflah, K. and Menanteau, J. (2001) Only fibres promoting a stable butyrate producing colonic ecosystem decrease the rate of aberrant crypt foci in rats. *Gut* **48**, 53–61.

Petuely, F. (1957) Bifidusflora bei flaschenkindern durch bifidogene substanzen (Bifidusfaktor). *Zeitschrift Kinderheilkunde* **79**, 174–179.

Phillips, J., Muir, J. G., Birkett, A., Lu, Z. X., Jones, G. P., O'Dea, K. and Young, G. P. (1995) Effect of resistant starch on fecal bulk and fermentation-dependent events in humans. *American Journal of Clinical Nutrition* **62**, 121–130.

Playne, M. J. and Crittenden, R. (1996) Commercially available oligosaccharides. *Bulletin of the International Dairy Federation* **313**, 10–22.

Playne, M. J. and Crittenden, R. G. (2004) Prebiotics from lactose, sucrose, starch, and plant polysaccharides. In: Neeser, J.-R. and German, J. B. (eds), *Bioprocesses and Biotechnology for Functional Foods and Neutraceuticals*. Marcel Dekker, New York, pp. 99–134.

Ponz de Leon, M. and Roncucci, L. (1997) Chemoprevention of colorectal tumours: role of lactulose and of other agents. *Scandinavian Journal of Gastroenterology* **32** (Suppl. 222), 72–75.

Rao, A. V. (1999) Dose–response effects of inulin and oligofructose on intestinal bifidogenesis effects. *Journal of Nutrition* **129** (Suppl. 7), 1442–1445.

Rautonen, N., Apajalahti, J., Kettunen, A. and Siikanen, O. (2004) Use of slowly fermented oligomeric or polymeric carbohydrate(s) for preparation of a composition for treating and/or preventing diseases and/or disorders caused by imbalanced fermentation in the colon e.g. food allergies and celiac disease. *Patent WO2004026316-A1.*

Rencurosi, A., Poletti, L., Guerrini, M., Russo, G. and Lay, L. (2002) Human milk oligosaccharides: an enzymatic protection step simplifies the synthesis of 3′- and 6′-*O*-sialyllactose and their analogues. *Carbohydrate Research* **337**, 473–483.

Rowland, I. R., Bearne, C. A., Fischer, R. and Pool-Zobel, B. L. (1996) The effect of lactulose on DNA damage induced by DMH in the colon of human flora-associated rats. *Nutrition and Cancer* **26**, 37–47.

Rumi, G., Tsubouchi, R., Okayama, M., Kato, S., Mozsik, G. and Takeuchi, K. (2004) Protective effect of lactulose on dextran sulfate sodium-induced colonic inflammation in rats. *Digestive Diseases and Sciences* **49**, 1466–1472.

Rycroft, C. E., Jones, M. R., Gibson, G. R. and Rastall, R. A. (2001) A comparative *in vitro* evaluation of the fermentation properties of prebiotic oligosaccharides. *Journal of Applied Microbiology* **91**, 878–887.

Salminen, S., Gorbach, S., Lee, Y.-K. and Benno, Y. (2004) Human studies on probiotics: What is scientifically proven today? In: Salminen, S., von Wright, A. and Ouwerhand A. (eds), *Lactic Acid Bacteria: Microbiological and Functional Aspects.* Marcel Dekker, New York, pp. 515–530.

Satouchi, M., Wakabayashi, S., Ohkuma, K. and Tsuji, K. (1996) Effect of depolymerised pyrodextrin on human intestinal flora. *Bioscience Microflora* **15**, 93–101.

Schell, M. A., Karmirantzou, M., Snel, B., Vilanova, D., Berger, B., Pessi, G., Zwahlen, M. C., Desiere, F., Bork, P., Delley, M., Pridmore, R. D. and Arigoni, F. (2002) The genome sequence of *Bifidobacterium longum* reflects its adaptation to the human gastrointestinal tract. *Proceedings of the National Academy of Sciences USA* **99**, 14422–14427.

Schumann, C. (2002) Medical, nutritional and technological properties of lactulose. An update. *European Journal of Nutrition* **41** (Suppl. 1), 17–25.

Shida, K., Takahashi, R., Iwadate, E., Takamizawa, K., Yasui, H., Sato, T., Habu, S., Hachimura, S. and Kaminogawa, S. (2002) *Lactobacillus casei* strain Shirota suppresses serum immunoglobulin E and immunoglobulin G1 responses and systemic anaphylaxis in a food allergy model. *Clinical and Experimental Allergy* **32**, 563–570.

Silvi, S., Rumney, C. J., Cresci, A. and Rowland, I. R. (1999) Resistant starch modifies gut microflora and microbial metabolism in human flora-associated rats inoculated with faeces from Italian and UK donors. *Journal of Applied Microbiology* **86**, 521–530.

Stevens, B. J. and Selvendran, R. R. (1988) Changes in composition and structure of wheat bran resulting from the action of human faecal bacteria *in vitro*. *Carbohydrate Research* **183**, 311–319.

Terada, A., Hara, H., Katoaka, M. and Mitsuoka, T. (1992) Effect of lactulose on the composition and metabolic activity of human faecal flora. *Microbial Ecology in Health and Disease* **5**, 43–50.

Teramoto, F., Rokutan, K., Kawakami, Y., Fujimura, Y., Uchida, J., Oku, K., Oka, M. and Yoneyama, M. (1996) Effect of 4G-beta-D-galactosylsucrose (lactosucrose) on fecal microflora in patients with chronic inflammatory bowel disease. *Journal of Gastroenterology* **31**, 33–39.

Topping, D. L. and Clifton, P. M. (2001) Short-chain fatty acids and human colonic function: roles of resistant starch and nonstarch polysaccharides. *Physiology Reviews* **81**, 1031–1064.

Tuohy, K. M., Ziemer, C. J., Klinder, A., Knobel, Y., Pool-Zobel, B. L. and Gibson, G. R. (2002) A human volunteer study to determine the prebiotic effects of lactulose powder on human colonic microbiota. *Microbial Ecology in Health and Disease* **14**, 165–173.

van Geel-Schutten, G. H. (2003) Novel glucan produced by glucosyltransferase activity of lactic acid bacterium on sucrose substrate, and having backbone consisting of alpha (1,3)- and alpha (1,6)-linked anhydroglucose units, useful as thickener. *Patent WO2003008618-A*.

Van Laere, K. M., Hartemink, R., Bosveld, M., Schols, H. A. and Voragen, A. G. (2000) Fermentation of plant cell wall derived polysaccharides and their corresponding oligosaccharides by intestinal bacteria. *Journal of Agricultural and Food Chemistry* **48**, 1644–1652.

Van Loo, J. (2004) The specificity of the interaction with intestinal bacterial fermentation by prebiotics determines their physiological efficacy. *Nutrition Research Reviews* **17**, 89–98.

Vercauteren, R. L. M. and Nguyen, V. S. (2004) Elongation of chain length of isomalto-oligosaccharides used for syrup useful in, e.g. food, comprises enzymatic transfer reaction between sucrose and isomalto-oligosaccharide present in carbohydrate syrup. *Patent WO2004068966-A1*.

Vincent, S. J., Faber, E. J., Neeser, J.-R., Stingele, F. and Kamerling, J. P. (2001) Structure and properties of the exopolysaccharide produced by *Streptococcus macedonicus* Sc136. *Glycobiology* **11**, 131–139.

Wada, K., Watabe, J., Mizutani, J., Tomoda, M., Suzuki, H. and Saitoh, Y. (1992) Effects of soybean oligosaccharides in a beverage on human fecal flora and metabolites. *Journal of the Agricultural Chemistry Society of Japan* **66**, 127–135.

Wang, B. and Brand-Miller, J. (2003) The role and potential of sialic acid in human nutrition. *European Journal of Clinical Nutrition* **57**, 1351–1369.

Wang, H. F., Lim, P. S., Kao, M. D., Chan, E. C., Lin, L. C. and Wang, N. P. (2001) Use of isomalto-oligosaccharide in the treatment of lipid profiles and constipation in hemodialysis patients. *Journal of Renal Nutrition* **11**, 73–79.

Wang, X., Brown, I. L., Evans, A. J. and Conway, P. L. (1999) The protective effects of high amylose maize (amylomaize) starch granules on the survival of *Bifidobacterium* spp. in the mouse intestinal tract. *Journal of Applied Microbiology* **87**, 631–639.

Wang, X., Brown, I. L., Khaled, D., Mahoney, M. C., Evans, A. J. and Conway, P. L. (2002) Manipulation of colonic bacteria and volatile fatty acid production by dietary high amylose maize (amylomaize) starch granules. *Journal of Applied Microbiology* **93**, 390–397.

Wilson, J. L. (2003) Nutritional composition for use as infant formula, follow-on formula, toddler's beverage, milk yogurt, fruit juice, fruit-based drink, candy, chewing gum and/or lozenge, comprises oligofructose and sialyllactose. *Patent US2003060445-A1*.

Yamada, H., Itoh, K., Morishita, Y. and Taniguchi, H. (1993) Structure and properties of oligosaccharides from wheat bran. *Cereal Foods World* **38**, 490–492.

Yoneyama, M., Mandai, T., Aga, H., Fujii, K., Sakai, S. and Katayama, Y. (1992) Effects of 4-β-D-galactosylsucrose (lactosucrose) intake on intestinal flora in healthy humans. *Journal of the Japanese Society of Nutrition and Food Science* **45**, 101–107.

Yoshida, T., Hirano, A., Wada, H., Takahashi, K. and Hattori, M. (2004) Alginic acid oligosaccharide suppresses Th2 development and IgE production by inducing IL-12 production. *International Archives of Allergy and Immunology* **133**, 239–247.

Younes, H., Coudray, C., Bellanger, J., Demigne, C., Rayssiguier, Y. and Remesy, C. (2001) Effects of two fermentable carbohydrates (inulin and resistant starch) and their combination on calcium and magnesium balance in rats. *British Journal of Nutrition* **86**, 479–485.

6

Molecular Microbial Ecology of the Human Gut

Kieran M. Tuohy and Anne L. McCartney

6.1 Introduction

The advent of molecular biology and the relative abundance of new technologies developed over the last decade has resulted in a barrage of genetic methodologies available to examine biological succession and/or microbial ecology of complex ecosystems. Initial applications of such techniques in the field of human gastrointestinal microflora has expanded the depth of our knowledge, both in terms of bacterial diversity and biological function. The recent advancement in genomics, proteomics and metabonomics assays, including high throughput systems, provides the necessary tools for 'systems biology' investigations. The imminent future of gut microbial ecology encompasses polyphasic analyses affording detailed profiles of the microbial complexity, dynamics, activity and host functions — in relation to genome, proteome, metabolome, diet and health.

The last decade has seen a flourish of molecular techniques developed and applied to elucidating the microbial diversity and ecological analysis of complex biosystems — including the human gastrointestinal (GI) tract. Such applications include investigating functionality, particularly changes in microbial composition, as a result of prebiotic fermentation and/or intervention. The aims of this chapter are to critically assess such approaches, assess the current state of knowledge of the microbial ecology of the human gut and address future avenues towards understanding the GI microflora and its impact upon host health and wellbeing.

Prebiotics: Development and Application Edited by G.R. Gibson and R.A. Rastall

Much of the early work investigating the microbial ecology of the human GI tract relied solely on cultivation work and subsequent characterisation of bacterial isolates (Finegold and Sutter, 1978; Moore *et al.*, 1978; Stark and Lee, 1982; McCartney *et al.*, 1996). Initially, phenotypic characterisation was employed to differentiate/identify isolates. Such assays included colonial and cellular morphology, growth requirements, resistance assessments and fermentation profiles of bacterial isolates. However, particular aspects of phenotypic characterisation do not lend themselves to accurate taxonomic classification. For example, certain phenotypic characteristics may be displayed by a range of genetically unrelated biotypes, assays often suffer reproducibility problems or provide ambiguous results due to bacterial metabolic plasticity. Furthermore, the logistics of measuring sufficient phenotypic parameters are prohibitive. The advent of the molecular age has afforded genotypic classification of cultured bacteria, which have generally proven to be taxonomically/phylogenetically superior to phenotypic identification systems (McCartney, 2002). In addition, molecular-based technology enables faster and more reliable classification of isolates and has spawned an array of genotyping assays for different applications (including differentiation) (Olive and Bean, 1999; O'Sullivan, 1999). One must not, however, completely discount cultivation work and phenotypic characterisation, as these tools provide essential information and knowledge of microbial physiology and function.

6.2 The Normal Microflora of the Human GI Tract

The human GI tract essentially comprises three distinct compartments, both in terms of anatomical regions and microbial colonisation. The distinct physicochemical environments of each region (stomach, small intestine and large intestine) impact the indigenous microbial populations (Tannock, 1995; Holzapfel *et al.*, 1998; Rabiu and Gibson, 2002). The healthy human stomach is minimally colonised by bacteria, due to an acidic pH of gastric contents and rapid transit time of digesta (Macfarlane and Cummings, 1991; Guarner and Malagelada, 2003). *Helicobacter pylori* is known to inhabit the stomach of humans, though it is debatable whether this is a commensal organism. *Helicobacter* evade or combat peristaltic movement with their flagellae and can burrow into the gastric mucosa away from the harsh acidity of the stomach contents (Marshall, 1994). Bile, pancreatic fluids and other small intestinal secretions, along with gastric juices, extend inhibition of microbial colonisation of the upper GI tract. Some aciduric lactobacilli and streptococci have been detected in the duodenum and jejunum, although population numbers are in the region of 100–10000 cells per millilitre of contents. The slowing of intestinal motility in distal regions of the small intestine, and less acidic pH, renders the ileum more favourable to microbial colonisation. Bacterial populations in the range of 10^6–10^8 bacterial cells per millilitre of contents have been demonstrated (Simon and Gorbach, 1984; Tannock, 1995). In addition, greater diversity has been seen in this region than in the upper GI tract (stomach, duodenum and jejunum), including obligate anaerobes, Gram-negative facultatively anaerobic bacteria, lactobacilli and enterococci. However, by far the greatest proportion of the human normal microflora inhabits the colon, where the typical transit time is 55–70 h (in health) (Macfarlane *et al.*, 1998; Bourlioux *et al.*, 2003). In the healthy situation, the ileocaecal valve restricts proximal movement of the colonic microflora (Conway, 1995). The slower flow rate, more neutral pH and plentiful supply of nutrients make the colon extremely favourable to microbial colonisation. Indeed, a vastly diverse and highly populated indigenous microflora, attaining levels of 10^{11}–10^{12} bacterial cells per gram of contents, resides in the colon of healthy adult humans (Guarner and Malagelada, 2003). The majority of members of the colonic microflora are strict anaerobes, such as *Atopobium* spp., *Bacteroides* spp., *Bifidobacterium* spp., the clostridia and other families within the *Clostridium* mega-genus (including *Butyrovibrio* spp., *Eubacterium* spp., *Fusobacterium* spp., *Peptostreptococcus* spp. and *Ruminococcus* spp.) and the peptococci. Facultatively anaerobic bacteria, such as lactobacilli and enterococci, generally form less dominant portions of the colonic microflora of healthy adults (typically 1000-fold lower than the predominant obligate anaerobes) (Tannock, 1995; Suau *et al.*, 1999).

The gut microbiota can be further differentiated in relation to distinctive micro-niches within the colon. Freter and colleagues highlighted four micro-habitat components of the microbial community in the distal GI tract; namely, the lumen, mucous layer, crypts, and surface of epithelial cells (Fanaro *et al.*, 2003). However, samples are extremely difficult to obtain from such areas and the majority of data pertaining to the gut microflora of humans relates to the faecal microflora of subjects. Cultivation studies of the luminal content of sudden death individuals by Moore and Holdeman (1975), and later by Macfarlane *et al.* (1998), allayed concerns over the accuracy of the faecal flora representation of the colonic microflora. More recently, advances in molecular assays and medical technology (including sampling methods), enable analysis of previously inaccessible regions of the GI tract of live subjects. A limited body of work with clinical samples, as well as those from healthy volunteers, have examined the microbial content of mucosal and/or biopsy material. One such study showed similarity in the microbial profiles of mucosally associated bacteria from the ascending, transverse and distal colon of healthy humans, which were distinct to the profile for faecal samples from the same volunteer (Zoetendal *et al.*, 2002a). Unfortunately, no analysis of the bacterial profiles of luminal samples from the different regions of the colon was performed.

Multiple factors have been shown to impact the composition and numerics of the human colonic flora, including host genetics, mode of delivery, gestational age at birth, diet, age, health, environment and cell-to-cell interactions (host–bacteria as well as bacteria–bacteria). The complexity and dynamics of the resident gut flora also play an important role in human health, providing a barrier to overgrowth of allocthonous micro-organisms (which may cause disease) and stimulating the immune system (Tannock, 1995).

The bacteriological succession of the gut flora commences during or shortly after birth (depending on the mode of delivery). Caesarean section delivery results in a delayed acquisition and development of the GI microflora and is associated with higher clostridial levels than vaginally birthed infants (Grönlund *et al.*, 1999; Mackie *et al.*, 1999). Preterm birth and/or extremely low birth weight (ELBW) is also associated with a disrupted acquisition and succession of the neonatal gut flora. Recent work suggests that there may be long-term effects of such delayed onset of, or altered composition of, the neonatal microflora. Indeed, 7-year-old children who had been born by Caesarean section harboured a higher clostridial population than an age and gender matched cohort of children delivered vaginally (Salminen *et al.*, 2004).

The neonatal microflora initially comprises mainly facultative anaerobes (phase 1: primary colonisers), which rapidly reduce the redox potential of the gut, enabling colonisation by more obligate anaerobes – such as bifidobacteria, bacteroides and clostridia (phase 2). The infant's feeding regime has been shown to play an integral part in the complexity and microbial components of the faecal flora during phase 2 acquisition and development (Mountzouris *et al.*, 2002). A large body of work has been published comparing the faecal microflora of breast- and formula-fed infants. Most of these studies encompassed cultivation data, although more recent approaches utilised molecular biological methods. Whilst there is some disparity between the different studies, it is generally accepted that breast- and formula-fed infants harbour distinctive faecal microbial populations. The inclusion of prebiotics in selected modern formulae adds a further distinction – eliciting a flora more closely resembling that of breast-fed infants than that associated with ingestion of standard formula (Boehm *et al.*, 2002).

The discrepancy seen between different studies most likely reflect methodological inconsistencies (collection and handling of samples, as well as cultivation and characterisation strategies), environmental factors (such as geographical, cultural or socioeconomic differences), or different formulae. The overriding findings, however, are that breast-feeding (or use of prebiotic-supplemented formulae) elicit a bifidobacterially predominated microflora during exclusive milk-feeding (phase 2), whilst infants fed standard formulae develop a more complex and diverse microflora (Conway, 1997). Interestingly, it appears that the levels of bacterial populations other than bifidobacteria (such as bacteroides, clostridia and enterobacteria) may be the key (Mitsuoka and Kaneuchi, 1977; Kleessen *et al.*, 1995).

Phase 3 in the bacterial succession of the infant microflora is associated with the introduction of complex foods (i.e. weaning). In general, this is associated with greater microbiological change in breast-fed infants – towards a more complex and diverse microbial community, similar to that seen in formula-fed infants during phase 2. This is often represented by an increased predominance of bacteroides (Mackie *et al.*, 1999). Phase 4 is essentially attainment of the climax or adult community. However, a fifth phase of succession of the human gut microflora has unravelled, that observed in latter life. Gorbach and colleagues (1967) demonstrated that elderly people harboured lower levels of bifidobacteria and higher levels of enterobacteria and yeasts than adults. Mitsuoka and Hayakawa's results corroborated these microbiological alterations associated with older age (Mitsuoka, 1992). More recent studies, employing direct 16S rRNA profiling has shown that the gut microflora of the elderly has increased bacterial diversity compared with younger adults (Saunier and Doré, 2002). Such observations may have direct implications for health and disease within ageing populations, especially in relation to increased risk of GI infections, the severity of such infections and chronic diseases (such as colon cancer, which has a direct correlation with increased age).

6.3 The Molecular Revolution

Whilst much of the initial work investigating the human gut microflora relied on cultivation work, including the use of selective growth media and/or conditions, a number of recent publications employing modern molecular technologies corroborate many of the findings. The advent of molecular biology, and an ever-increasing catalogue of genetic applications, has highlighted the under-representation of microbial diversity achieved by cultivation (Zoetendal *et al.*, 1998; Suau *et al.*, 1999; Harmsen *et al.*, 2000; Favier *et al.*, 2002). Furthermore, molecular profiling (including genetic fingerprinting, species-specific probing strategies and community analysis assays) affords reliable differentiation to lower phylogenetic levels (for example, species or sub-species).

Formerly, molecular assays were developed to discriminate bacterial isolates obtained during cultivation studies, generally involving genetic fingerprinting based on restriction fragment length polymorphism (RFLP) (McCartney, 2002). Sequencing methods were developed and the database of sequences for bacterial ribosomal RNA genes rapidly expanded, affording a relatively quick and accurate method for identification based on genotype. Probing techniques also evolved, and continue to increase with the expansion of sequencing information. Similarly, the emergence and subsequent prevalence of specific amplification techniques (using polymerase chain reaction, PCR), community profiling strategies, such as phylogenetic information from sequencing clone libraries and denaturing gradient gel electrophoresis (DGGE) profiling, have revolutionised our knowledge and understanding of the human GI microflora.

To date, molecular techniques have been used in gut microbiology to (a) identify bacterial isolates, (b) enumerate specific phylogenetic groups, (c) track organisms of particular interest (such as probiotics) and (d) characterise the microbial diversity and dynamics over time and between different subjects or study groups. Current developments in molecular ecology of the human gut include genetic profiling strategies, high throughput systems (including micro-arrays and micro-chip technology) and functional genomics (such as metabonomics, proteomics and nutrigenomics).

6.4 Molecular Probing Strategies

Molecular probing assays can be employed to accurately monitor the predominant bacterial groups of mixed microbial populations – provided appropriate probing strategies are available. A number of probing methodologies exist, including colony hybridisations, dot blot assays and *in situ* hybridisations. The heart of all probing technology is the hybridisation of oligonucleotides (or probes), often synthetically manufactured, targeting specific sequences on bacterial DNA. In theory, any DNA sequence could be used. However, specificity and selectivity is crucial. Furthermore, the physicochemical nature of the bacterial genome, including secondary and tertiary structure, negates the application of some regions. The specificity of oligonucleotide probes is reliant on the target sequence, hybridisation temperature and stringency of both hybridisation and washing conditions (Charteris *et al.*, 1997; O'Sullivan, 1999).

The greater proportion of probing strategies used to examine mixed bacterial populations concentrate on the bacterial ribosome, predominantly 16S rDNA probes. The nature of the 16S rRNA gene, namely the availability of highly conserved, variable and highly variable regions (enabling selection of oligonucleotide sequences of varying taxonomic specificity), and the extensive databases of known sequences are largely responsible for the unparalleled application of this house-keeping gene. Highly conserved regions provide universal target sequences (e.g. the domain probe, Bact 338), variable regions for group level sequences (such as Erec 482 probe for the *Clostridium coccoides/Eubacterium rectale* subgroup), and hypervariable regions for genus- and/or species-specific targets (e.g. Bif 164 and Bdis 656, for *Bifidobacterium* spp. and *Bacteroides distasonis*, respectively) (Franks *et al.*, 1998; Blaut *et al.*, 2002; Harmsen *et al.*, 2002). A limited amount of work has also investigated the potential of other bacterial genes, for example the fructose 6-phosphate phosphoketolase enzyme specific to *Bifidobacterium* and *Gardnerella* species (O'Sullivan, 1999). Specificity levels and detection limits are the major limitations of genetic probes. As our understanding of gut microbiology expands, through identification of novel bacterial species and/or clones (during molecular community analysis studies), so the repertoire of oligonucleotide probes (especially 16S rRNA-targeted sequences) also expands.

Dot-blot hybridisations and fluorescence *in situ* hybridisations are the most common probing strategies employed to date in studying the microbiology of the human GI tract. In general, dot-blot assays probe DNA extracts (either from bacterial isolates or mixed samples directly) and provide an index of the percentage of total 16S rRNA each specific population comprises. An array of oligonucleotide probes and/or samples can be simultaneously examined during dot-blot hybridisation using a checkerboard system (or probe grid), provided the stringency conditions of each probe are similar (McCartney, 2002). Inter-individual variation was demonstrated in two separate studies using dot-blot hybridisations monitoring the predominant faecal microflora of healthy humans (Sghir *et al.*, 2000; Marteau *et al.*, 2001). The earlier work, by Sghir and colleagues, demonstrated a predominance of the *Bacteroides* group, *Clostridium leptum* subgroup and *Clostridium coccoides/Eubacterium rectale* group, accounting for 37, 16 and 14 % of the total rRNA, respectively (Sghir *et al.*, 2000). Whilst different proportions were seen in the second study, by Marteau and colleagues (2001), the same three bacterial groups formed the predominant population – accounting for 44 % of the total rRNA.

Hopkins *et al.* (2001) employed both dot-blot analysis and traditional cultivation assays to investigate the microbial composition of the human microflora across four cohorts [children, adults, healthy elderly people and elderly patients suffering from *Clostridium difficile* associated diarrhoea (CDAD)]. Interestingly, significantly higher levels of enterobacteria were seen in children, compared with adults. Age-related variation was evidenced for the bifidobacterial levels harboured by humans, with lower indices seen for both elderly groups compared with children and adult cohorts. Great inter-individual variation was also seen in relation to the bifidobacterial content of elderly subjects (with some showing relatively high levels and others extremely low levels) (Hopkins *et al.*, 2001). However, the elderly CDAD group had the lowest bifidobacterial populations, of all groups. Furthermore, this cohort of subjects displayed reduced bacteroides and elevated clostridial levels compared with all other groups. Additional studies have also reported distinctions in faecal 16S rRNA species levels in elderly subjects compared with healthy adults (Saunier and Doré, 2002). Most notably, reduced rRNA indices for bacteroides, bifidobacteria and *Clostridium leptum*, and increased rRNA indices for lactobacilli were seen. Furthermore, these authors demonstrated that bacterial diversity increased with age, as the probes which cover 80 % of the total rRNA of healthy adults accounted for only 50 % of the total rRNA of elderly subjects (Saunier and Doré, 2002).

Fluorescence *in situ* hybridisation (FISH) affords direct enumeration of fixed bacterial cells in mixed populations (such as the human gut flora) (Langendijk *et al.*, 1995). Epifluorescence microscopy or flow cytometry is used to detect the hybridised cells in the sample and allow quantification of the target population (either as number of cells per sample or relative proportion of total rRNA). FISH analysis has become one of the methods of choice in monitoring the predominant bacteria of the human GI tract, with an abundance of published work applying this technique to investigations of the microflora in relation to health and dietary interventions (Franks *et al.*, 1998; Harmsen *et al.*, 2000; Suau *et al.*, 2001; Tuohy *et al.*, 2001; Harmsen *et al.*, 2002; Zoetendal *et al.*, 2002b).

One longitudinal study employed FISH to monitor the faecal microflora of nine healthy subjects (Franks *et al.*, 1998). Overall, greater than 90 % of 4′,6-diamidino-2-phenylindol dihydrochloride (DAPI)-stained cells were hybridised by the universal bacterial probe (Bact 338). As with dot-blot data, the *Bacteroides* and *Clostridium coccoides/Eubacterium rectale* groups comprised the predominant flora (making up nearly 50 % of the total bacteria). Other dominant populations included the Low G+C #2 group and bifidobacteria (Franks *et al.*, 1998). Variations were observed in the faecal microflora over time, with the greatest fluctuations seen in the bifidobacterial component.

Recent studies using a greater barrage of probes confirmed the predominance of the *Bacteroides* and *Clostridium coccoides/Eubacterium rectale* groups, 28 % and 23 %, respectively (Harmsen *et al.*, 2002). The additional probes highlighted the relative dominance of the *Atopobium* (12 %), *Eubacterium* low G+C/*Fusobacterium prausnitzii* (reclassified as *Faecalibacterium prausnitzii*) (11 %) and *Ruminococcus* groups (10 %). *Bifidobacterium* again represented approximately 5 % of the total bacterial load, whilst the *Eubacterium hallii*, *Lachnospira* and *Eubacterium cylindroides* groups were also members of the dominant faecal flora of healthy humans (Harmsen *et al.*, 2002). Subdominant levels (<1 %) of other bacterial groups were also identified.

Knowledge gathered from other molecular strategies has highlighted potentially important bacterial groups within the faecal microflora of humans. For example, ribosomal clonal libraries of human stool samples often contain sequences related to *Ruminococcus obeum*-like bacteria. The question arises as to whether this is due to their relative predominance in the original samples or a result of biasing/artefact of the cloning assay. To investigate this, Zoetendal *et al.* (2002b) developed the Urobe 63 probe, specific for the *Ruminococcus obeum* group. Preliminary investigations, using single samples from nine Dutch subjects (five males and four females), demonstrated the *Ruminococcus obeum* group was relatively dominant (comprising ~2.5 % of cells hybridised with Bact 338 – the universal bacterial probe; and ~16 % of the *Clostridium coccoides/Eubacterium rectale* group) (Zoetendal *et al.*, 2002b). Similar developmental work has been done for the *Faecalibacterium prausnitzii* cluster (Suau *et al.*, 2001).

Microarray/membrane array techniques probing extensive libraries of oligonucleotides simultaneously, are the new generation of probing strategies, affording high throughput analysis of mixed microflora systems. To date, two studies from the same group have been published using such technology examining 60 oligonucleotide probes covering 20 bacterial species (Wang *et al.*, 2002a,b). Overall, the data from healthy humans were consistent with previous findings, with *Bacteroides* species, *Clostridium clostridioforme, Clostridium leptum, Faecalibacterium prausnitzii, Ruminococcus* species and *Bifidobacterium* species forming the predominant microflora. Inter-individual variation was again seen between healthy subjects. However, the microflora of an individual suffering from long-term diarrhoea was shown to lack a number of the common bacterial species predominant in healthy samples (Wang *et al.*, 2002a).

Overall, genetic probing strategies afford reasonable coverage of the predominant bacteria of the human gut microflora, with new probe development ongoing. Detection thresholds are a limiting factor, particularly in relation to monitoring less dominant populations in the mixed microbiota (Franks *et al.*, 1998). However, technology is ever improving and, as well as a continually increasing library of probes, scientists and manufacturers strive for instrumentation enabling higher sample throughput and greater coverage and/or sensitivity.

6.5 PCR-based Community Analyses

The presence of particular bacterial groups (at either genus or species level) can also be identified using PCR technology. However, this can quickly develop into a logistically intense mode of examining the predominant members of a complex microbial community, such as the human gut flora. Community profiling techniques, such as DGGE, may be applied to such samples to examine the predominant components by way of a snap-shot. Selecting the focus of such snap-shots can be adjusted through primer selection. For example, panoramic views can be obtained using universal primers, whilst portraits or even highly magnified (detailed) pictures can be taken with group-, genus- or even species-specific amplification conditions. As such, DGGE can generate a map (at selected scale/level of detail) of the microbial diversity present in environmental samples including faeces, biopsies and biopsy fluids. Amplicons are separated according to the chemical stability or melting temperature of the gene sequence (Muyzer *et al.*, 1993; Muyzer, 1999). One limitation is that phylogenetically unrelated heterologous sequences may migrate similarly. However, narrowing of the gradient may alleviate this. Another limitation is the detection limit, with poor detection sensitivity to less dominant members of mixed ecosystems (Zoetendal *et al.*, 1998). The power of DGGE is the ability to examine and compare the bacterial diversity of multiple samples relatively quickly. Bands of particular interest, such as those which are differentially represented across different samples, may be examined further (for example by excision and/or cloning, and sequencing).

Much work is accumulating which employed DGGE to monitor the faecal microflora across numerous studies (including feeding trials) (Zoetendal *et al.*, 1998; Satokari *et al.*, 2001; Walter *et al.*, 2001; Zoetendal *et al*, 2001; Favier *et al.*, 2002; Heilig *et al.*, 2002; Zoetendal *et al.*, 2002a). In general, inter-individual variations have been seen in DGGE profiles, as has intra-individual stability (i.e. DGGE profiles of an individual over time) (Table 6.1). Common bands have also been demonstrated in the DGGE profiles of different subjects, indicating the presence of dominant bacteria across samples/subjects. A positive linear relationship was shown between host genetic relatedness and the similarity index of DGGE profiles (Zoetendal *et al.*, 2001). However, no such correlation was seen between the DGGE profile similarity indices and either gender or living arrangements of unrelated subjects. As such, DGGE analysis has clearly demonstrated that host genotype factors are correlated to the bacterial composition of the human gut flora (Zoetendal *et al.*, 2001).

Favier *et al.* (2002) employed PCR-DGGE to examine the microbial diversity and dynamics in two baby boys during the first 10–12 months of life. Initial profiles were

Table 6.1 *Overall findings of published studies employing DGGE profiling of the human gut microflora*

Target	Subjects	Investigation	Summary of results	Reference
Bifidobacteria (16S rRNA)	Finnish adults ($n=6$; 3 males)	Stability over 4 weeks	Inter-individual variation Multiple bifidobacterial biotypes Relatively stable	Satokari *et al.*, 2001
Lactobacilli (16S rRNA)	NZ adults ($n=4$; 2 males)	Validation of developed group-specific primers	Inter-individual variation Intra-individual variation over 6 months	Walter *et al.*, 2001
Total community (16S rRNA)	Adults residing in The Netherlands ($n=50$)	Impact of genetic relatedness on the faeal flora	Inter-individual variation Positive linear relationship between host genetic factors and similarity index of profiles	Zoetendal *et al.*, 2001
Total community (16S rRNA)	Dutch babies ($n=2$; both males)	Diversity and stability	Inter-individual variation Initial DGGE profiles were simple (first few days of life) Bacterial diversity increased with time (bacterial succession) Dietary challenge correlated to DGGE profile shifts	Favier *et al.*, 2002
Lactobacilli (16S rRNA)	Adults residing in The Netherlands ($n=12$)	Diversity and stability	Inter-individual variation Relatively stable DGGE profiles (total bacteria) *Lactobacillus* stability variable across the group (stable for some, dynamic for others)	Heilig *et al.*, 2002
	Baby boy	Diversity and stability	No lactobacilli profile for initial 55 days Relative stable from 2 to 3 months of age Bacterial succession due to dietary challenge	

Table 6.1 *(Continued)*

Target	Subjects	Investigation	Summary of results	Reference
Bifidobacteria (transadolase)	NZ adults (n=10)	Diversity	Inter-individual variation: six adults harboured two amplicons (one being *Bifidobacterium adolescentis*), three harboured one amplicon and one gave no PCR-DGGE product	Requena *et al.*, 2002
	NZ babies ($n=2$; total of 10 samples)	Diversity	Relatively similar profiles: eight samples harboured *Bifidobacterium bifidum* (one of which contained a second amplicon), one harboured three amplicons and one gave no PCR-DGGE product	
Total community and lactobacilli (16S rRNA)	Dutch adults ($n=10$; 5 males)	Mucosally associated flora	Total community PCR-DGGE Large inter-individual variation (both faecal and mucosal samples) High similarity index of biopsy samples from same donor Mucosal and faecal profiles significantly different Lactobacillus PCR-DGGE Good similarity between mucosal and faecal profiles of 6 subjects, minor variations in profiles of different biopsy samples of 3 subjects	Zoetendal *et al.*, 2002a

rather simple in both infants, although inter-individual variation was also noted. Bacterial succession was clearly demonstrated, with increased diversity seen over time, yet both infants harboured relatively stable microbial communities during certain weeks. Dietary challenges [either inclusion of milk formula or introduction of solid foods (weaning)] resulted in microbiological shifts and increased complexity of the DGGE profiles (Favier *et al.*, 2002). Interestingly, weaning had a more pronounced impact on the diversity of the microflora of infant D, who was exclusively breast-fed prior to weaning. This observation was largely associated with the increased complexity and relative stability of multiple dominant amplicons in the mixed-fed infant prior to the introduction of solid foods in the diet. Both infants harboured profiles containing relatively dominant bifidobacterial components during the first 6 months of life (Favier *et al.*, 2002).

Investigations of the lactic acid bacterial component of the human faecal flora have also been successfully achieved using DGGE analysis (Satokari *et al.*, 2001; Walter *et al.*, 2001; Heilig *et al.*, 2002). Multiple bifidobacterial biotypes were seen in five out of six Finnish adults, no bifidobacterial PCR amplification was observed for the last individual (Satokari *et al.*, 2001). Inter-individual variation was evident, but bifidobacterial PCR-DGGE demonstrated a relatively stable population over the 4 weeks monitoring (with minor changes in patterns seen for one of the subjects). *Lactobacillus* group-specific primers coupled with DGGE showed fluctuations in profiles for this group over 6 months in four healthy adults from New Zealand (two males and two females), as well as inter-individual variations (Walter *et al.*, 2001). A study of 12 adults living in The Netherlands also examined the *Lactobacillus* component of human faecal samples (using a separate set of primers to the work of Walter and colleagues) (Heilig *et al.*, 2002). Interestingly, variable intra-individual stability was identified, with some individuals harbouring relatively stable lactobacilli profiles and other harbouring more dynamic ones. Inter-individual variations were again evident (Heilig *et al.*, 2002). The same PCR-DGGE assay was used to follow the lactobacilli population of one baby boy from 1 day old to 5 months old. No amplification was seen with this system prior to day 55. Subsequently, a *Lactobacillus* DGGE profile comprising two prominent amplicons was displayed until weaning, when a third prominent amplicon was also observed (suggesting bacterial succession of the lactobacilli component as a result of dietary modulation) (Heilig *et al.*, 2002).

One study investigated the application of PCR-DGGE targeting the transadolase gene for identification and detection of faecal bifidobacteria (Requena *et al.*, 2002). Of the nine bifidobacterial species investigated, only *Bifidobacterium angulatum* and *Bifidobacterium catenulatum* could not be differentiated by this assay. Investigation of the diversity of the bifidobacterial flora of 10 healthy adults demonstrated four individuals harboured *Bifidobacterium adolescentis* and *Bifidobacterium longum*, one contained two distinct *Bifidobacterium adolescentis* amplicons, one *Bifidobacterium adolescentis* and an unidentified amplicon, two a single *Bifidobacterium longum* amplicon, one a single *Bifidobacterium bifidum* amplicon and one produced no PCR-DGGE product (Requena *et al.*, 2002). A similar examination (by the same research group) of the bifidobacterial flora of 10 samples from two healthy babies showed eight samples contained *Bifidobacterium bifidum* (one of which also harboured a second, unidentified amplicon), one sample harboured *Bifidobacterium infantis, Bifidobacterium longum* and an unidentified amplicon, whilst the last baby gave no PCR-DGGE product. Confirmation of the bifidobacterial diversity of all 20 faecal samples (10 from adults and 10 from babies), using Matsuki and colleagues' species-specific 16S rRNA-targeted PCR primers (using conventional gel electrophoresis), highlighted some discrepancies between the two assays. In general, greater diversity was observed using the 16S rRNA-targeted PCR method – although this was not coupled with DGGE.

Examination of the mucosally associated and faecal microflora of 10 subjects (five male and five female) showed high similarity indices for mucosal DGGE profiles from different colonic regions of the same donor (Zoetendal *et al.*, 2002a). Furthermore, the mucosally associated microflora of an individual was distinctly different from that of the faecal sample from the same subject. *Lactobacillus*-specific DGGE profiles of mucosal and faecal samples were very similar in 6 of the 10 subjects (with a single prominent amplicon). Large inter-individual variation was, however, observed with total community DGGE for both faecal and mucosal samples.

An alternative to DGGE profiling is sequencing the clonal libraries, rather than separating the amplicons using a denaturing gradient. DGGE affords rapid comparative analysis of multiple samples, which requires additional work in relation to characterisation of amplicons contained within each profile. Clone libraries enable detailed analysis of the predominant species based on the sequence data from each amplicon. Both methods suffer from inherent biasing associated with PCR-based techniques, particularly concerning insufficient or preferential cell lysis, PCR inhibition and differential amplification (whether a result of amplification efficiency or differing rRNA gene copy numbers) (Wintzingerode *et al.*, 1997). Overall, however, PCR-based technology has revolutionised microbial ecology and provides a rapid and powerful tool to investigate diversity within mixed microbial populations (including complex systems such as the human gut).

Community analysis using clone libraries, as with DGGE, first requires extraction of total DNA from the samples and amplification of the targeted genes. Individual amplicons are then separated by cloning and subsequently sequenced (Suau *et al.*, 1999). The number of clones obtained from complex microbial samples, such as the human gut, limit the use of cloning libraries. Although the advent of capillary sequencing machinery and robotics somewhat reduces the logistics of such analyses. Critically, such investigations fully alerted the microbiological community to the relative anonymity of the greater proportion of bacterial diversity of the human GI microflora, and the extent to which our understanding of its components was biased or blinded due to utility of cultivation strategies.

To date, three studies have been reported which sequenced partial cloning libraries to examine the predominant bacterial microflora of humans (Wilson and Blitchington, 1996; Suau *et al.*, 1999; Blaut *et al.*, 2002). Three major monophyletic groups contained the majority of clones sequenced in the first two studies: namely, the *Bacteroides* group, *Clostridium coccoides* group and *Clostridium leptum* group (Wilson and Blitchington, 1996; Suau *et al.*, 1999). Disparity was noted between the relative proportions seen in each study, but may simply reflect inter-individual variation. Notably, bifidobacterial sequences were lacking in both cloning libraries, even though 16S rRNA dot-blot analysis demonstrated their presence (Suau *et al.*, 1999). Investigations performed as part of the European Union human gut flora project showed greater microbial diversity in cloning libraries of older subjects (Blaut *et al.*,

2002). Furthermore, a larger proportion of operational taxonomic units in the cloning library of an elderly subject were unknown, compared with that seen with cloning libraries of the infant and adult subjects. Such research highlighted important components of the faecal flora of humans, providing essential direction for the expansion of probe and primer development, with the objective of fully elucidating the predominant microbial composition of the human GI tract.

In addition to PCR-cloning and PCR-DGGE community profiling assays, standard PCR techniques have been used to determine the presence/absence and/ or activity of particular bacterial groups of interest. Whilst initially developed to afford rapid and reliable characterisation, PCR technology was applied to gut microbiology to detect qualitative changes between distinct samples or population groups. Recent technology couples PCR amplification with fluorescence emission to afford quantitative analysis of the initial copy number of the target sequence in the template (i.e. real-time PCR) (Sharkey *et al.*, 2004). This is achieved by incorporating chemistries, such as nonspecific DNA-binding fluoropores (e.g. SYBR Green) or fluorescently labelled oligonucleotide probes, to measure the yield of PCR amplicons after each cycle (Mackay, 2004; McKillip and Drake, 2004; Sharkey *et al.*, 2004). Quantitative PCR (qPCR) can also be achieved through competitive qPCR strategies (CqPCR), which incorporate an internal standard (competitor DNA of known concentration), calculate the ratios of target amplicons to competitor amplicons and determines target DNA in template from the calibration curve (Lim *et al.*, 2001).

Both qPCR methodologies have their limitations, largely inherent to all PCR strategies, especially in relation to examining the microbial diversity of mixed populations (which is the primary objective of microbial ecological studies). One advantage CqPCR has over real-time PCR is the ability to correct for sample-to-sample variation resulting from PCR inhibitors or extensive background DNA, due to the co-amplification of internal standard and target DNA (Lim *et al.*, 2001). However, the selection of competitor DNA is essential (in practice it should have the same amplification efficiency as, yet be readily distinguishable from, target DNA). Lim *et al.* (2001) demonstrated that combining CqPCR with constant-denaturant capillary electrophoresis (CDCE) alleviated these limitations, as it afforded differentiation of competitor DNA virtually identical to target DNA. Real-time PCR, on the other hand, affords measurement of PCR products as they accumulate and does not require post-PCR handling (Klein, 2002). A number of real-time PCR systems and chemistries are available, including 5′-nuclease assays (TaqMan), DNA binding dyes (SYBR Green I assay), fluorogenic PCR primers (scorpions), molecular beacons, fluorescently labelled oligonucleotide probe assays and light up extension (LUX) fluorogenic primers (Sharkey *et al.*, 2004).

The advantages of PCR-based assays, once optimised, include their simplicity, speed, selectivity (in relation to both the phylogenetic level of investigation and specificity) and sensitivity. To date, the application of qPCR has largely concentrated on pathogenic organisms and potential diagnostic tools (Nogva *et al.*, 2000; Ge *et al.*, 2001) or quantification of total bacteria (Suzuki *et al.*, 2000; Bach *et al.*, 2002; Nadkarni *et al.*, 2002). In addition, qPCR strategies have been employed in a number of environmental studies, including soil microbial ecology and aquatic/marine microbiological samples (Leser *et al.*, 1995; Lee *et al.*, 1996; Möller and Jansson, 1997; Johnsen *et al.*, 1999; Tay *et al.*, 2001). However, the potential of the technique is clearly apparent and, with the array of chemistries and systems increasing, it is envisaged the next decade will unfold a plethora of information using qPCR.

The extensive interest in the microbial ecology of the human gut, together with the vast array of PCR primers and oligonucleotide probes targeting indigenous organisms of interest (at all phylogenetic levels), suggests it is only a matter of time before optimised qPCR is available for such investigations. Indeed, published work has begun to slowly filter out – of initial investigations of such applications (Huijsdens *et al*., 2002; Malinen *et al*., 2003; Rousselon *et al*., 2004; Bartosch *et al*., 2005). Furthermore, Furret *et al*. (2004) demonstrated the use of qPCR to analyse the lactic acid bacterial content of fermented milk products, whilst Vitali *et al*. (2003) used real-time PCR to quantify specific *Bifidobacterium* strains in a mixed probiotic product.

Huijsdens *et al*. (2002) examined the application of real-time qPCR to investigate the adhesion of gut commensals (namely, *Escherichia coli* and *Bacteroides vulgatus*) to GI mucosa. Whilst preliminary in nature, the research demonstrated the potential of the 5′-nuclease PCR technique for studying the microbial composition of the GI microflora, including biopsy samples. A subsequent study by Malinen *et al*. (2003) compared the 5′-nuclease assay with the SYBR Green I assay and dot-blot hybridisations, with respect to quantifying specific faecal organisms. In this case, the target bacterial populations were *Bacteroides fragilis, Ruminococcus productus, Bifidobacterium longum, Escherichia coli, Lactobacillus acidophilus* and *Bifidobacterium lactis* (a probiotic bacterium). Overall, they found that the two real-time qPCR assays were comparable for all target populations except *R. productus*. The 5′-nuclease assay (TaqMan) identified a higher quantity of *R. productus* in the original stool samples than the SYBR Green I assay. Though the authors did not conclude whether this was due to underestimation by the SYBR Green I assay or overestimation by the TaqMan technique. However, real-time PCR was shown to provide higher sensitivity than dot-blot hybridisations (Malinen *et al*., 2003).

Rousselon *et al*. (2004) developed a qPCR system to detect the Gram-positive low G+C cluster of phylotypes (Fec1) belonging to the *Clostridium coccoides* group. The aim of the work was to develop a system for assessing faecal contamination of environmental samples. However, the assay was used, in conjunction with qPCR measuring the total bacterial load of human faecal samples ($n = 5$), and showed Fec1 16S rDNA comprised between 0.57 % and 4.08 % of total 16S rDNA (Rousselon *et al*., 2004).

More recently, real-time qPCR has been used to examine the effects of synbiotic administration on the faecal bifidobacterial population of elderly subjects (Bartosch *et al*., 2005). The synbiotic contained *Bifidobacterium bifidum* BB-02 and *Bifidobacterium lactis* BL-01, in conjunction with Synergy 1 (an inulin-based prebiotic). Bifidobacterial DNA was detected, by qPCR, in all stool samples from both the test ($n=9$) and control ($n=9$) cohorts. Real-time PCR showed significantly higher levels of bifidobacterial DNA in samples from the test group during and after synbiotic administration, compared with the control group. Species-specific primers detected *B. bifidum* and *B. lactis* in all stool samples from the synbiotic group during synbiotic feeding. However, only five subjects in the synbiotic group (and one of the nine control subjects) harboured *B. bifidum* during the baseline period of the study (Bartosch *et al*., 2005). Overall, significantly higher titres of total *B. bifidum* DNA were identified in the samples of the synbiotic group throughout the study (including baseline) compared with the control group. Synbiotic feeding elicited significantly higher copy numbers of *B. bifidum* rRNA genes in the test group, compared with baseline quantities ($P<0.01$). Significantly higher copy numbers of *B. lactis* rRNA genes were seen in the synbiotic group, compared with control group, during synbiotic feeding. Furthermore, synbiotic administration resulted in a highly significant increase in *B. lactis* rRNA genes in the synbiotic group, compared with baseline levels ($P<0.001$).

These data demonstrate that qPCR can be applied to investigations of the microbial ecology of the human GI tract. Although differences with respect to rRNA gene copy numbers between different bacterial species and genera, complicate the situation, as does the indiscriminate detection of genes from bacterial chromosomal DNA, whether cells are viable or not. One avenue would be to utilise real-time reverse transcription PCR (real-time RT-PCR), which provides qPCR analysis of gene expression (as opposed to qPCR analysis of copy numbers of genes). Another alternative is microarray analysis, which provides expression patterns of multitudes of genes simultaneously (Sharkey *et al.*, 2004). The difficulty of both methods (real-time RT-PCR and microarrays) is accurate isolation of total mRNA from samples, and subsequent concerns that any observed differences in gene expression between two samples are merely an artefact of mRNA isolation rather than expression levels.

6.6 DNA Microarrays

The ability to construct DNA microarrays comprising thousands of distinct oligonucleotide probes on a single microchip or glass slide, opens up an exciting array of opportunity for monitoring the microbial ecology of environmental or human GI samples (Kuipers *et al.*, 1999). In addition, DNA microarrays may be used to investigate host–microbe communication (so-called cross-talk) in the GI tract, including monitoring the expression of genes in response to introduced bacteria (probiotics or pathogens), as well as those expressed in relation to commensal organisms (Cummings and Relman, 2000; Hooper *et al.*, 2001). Alternatively, one could construct microarrays targeting genes encoding important bacterial traits, for example those associated with probiotic characteristics (Klaenhammer *et al.*, 2002) or those associated with virulence or pathogenesis.

Microarrays developed with oligonucleotide probes targeting specific 16S rRNA genes would afford a high throughput system for rapid characterisation of mixed microbial populations (Guschin *et al.*, 1997; Rudi *et al.*, 2002). As peviously discussed, Wang and colleagues (2002a,b) utilised such an approach, although they employed membrane arrays, to monitor the presence of predominant bacteria using 60 oligonucleotide probes.

6.7 Functional Investigations, Introducing the 'Omics' Era

Microarrays and qPCR techniques provide assays for comparing gene expression levels, or gene copy numbers, of different samples. However, such methods generally do not provide absolute levels of expression, as actual expression or activity often does not reflect transcript levels (McAuliffe and Klaenhammer, 2002). As such, modern omics assays [e.g. proteomics using matrix-assisted laser desorption/ionisation time-of-flight

(MALDI-TOF) mass spectrometry to analyse the protein composition of cells; and metabonomics, which quantifies low molecular weight molecules (metabolites) in biological samples] are essential for functional investigations of biological systems (McAuliffe and Klaenhammer, 2002; Nicholson *et al.*, 2005).

In gut microbial ecology, proteomics has the potential to connect genetic information encoded by the bacterial community and their expressed phenotypes. Furthermore, MALDI-TOF analysis may allow identification of important gut microbial attributes (such as receptor sites for adhesion, including those necessary for mucosal attachment). The combined use of proteomic and genomic approaches will further facilitate detailed analysis of the microbial ecology of the human gut. Metabonomics, using nuclear magnetic resonance (NMR) spectroscopy or liquid chromatrography-mass spectrometry (LC-MS), may be employed to monitor microbial metabolites as biomarkers of health status, or potential drug efficacy. Metabonomics is the study of the whole metabolic output of an organism or living system. In the case of humans, this is the sum of genomically encoded activities coupled with metabolic activities of our resident gut microflora (Nicholson and Wilson, 2003). The use of post-genomic principles, including proteomics and metabonomics, in microbial ecological studies will expand our understanding of the global function and activity of the human gut microflora (beyond that of bacterial succession and compositional dynamics). Such analyses will afford detailed assessment of the impact of diet and health, including prebiotic administration, on the GI microflora and host interactions. Indeed, the complex nature of the symbiotic relationship between human host and gut microflora led Nicholson *et al.* (2005) to propose 'a "conditional metabolic phenotype" of the host, dependent on the host genome, gut microbiome and other environmental factors, including nutrition'. Furthermore, they alluded to the fact that this 'conditional metabolic phenotype' is essential to health outcome and drug efficacy. As such, tampering with the symbiotic microbiome (whether by antibiotics or dietary means) may form uncontrolled evolution of the GI microbiota, the long-term effects of which are unknown (Nicholson *et al.*, 2005).

Genomics, proteomics and metabonomics will contribute to our understanding of the molecular mechanisms involved in relationships between food components (such as prebiotics), the GI microflora and the host (including bacterial interactions in the GI tract). The ultimate endpoint being the potential to ensure the safety of functional foods and provide evidence of the health benefits of consuming such products (McAuliffe and Klaenhammer, 2002).

6.8 Conclusion

Molecular microbial technology has advanced incredibly in the last decade and high throughput systems are now available to provide rapid and reliable analysis of the microbial composition and activity, even of complex ecosystems – such as the human GI tract. The scene is set for detailed profiling of the mass array of genomic, proteomic and metabonomic entities of the human gut, in relation to host genome, nutrition and health. Indeed, the future potential of personalised dietary and pharmacological health care, leading towards improved quality of life looks attainable in the light of such breakthroughs.

References

Bach, H.-J., Tomanova, J., Schloter, M. and Munch, J. C. (2002) Enumeration of total bacteria and bacteria with genes for proteolytic activity in pure cultures and in environmental samples by quantitative PCR mediated amplification. *Journal of Microbiological Methods* **49**, 235–245.

Bartosch, S., Woodmansey, E. J., Paterson, J. C. M., McMurdo, M. E. T. and Macfarlane, G. T. (2005) Microbiological effects of consuming a synbiotic containing *Bifidobacterium bifidum, Bifidobacterium lactis*, and oligofructose in elderly persons, determined by real-time polymerase chain reaction and counting of viable bacteria. *Clinical Infectious Diseases* **40**, 28–37.

Blaut, M., Collins, M. D., Welling, G. W., Doré, J., van Loo, J. and de Vos, W. (2002) Molecular biological methods for studying the gut microbiota: the EU human gut flora project. *British Journal of Nutrition* **87**(Suppl. 2), S203–S211.

Boehm, G., Lidestri, M., Casetta, P., Jelinek, J., Negretti, F., Stahl, B. and Marini, A. (2002) Supplementation of a bovine milk formula with an oligosaccharide mixture increases counts of faecal bifidobacteria in preterm infants. *Archives of Disease in Childhood. Fetal and Neonatal Edition* **86**, F178–F181.

Bourlioux, P., Koletzko, B., Guarner, F. and Braesco, V. (2003) The intestine and its microflora are partners for the protection of the host: report on the Danone Symposium The Intelligent Intestine, Paris, 14 June 2002. *American Journal of Clinical Nutrition* **78**, 675–683.

Charteris, W. P., Kelly, P. M., Morelli, L. and Collins, J. K. (1997) Review article: selective detection, enumeration and identification of potential probiotic *Lactobacillus* and *Bifidobacterium* species in mixed culture populations. *International Journal of Food Microbiology* **35**, 1–27.

Conway, P. L. (1995) Microbial ecology of the human large intestine. In *Human Colonic Bacteria: Role in Nutrition, Physiology and Pathology*, pp. 1–24 (G. R. Gibson and G. T. Macfarlane, eds). Boca Raton, FL: CRC Press.

Conway, P. (1997). Development of intestinal microbiota. In *Gastrointestinal Microbiology* (vol. 2), pp. 3–38 (R. I. Mackie, B. A. White, and R. E. Isaacson, eds). New York: Chapman and Hall.

Cummings, C. A. and Relman, D. A. (2000) Using DNA microarrays to study host:microbe interactions. *Emerging Infectious Diseases* **6**, 513–525.

Fanaro, S., Chierici, R., Guerrini, P. and Vigi, V. (2003) Intestinal microflora in early infancy: composition and development. *Acta Paediatrica Supplement* **441**, 48–55.

Favier, C. F., Vaughan, E. E., de Vos, W. M. and Akkermans, A. D. L. (2002) Molecular monitoring of succession of bacterial communities in human neonates. *Applied and Environmental Microbiology* **68**, 219–226.

Finegold, S. M. and Sutter, V. L. (1978) Fecal flora in different populations, with special reference to diet. *American Journal of Clinical Nutrititon* **31**, S116–S122.

Franks, A. H., Harmsen, H. J. M., Raangs, G. C., Jansen, G. J., Schut, F. and Welling, G. W. (1998) Variations of bacterial populations in human feces measured by fluorescence in situ hybridization with group-specific 16S rRNA-targeted oligonucleotide probes. *Applied and Environmental Microbiology* **64**, 3336–3345.

Furret, J. P., Quence, P. and Tailliez, P. (2004) Molecular quantification of lactic acid bacteria in fermented milk products using real-time quantitative PCR. *International Journal of Food Microbiology* **15**, 197–207.

Ge, Z., White, D. A., Whary, M. T. and Fox, J. G. (2001) Fluorogenic PCR-based quantitative detection of a murine pathogen, *Helicobacter hepaticus. Journal of Clinical Microbiology* **39**, 2598–2602.

Gorbach, S. L., Nahas, L., Lerner, P. I. and Weinstein, L. (1967) Studies of intestinal microflora. I. Effects of diet, age, and periodic sampling on numbers of fecal microorganisms in man. *Gastroenterology* **53**, 845–855.

Grönlund, M.-M., Lehtonen, O. P., Eerola, E. and Kero. P. (1999) Fecal microflora in healthy infants born by different methods of delivery: permanent changes in intestinal flora after Cesarean delivery. *Journal of Pediatric Gastroenterology and Nutrition* **28**, 19–25.

Guarner, F. and Malagelada, J.-R. (2003) Gut flora in health and disease. *Lancet* **361**, 512–519.

Guschin, D. Y., Mobarry, B. K., Proudnikov, D., Stahl, D. A., Rittmann, B. E. and Mirzabekov, A. D. (1997) Oligonucleotide microchips as genosensors for determinative and environmental studies in microbiology. *Applied and Environmental Microbiology* **63**, 2397–2402.

Harmsen, H. J. M., Raangs, G. C., He, T., Degener, J. E. and Welling, G. W. (2002) Extensive set of 16S rRNA-based probes for detection of bacteria in human feces. *Applied and Environmental Microbiology* **68**, 2982–2990.

Harmsen, H. J. M., Wildeboer-Veloo, A. C. M., Raangs, G. C., Wagendorp, A. A., Klijn, N., Bindels, J. G. and Welling, G. W. (2000) Analysis of intestinal flora development in breast-fed and formula-fed infants by using molecular identification and detection methods. *Journal of Pediatric Gastroenterology and Nutrition* **30**, 61–67.

Heilig, H. G. H. J., Zoetendal, E. G., Vaughan, E. E., Marteau, P., Akkermans, A. D. L. and de Vos, W. M. (2002) Molecular diversity of *Lactobacillus* spp. and other lactic acid bacteria in the human intestine as determined by specific amplification of 16S ribosomal DNA. *Applied and Environmental Mircobiology* **68**, 114–123.

Holzapfel, W. H., Haberer, P., Schillinger, U. and Huis in't Veld, J. H. J. (1998) Overview of gut flora and probiotics. *International Journal of Food Microbiology* **41**, 85–101.

Hooper, L. V., Wong, M. H., Thelin, A., Hansson, L., Falk, P. G. and Gordon, J. I. (2001) Molecular analysis of commensal host–microbial relationships in the intestine. *Science* **291**, 881–884.

Hopkins, M. J., Sharp, R. and Macfarlane, G. T. (2001) Age and disease related changes in intestinal bacterial populations assessed by cell culture, 16S rRNA abundance, and community cellular fatty acid profiles. *Gut* **48**, 198–205.

Huijsdens, X. W., Linskens, R. K., Mak, M., Meuwissen, S. G. M., Vandenbroucke-Grauls, C. M. J. E. and Savelkoul, P. H. M. (2002) Quantification of bacteria adherent to gastrointestinal mucosa by real-time PCR. *Journal of Clinical Microbiology* **40**, 4423–4427.

Johnsen, K., Enger, Ø., Jacobsen, C. S., Thirup, L. and Torsvik, V. (1999) Quantitative selective PCR of 16S ribosomal DNA correlates well with selective agar plating in describing population dynamics of indigenous *Pseudomonas* spp. in soil hot spots. *Applied and Environmental Microbiology* **65**, 1786–1789.

Klaenhammer, T., Altermann, E., Arigoni, F., Bolotin, A., Breidt, F., Broadbent, J., Cano, R., Chaillou, S., Deutscher, J., Gasson, M., van de Guchte, M., Guzzo, J., Hartke, A., Hawkins, T., Hols, P., Hutkins, R., Kleerebezen, M., Kok, J., Kuipers, O., Lubbers, M., Maguin, E., McKay, L., Mills, D., Nauta, A., Overbeek, R., Pel, H., Pridmore, D., Saier, M., van Sinderen, D., Sorokin, A., Steele, J., O'Sullivan, D., de Vos, W., Weimer, B., Zagorec, M. and Siezen, R. (2002) Discovering lactic acid bacteria by genomics. *Antonie van Leeuwenhoek* **82**, 29–58.

Kleessen, B., Bunke, H., Tovar, K., Noack, J. and Sawatzki, G. (1995) Influence of two infant formulas and human milk on the development of the faecal flora in newborn infants. *Acta Paediatrica* **84**, 1347–1356.

Klein, D. (2002) Quantification using real-time PCR technology: applications and limitations. *Trends in Molecular Medicine* **8**, 257–260.

Kuipers, O. P., de Jong, A., Holsappel, S., Bron, S., Kok, J. and Hamoen, L. W. (1999) DNA-microarrays and food-biotechnology. *Antonie van Leeuwenhoek* **76**, 353–355.

Langendijk, P. S., Schut, F., Jansen, G. J., Raangs, C. J., Kamphuis, G. R., Wilkinson, M. H. F. and Welling, G. W. (1995) Quantitative fluorescence *in situ* hybridisation of *Bifidobacterium* spp. with genus-specific 16S rRNA-targeted probes and its application in fecal samples. *Applied and Environmental Microbiology* **61**, 3069–3075.

Lee, S.-Y., Bollinger, D., Bezdicek, D. and Ogram, A. (1996) Estimation of the abundance of an uncultivated soil bacterial strain by a competitive quantitative PCR method. *Applied and Environmental Microbiology* **62**, 3787–3793.

Leser, T. D., Boye, M. and Hendriksen, N. B. (1995) Survival and activity of *Pseudomonas* sp. strain B13(FRI) in a marine microcosm determined by quantitative PCR and an rRNA-targeting probe and its effect on the indigenous bacterioplankton. *Applied and Environmental Microbiology* **61**, 1201–1207.

Lim, E. L., Tomita, A. V., Thilly, W. G., and Polz, M. (2001) Combination of competitve quantitative PCR and constant-denaturant capillary electrophoresis for high-resolution detection and enumeration of microbial cells. *Applied and Environmental Microbiology* **67**, 3897–3903.

Macfarlane, G. T. and Cummings, J. H. (1991) The colonic flora, fermentation, and large bowel digestive function. In *The Large Intestine: Physiology, Pathophysiology, and Disease*, pp. 51–92 (S. F. Phillips, J. H. Pemberton and R. G. Shorter, eds). New York: Raven Press Ltd.

Macfarlane, G. T., Macfarlane, S. and Gibson, G. R. (1998) Validation of a three-stage continuous culture system for investigating the effect of retention time on the ecology and metabolism of bacteria in the human colon. *Microbial Ecology* **35**, 180–187.

Mackay, I.M. (2004) Real-time PCR in the microbiology laboratory. *Clinical Microbiology and Infection* **10**, 190–212.

Mackie, R. I., Sghir, A. and Gaskins, H. R. (1999) Developmental microbial ecology of the neonatal gastrointestinal tract. *American Journal of Clinical Nutrition* **69**(Suppl.), 1035S–1045S.

Malinen, E., Kassinen, A., Rinttilä, T. and Palva, A. (2003) Comparsion of real-time PCR with SYBR Green I or 5'-nuclease assays and dot-blot hybridization with rDNA-targeted oligonucleotide probes in quantification of selected faecal bacteria. *Microbiology* **149**, 269–277.

Marshall, B. J. (1994) *Helicobacter pylori. American Journal of Gastroenterology* **89**(Suppl.), S116–S128.

Marteau, P., Pochart, P., Doré, J., Béra-Maillet, C., Bernalier, A. and Corthier, G. (2001) Comparative study of bacterial groups within the human cecal and fecal microbiota. *Applied and Environmental Microbiology* **67**, 4939–4942.

McAuliffe, O.E. and Klaenhammer, T. R. (2002) Genomic perspectives on probiotic and the gastrointestinal microflora. In *Probiotics and Prebiotics: Where are we going?*, pp. 263–309 (G. W. Tannock, ed.). Norfolk, UK: Caister Academic Press.

McCartney, A. L. (2002) Application of molecular biological methods for studying probiotics and the gut flora. *British Journal of Nutrition* **88**(Suppl. 1), S29–S37.

McCartney, A. L., Wang, W. and Tannock, G. W. (1996) Molecular analysis of the composition of the bifidobacterial and lactobacillus microflora of humans. *Applied and Environmental Microbiology* **62**, 4608–4613.

McKillip, J. L. and Drake, M. (2004) Real-time nucleic acid-based detection methods for pathogenic bacteria in food. *Journal of Food Protection* **67**, 823–832.

Mitsuoka, T. (1992) Intestinal flora and aging. *Nutrition Reviews* **50**, 438–446.

Mitsuoka, T. and Kaneuchi, C. (1977) Ecology of the bifidobacteria. *American Journal of Clinical Nutrition* **30**, 1799–1810.

Möller, A. and Jansson, J. K. (1997) Quantification of genetically tagged cyanobacteria in Baltic Sea sediment by competitive PCR. *Biotechniques* **22**, 512–518.

Moore, W. E. C. and Holdeman, L. V. (1975). Discussion of current bacteriological investigations of the relationships between intestinal flora, diet, and colon cancer. *Cancer Research* **35**, 3418–3420.

Moore, W. E. C., Cato, E. P. and Holdeman, L. V. (1978) Some current concepts in intestinal bacteriology. *American Journal of Clinical Nutrititon* **31**, S33–S42.

Mountzouris, K. C., McCartney, A. L. and Gibson, G. R. (2002) Intestinal microflora of human infants and current trends for its nutritional modulation. *British Journal of Nutrition* **87**, 405–420.

Muyzer, G. (1999) DGGE/TGGE a method for identifying genes from natural ecosystems. *Current Opinions in Microbiology* **2**, 317–322.

Muyzer, G., de Waal, E. C. and Uitterlinden, A. G. (1993) Profiling of complex microbial populations by DGGE analysis of PCR-amplified genes coding for 16S rRNA. *Applied and Environmental Microbiology* **59**, 695–700.

Nadkarni, M. A., Martin, F. E. and Jacques, N. A. (2002) Determination of bacterial load by real-time PCR using a broad-range (universal) probe and primer set. *Microbiology* **148**, 257–266.

Nicholson, J. K. and Wilson, I. D. (2003) Opinion: understanding 'global' systems biology: metabonomics and the continuum of metabolism. *Nature Reviews Drug Discovery* **2**, 668–676.

Nicholson, J. K., Holmes, E. and Wilson, I. D. (2005) Gut microorganisms, mammalian metabolism and personalized health care. *Nature Reviews Microbiology* (Advance Online Publication: www.nature.com/reviews/micro). Published online 8 April 2005; doi: 10.1038/nrmicro1152.

Nogva, H. K., Bergh, A., Holck, A. and Rudi, K. (2000) Application of the 5'-nuclease PCR assay in the evaluation and development of methods for quantitative detection in *Campylobacter jejuni. Applied and Environmental Microbiology* **66**, 4029–4036.

Olive, D. M. and Bean, P. (1999) Minireview: principles and applications of methods for DNA-based typing of microbial organisms. *Journal of Clinical Microbiology* **37**, 1661–1669.

O'Sullivan, D. J. (1999) Methods for analysis of the intestinal microflora. In *Probiotics. A Critical Review*, pp. 23–44 (G. W. Tannock, ed.). Norfolk, UK: Horizon Scientific Press.

Rabiu, B. A. and Gibson, G. R. (2002) Carbohydrates: a limit on bacterial diversity within the colon. *Biological Reviews of the Cambridge Philosophical Society* **77**, 443–453.

Requena, T., Burton, J., Matsuki, T., Munro, K., Simon, M.A., Tanaka, R., Watanabe, K. and Tannock, G. W. (2002) Identification, detection, and enumeration of human *Bifidobacterium* species using PCR targeting the transaldolase gene. *Applied and Environmental Microbiology* **68**, 2420–2427.

Rousselon, N., Delgenès, J.-P. and Godon, J.-J. (2004) A new real time PCR (TaqMan(®)PCR) system for detection of the 16S rDNA gene associated with fecal bacteria. *Journal of Microbiological Methods* **59**, 15–22.

Rudi, K., Flateland, S. L., Hanssen, J. F., Bengtsson, G. and Nissen, H. (2002) Development and evaluation of a 16S ribosomal DNA array-based approach for describing complex microbial communities in ready-to-eat vegetable salads packed in a modified atmosphere. *Applied and Environmental Microbiology* **68**, 1146–1156.

Salminen, S., Gibson, G. R., McCartney, A. L. and Isolauri, E. (2004) Influence of mode of delivery on gut microbiota composition in seven year old children. *Gut* **53**, 1388–1389.

Satokari, R. M., Vaughan, E. E., Akkermans, A. D. L., Saarela, M. and de Vos, W. M. (2001) Bifidobacterial diversity in human feces detected by genus-specific PCR and denaturing gradient gel electrophoresis. *Applied and Environmental Microbiology* **67**, 504–513.

Saunier, K. and Doré, J. (2002) Gastrointestinal tract and the elderly: functional foods, gut microflora and healthy ageing. *Digestive and Liver Diseases* **34**, S19–S24.

Sghir, A., Gramet, G., Suau, A., Rochet, V., Pochart, P. and Doré, J. (2000) Quantification of bacterial groups within human fecal flora by oligonucleotide probe hybridization. *Applied and Environmental Microbiology* **66**, 2263–2266.

Sharkey, F. H., Banat, I. M. and Marchant, R. (2004) Detection and quantification of gene expression in environmental bacteriology. *Applied and Environmental Microbiology* **70**, 3795–3806.

Simon, G. L. and Gorbach, S. L. (1984) Intestinal flora in health and disease. *Gastroenterology* **86**, 174–193.

Stark, P. L. and Lee, A. (1982) The microbial ecology of the large bowel of breast-fed and formula-fed infants during the first year of life. *Journal of Medical Microbiology* **15**, 189–203.

Suau, A., Bonnet, R., Sutren, M., Godon, J.-J., Gibson, G. R., Collins, M. D. and Doré, J. (1999) Direct analysis of genes encoding 16S rRNA from complex communities reveals many novel molecular species within the human gut. *Applied and Environmental Microbiology* **65**, 4799–4807.

Suau, A., Rochet, V., Sghir, A., Gramet, G., Brewaeys, S., Sutren, M., Rigottier-Gois, L. and Doré, J. (2001) *Fusobacterium prausnitzii* and related species represent a dominant group within the human fecal flora. *Systematic and Applied Microbiology* **24**, 139–145.

Suzuki, M. T., Taylor, L. T. and DeLong, E. F. (2000) Quantitative analysis of small-subunit rRNA gene in mixed microbial populations via 5'-nuclease assays. *Applied and Environmental Microbiology* **62**, 625–630.

Tannock, G. W. (1995) *Normal Microflora. An Introduction to Microbes Inhabiting the Human Body.* London: Chapman & Hall.

Tay, T.-L., Hemond, H., Krumholz, L., Cavanaugh, C. M. and Polz, M. F. (2001) Population dynamics of two toluene degrading species in a contaminated stream assessed by quantitative PCR. *Microbial Ecology* **41**, 124–131.

Tuohy, K. M., Kolida, S., Lustenberger, A. M. and Gibson, G. R. (2001) The prebiotic effects of biscuits containing partially hydrolysed guar gum and fructo-oligosaccharides – a human volunteer study. *British Journal of Nutrition* **86**, 341–348.

Vitali, B., Candela, M., Matteuzzi, D. and Brigidi, P. (2003) Quantitative detection of probiotic *Bifidobacterium* strains in bacterial mixtures by using real-time PCR. *Systematic and Applied Microbiology* **26**, 269–276.

Walter, J., Hertel, C., Tannock, G. W., Lis, C. M., Munro, K. and Hammes, W. P. (2001) Detection of *Lactobacillus, Pedicoccus, Leuconostoc and Weissella* species in human feces by using group-specific PCR primers and denaturing gradient gel electrophoresis. *Applied and Environmental Microbiology* **67**, 2578–2585.

Wang, R.-F., Beggs, M. L., Robertson, L. H. and Cerniglia, C. E. (2002a) Design and evaluation of oligonucleotide-microarray method for the detection of human intestinal bacteria in fecal samples. *FEMS Microbiological Letters* **213**, 175–182.

Wang, R.-F., Kim, S-J., Robertson, L. H. and Cerniglia, C. E. (2002b) Development of a membrane-assay method for the detection of intestinal bacteria in fecal samples. *Molecular and Cellular Probes* **16**, 341–350.

Wilson, K. H. and Blitchington, R. B. (1996) Human colonic biota studied by ribosomal DNA sequence analysis. *Applied and Environmental Microbiology* **62**, 2273–2278.

Wintzingerode, F. V., Gobel, U. B. and Stackerbrandt, E. (1997) Determination of microbial diversity in environmental samples: pitfalls of PCR-based rRNA analysis. *FEMS Microbiology Reviews* **21**, 213–229.

Zoetendal, E. G., Akkermans, A. D. L. and de Vos, W. M. (1998) Temperature gradient gel electrophoresis analysis of 16S rRNA from human fecal samples reveals stable and host-specific communities of active bacteria. *Applied and Environmental Microbiology* **64**, 3854–3859.

Zoetendal, E. G., Akkermans, A. D. L., Akkermans-van Vliet, W. M., de Viosser, J. A. G. M. and de Vos, W. M. (2001) The host genotype affects the bacterial community in the human gastro-intestinal tract. *Microbial Ecology in the Health and Disease* **13**, 129–134.

Zoetendal, E. G., von Wright, A., Vilpponen-Salmela, T., Ben-Amor, K., Akkermans, A. D. L. and de Vos, W. M. (2002a) Mucosa-associated bacteria in the human gastrointestinal tract are uniformly distributed along the colon and differ from the community recovered from feces. *Applied and Environmental Microbiology* **68**, 3401–3407.

Zoetendal, E. G., Ben-Amor, K., Harmsen, H. J. M., Schut, F., Akkermans, A. D. L. and de Vos, W. M. (2002b) Quantification of uncultured *Ruminococcus obeum*-like bacteria in human fecal samples by fluorescent in situ hybridization and flow cytometry using 16S rRNA-targeted probes. *Applied and Environmental Microbiology* **68**, 4225–4232.

7

Dietary Intervention for Improving Human Health: Acute Disorders

Wolfram M. Brück

7.1 The Human Colon

Humans live in close association with a vast number of microorganisms present on the skin, in the mouth and in the gastrointestinal tract. In our bodies, we carry about 1 kg of microorganisms, which constitutes about 20 times more bacterial cells than mammalian. The vast majority (ca. 90 %) are found in the lower gastrointestinal tract (colon).

In contrast to the historic belief that the main roles of the colon were to absorb water and salt as well as store and dispose of waste materials, it is now believed that it has a variety of other important biological functions. It actively transports sodium and chloride ions, probably by the exchange of sodium–hydrogen and chloride–bicarbonate (Willis and Gibson, 1999). Additionally, it also absorbs many products of bacterial fermentation such as short-chain fatty acids (SCFA), ammonia and other metabolites. The colonic mucosa may also secrete fluids as a potential source of hormones or neuropeptides influencing colonic function through acting as local messengers (Stalheim-Smith and Fitch, 1993).

The large intestine and its indigenous microflora contribute towards the digestion of dietary compounds, such as dietary starches, plant polysaccharides and proteins which escape breakdown and absorption in the small intestine (Willis and Gibson, 1999).

Water, undigested food, and endogenous sources of nutrients enter the caecum from the small intestine thus providing a rich nutrient source for bacteria present in the proximal colon. Growth of bacteria is rapid and SCFA produced by fermentation lowers the gut pH. As contents progress through the large intestine, carbohydrates and other

Prebiotics: Development and Application Edited by G.R. Gibson and R.A. Rastall

bacterial substrates become depleted resulting in reduced acid levels in the distal colon and a pH, which approaches neutrality (Salminen *et al.*, 1998).

Many different types of microorganisms can flourish in the large bowel, mainly because of the wide variety of nutrients available and its heterogenous nature (Willis and Gibson, 1999). The normal adult flora consists of about 10^{12} colony forming units g^{-1} (CFU g^{-1}) dry weight of contents (Salminen *et al.*, 1998). Fifty different bacterial genera with well over 400 species are generally represented (Ziemer and Gibson, 1998). Most bacteria growing in the colon are nonsporing obligate anaerobes (Gibson and Roberfroid, 1995). Of these, the *Bacteroides*, which comprise about 30 % of all culturable bacteria in the large intestine, and eubacteria and bifidobacteria are most numerous (Willis and Gibson, 1999). Other resident microorganisms include other obligate anaerobes as well as factultative anaerobes such as peptostreptococci, fusobacteria, lactobacilli, enterobacteria, enterococci, methanogens and dissimilatory sulfate-reducing bacteria (Tuohy *et al.*, 2001). The facultatively anaerobic, gram-positive streptococci are represented by many species from Lancefield group D, including *Streptococcus faecalis*, *S. bovis* and *S. equinus*, and some from group K, such as *S. salivarius*, which is usually associated with the mouth (Gibson and Roberfroid, 1995). Gram-negative anaerobic cocci present include *Veillonella* and *Acidaminococcus* (Gibson and Roberfroid, 1995). Several types of spore-forming rods also inhabit the colon. Of these, *Clostridium* is probably the most common, principally *C. perfringens* (Gibson and Roberfroid, 1995). Although they are present only in low numbers, they are of considerable significance along with other facultative but nonsporing rods since the genus includes a number of important enteric pathogens. Members of the family Enterobacteriaceae, in particular *Escherichia coli*, are usually thought of as typical intestinal bacteria but also include a number of the most important pathogens found in human medicine (Steer *et al.*, 2000).

The microflora as a whole has marked influences on host homeostasis as observed in experiments whereby the characteristics of germfree and conventional animals have been compared (Tannock, 1998). Such assessments have shown that biochemical, physiological and immunological characteristics can be strongly influenced by the type of microflora present in the gut. Therefore, dietary factors such as pro- and prebiotics that influence and regulate the composition of the intestinal microflora and help to suppress the establishment of pathogenic bacteria through competitive exclusion, may also aid other host functions and systems by fortifying the immune system and homeostasis in general (Roberfroid, 2001).

7.2 General Aspects of Prebiotics

A prebiotic is 'a nondigestible food ingredient that beneficially affects the host by selectively stimulating the growth and/or the activity of one or a limited number of bacteria in the colon' (Gibson and Roberfroid, 1995). Specifically, such a food component must fulfil the following criteria (Roberfroid, 2001):

(1) stability under acidic conditions in the stomach and the small colon;
(2) resistance to enzymatic digestion;

(3) hydrolysis and fermentation by colonic bacteria; and
(4) selective stimulation of growth of a limited number of beneficial colonic microorganisms.

The concept of prebiotics is based on the generalised hypothesis that the microflora of the human large colon may be divided into bacteria that are either beneficial or detrimental to health (Gibson and Roberfroid, 1995). Thus, as for probiotics, lactic acid bacteria and bifidobacteria are the main targets because of their associated functions in the gut and stimulation of the immune system (Fuller and Gibson, 1997).

Any dietary material that enters the large intestine is a candidate prebiotic (Collins and Gibson, 1999). This includes carbohydrates such as resistant starch and dietary fibre as well as certain proteins and lipids (Ziemer and Gibson, 1998). However, several attributes also need to be fulfilled in order for a prebiotic to be effective (Table 7.1). Thus, in reality most prebiotics are confined to nondigestible oligosaccharides, many of which seem to fulfil the degree of fermentation selectivity that is required (Gibson *et al.*, 1995). Nevertheless, future aspects on the study of prebiotics may focus more on the immediate aspects of their effect on the pathogenic flora components, as is the case with cellobiose and its ability to attenuate virulence in *Listeria monocytogenes* (Collins and Gibson, 1999).

Bllongue *et al.* (1997) carried out a volunteer trial to confirm the prebiotic nature of lactulose. The feeding was a parallel group, randomised, double blind and placebo-controlled study with 12 healthy volunteers per group. Two weeks baseline was followed by 4 weeks of treatment and a 3-week follow-up. Subjects were fed $2 \times 10\,\mathrm{g\,day}^{-1}$ lactulose or placebo of 50:50 glucose and lactose. Faecal samples were taken throughout and bacteria determined by selective agars. The target populations were *Bacteroides, Bifidobacterium, Clostridium*, coliforms, *Eubacterium, Lactobacillus* and *Streptococcus*. The authors found significant increases in bifidobacteria, lactobacilli and streptococci, whilst bacteroides clostridia, coliforms and eubacteria all decreased during the test period. This would suggest prebiotic activity, although a reliance on selective agars for bacterial discrimination is not a wholly reliable approach.

Fructooligosaccharides (FOS) and inulins naturally occur in numerous different plants (Van Loo *et al.*, 1995) with garlic, onion, asparagus, chicory, artichoke, banana, wheat

Table 7.1 *Design parameters for prebiotics*

Desirable attribute in prebiotic	Properties of oligosaccharides
Good storage and processing stability	Possess 1-6 linkages and pyranosyl sugar rings
Persistence through the colon	High molecular weight
Lack of side effects	Selectively metabolised by 'beneficial' bacteria but not by gas producers, putrefactive organisms, etc.
Active at low dosage	Selectively and efficiently metabolised by bifidobacteria and/or lactobacilli
Control of microflora modulation	Selectively metabolised by restricted species
Inhibit adhesion of pathogens	Possess receptor sequence
Varying sweetness	Different monosaccharide composition
Varying viscosity	Available in different molecular weights and linkages

and leek being especially rich (Gibson and Wang, 1994). Several studies, confirming their prebiotic activity have been conducted using human subjects although the dose, substrate, duration and volunteers have varied. Gibson *et al.* (1995) reported a volunteer trial with adult subjects on strictly controlled diets supplemented with 15 g day^{-1} FOS. Sucrose was used as the control and faecal samples were processed, in a blind manner, within 30 min of passage. Most importantly, the study used follow-up phenotypic characterisation techniques to fully identify the microflora that developed during the feeding regimes. These studies showed that the intake of FOS significantly modified composition of the faecal microflora by stimulating the growth of bifidobacteria, which, after 2 weeks of the feeding period, became the most numerically predominant bacterial group. Additionally, FOS significantly reduced counts of bacteroides, fusobacteria and clostridial populations. These effects lasted for as long as the prebiotic was consumed. However, at the end of a 2-week follow up on the control diet, the faecal flora of all volunteers still had higher bifidobacterial numbers than when recruitment started.

Prebiotics that may exhibit anti-adhesive properties against pathogenic organisms would also confer enhanced functionality by altering gut pathogenesis (Steer *et al.*, 2000). Intestinal pathogens commonly use carbohydrate-binding proteins in order to attach to cells and initiate disease (Kawasaki *et al.*, 1993). Exogenous substances containing the same carbohydrate residues that are required for a pathogen to initiate invasion might competitively inhibit bacterial adhesion to intestinal cells and thereby inhibit colonisation (Kawakami, 1997). Thus, prebiotics bearing such sequences may act as blocking factors through steric hindrance (Kawakami *et al.*, 1992). Several of these saccharides have previously been identified in human breast milk, which is often regarded as a complete food and the ultimate stimulant of bifidobacterial growth (Collins and Gibson, 1999; Kelleher and Lönnerdal, 2001).

7.3 Acute Gastroenteritis

Acute gastroenteritis is something that probably affects everyone at one time or another and due to its widespread occurrence the economic costs and medical aspects are enormous. Usually, acute gastroenteritis involves the ingestion of food or water contaminated with pathogenic microorganisms and/or their toxins. Infections are usually self-limiting, and characterised by diarrhoea and often vomiting (Salminen *et al.*, 1998). The principal pathogens are viruses (e.g. rotavirus) and bacteria such as *E. coli*, *Campylobacter* sp., *Vibrio cholerae*, *Staphylococcus aureus*, *Bacillus cereus*, *C. perfringens*, *Salmonella* sp. *Shigella* sp., *Yersinia* sp. and a number of protozoa (Salminen *et al.*, 1998). Human intestinal bacterial pathogens can be characterised according to the virulence factors that enable them to overcome host defences. These include invasion, which enables bacterial multiplication within enterocytes or colonocytes, for example, *E. coli*, *Shigella* sp., salmonellae and yersinae. Cytotoxic bacteria, which include enteropathogenic and enterohaemorrhagic strains of *E. coli* as well as some shigellae are able to produce substances, which directly cause cell injury. Toxigenic bacteria such as *V. cholerae* and some shigellae are capable of producing enterotoxins, which affect salt and water secretion in the host. Enteroaggregative *E. coli* have the ability to tightly adhere to the colonic mucosa. Such mechanisms enable potentially pathogenic bacteria to establish infections in the gastrointestinal tract, evade the immune system and defeat colonisation resistance afforded by the indigenous gut microflora. Infants, children and the elderly in particular face severe health threats from acute diarrhoea, causing problems in both industrial and developing nations. Many return to hospital after initial treatment, since preventive care and instruction is rarely provided (Ribeiro, 2000).

7.4 Enteropathogenic *Escherichia coli* (EPEC) Gastroenteritis

EPEC is an important category of diarrhoeagenic *E. coli*, which has been linked with infantile diarrhoea in the developing world, particularly in countries lacking adequate sanitation. It can be defined on the basis of pathogenic characteristics.

The hallmark of EPEC infections are attaching and effacing (A/E) histopathology. They are characterised by effacement of microvilli and intimate adherence between bacteria and the epithelial cell membranes. The epithelial cell cytoskeleton changes to include dense accumulations of polymerised filamentous actin (F-actin), beneath the adherent bacteria, which sometimes sit on pseudopod-like pedestal structures that can extend up to 10 μm out from the epithelial cell. Some attached EPEC may also move along the epithelial cell surface in a process driven by polymerisation of actin beneath the pedestal base resembling motility seen in *Listeria* sp. Similar lesions may also be found in animal and cell culture models of enterohaemorrhagic *E. coli* (EHEC) and *Hafnia alvei* isolated from children with diarrhoea. In addition to EPEC and EHEC, a variety of other *E. coli* strains are able to produce A/E lesions, making EPEC the prototype of an entire family of enteric pathogens that produce such lesions on epithelial cells (Nataro and Kaper, 1998).

An age related distribution seen in persons infected with EPEC is one of its most notable epidemiological features. Infection primarily occurs in infants under 2 years of age and numerous case-control studies have demonstrated a strong correlation between the isolation of EPEC from infants with diarrhoea when compared with healthy children (Levine and Edelman, 1984). EPEC can also cause diarrhoea in adults if high inocula (10^8–10^{10}) are given after gastric acid has been neutralised with bicarbonate. The infectious dose in naturally transmitted infection in infants is presumed to be much lower (Levine *et al.*, 1978).

As with other diarrhoeagenic *E. coli* strains, transmission of EPEC is faecal-oral, with contaminated hands, weaning foods or formulae, amongst the vehicles (Levine and Edelman, 1984). The reservoir of EPEC infection is thought to be asymptomatic children and adult carriers. Numerous studies have documented the spread of infection through hospitals, nurseries and day-care centres (Nataro and Kaper, 1998). In symptomatic patients, EPEC can be isolated from stools up to 2 weeks after the cessation of symptoms (Hill *et al.*, 1991). The infection is no longer an important cause of diarrhoea in the developed world even though sporadic cases still exist. In contrast, EPEC is a major cause of infantile diarrhoea in developing countries (Donnenberg, 1995). Particularly in the 0–6 months age group, EPEC strains are the most isolated bacterial diarrhoea-causing pathogen and sometimes exceed rotaviral infections in incidence (Robins-Browne *et al.*, 1980; Craviota *et al.*, 1988). Several studies have shown that breast feeding can be protective against diarrhoea from EPEC since both human colostrum and milk inhibit the adhesion of EPEC to Hep-2 *in vitro* (Robins-Browne *et al.*, 1980). This inhibitory activity has also been detected in sIgA and oligosaccharide fractions excreted in breast milk (Camara *et al.*, 1994).

Infection with EPEC can be severe, with recent outbreaks in developing countries reporting up to 30 % mortality. In addition to acute watery diarrhoea, vomiting and low-grade fever are common symptoms and the primary treatment goal is to prevent dehydration by correcting fluid and electrolyte imbalances. A variety of antibiotics have proven helpful in many cases (Donnenberg, 1995) but resistance is common and alternative therapies such as bismuth subsalicylate (Figueroa-Quintanilla *et al.*, 1993) and specific bovine anti-EPEC milk immunoglobulins (Nzegwu and Levin, 1994) have been used successfully.

Because of the problem of resistance and toxin production, a variety of compounds with prebiotic properties may provide a useful alternative to the traditional treatment with antibiotics. Liehr *et al.* (1980) found that lactulose (β-galactosido-fructose) has anti-endotoxin properties and 670 mg lactulose abolished the gelating activity of *E. coli* endotoxin on Limulus lysate *in vitro*. This suggests that lactulose might offer a therapeutic basis in clinical situations in which endotoxemia is of pathogenetic significance. However, no other study has examined this anti-endotoxin effect making it difficult to assess whether a similar effect is observed *in vivo*. In any case, the well documented (positive) effects of lactulose on the lactic flora of human should provide some level of protection from *E. coli* infection by increasing colonisation resistance.

α-Lactalbumin, and glycomacropeptide, two milk proteins with prebiotic properties, may also prove useful for preventing and treatment of *E. coli* infection while stimulating the natural lactic acid flora.

α-Lactalbumin, a regulatory component of lactose synthase, shares a high level of sequence identity to c-type lysozymes and has a similar three-dimensional structure suggesting that they came from a common ancestral gene (Wold and Adlerberth, 2000). Several studies have been performed to suggest that α-lactalbumin might have similar effects on pathogens (Xue *et al.*, 2001). Glycomacropeptide, a derivative of κ-casein, has been extensively studied *in vitro*. It is thought to inhibit the adhesion of various bacteria, viruses and toxins as it contains a carbohydrate chain containing *N*-acetylneuraminic acid which might act as a possible anti-adhesive prebiotic (Brück *et al.*, 2002). Pihlanto-Leppälä *et al.*(1999) demonstrated that α-lactalbumin hydrolysed with pepsin or trypsin could lower the metabolic activity of *E. coli* JM103 to just 21 % of normal after 6 h incubation. The undigested protein did not inhibit bacterial growth nor metabolism. This bacteriostatic effect of the hydrolysates was found at a high concentration (25 mg ml^{-1}) when compared with a lactoferrin hydrolysate (10 μg ml^{-1}) (Saito, *et al.* 1991). However, the study was carried out under optimum growth conditions and metabolism for *E. coli*, which does not necessarily represent the environment of the colon. Pelligrini *et al.* (1999) found that proteolytic digestion of α-lactalbumin by trypsin and chymotrypsin yielded three peptide fragments with bactericidal properties against gram-positive bacteria and a limited function against gram-negative organisms. Hence, it may be assumed that digestion by endopeptidases allows α-lactalbumin some antimicrobial function. *In vitro* studies on the influence of glycomacropeptide and α-lactalbumin on the gut microflora have been confirmed in several recent studies. Using a two-stage continuous culture system containing fresh faecal specimens of 1, 3 and 6 months old human infants, it was shown that vessels containing breast milk, α-lactalbumin and glycomacropeptide supplemented formula had stable total counts of bifidobacteria. Bacteroides, clostridia and *E. coli* decreased significantly (Brück *et al.*, 2002). In an *in vitro* infection study, a similar effect was observed, while enteropathogenic *E. coli* counts decreased in vessels containing α-lactalbumin and breast milk (Brück *et al.*, 2003a). The influence of α-lactalbumin and glycomacropeptide was further examined on the gut microflora of Rhesus monkeys challenged with 10^8 CFU enteropathogenic *E. coli*. While monkeys fed a α-lactalbumin and glycomacropeptide-free control formula developed acute diarrhoea, those fed breast milk and formula supplemented with α-lactalbumin had no diarrhoea. Infant monkeys fed glycomacropeptide supplemented formulae also developed diarrhoea even though episodes were of shorter duration and associated symptoms such as fever and dehydration were less severe (Brück *et al.*, 2003b).

7.5 *Salmonella*–associated Gastroenteritis

Rates of *Salmonella* infections may be significantly higher in patients with acute diarrhoea than with many other enteric pathogens (Mathew *et al.*, 1991). Children, especially those less than 1 year of age, are susceptible and tend to experience severe infections (Gomez and Cleary, 1998).

Like *E. coli*, *Salmonella* is a member of the family Enterobacteriaceae and exhibits the typical faecal-oral route of transmission through contaminated foods and formulae. However, spread by contaminated hands is also common (Wilson *et al.*, 1982) and controlling outbreaks is difficult since *Salmonella* may be shed asymptomatically for up to 20 weeks after infection in children under 5 years of age and for 8 weeks by adults, becoming healthy excretors (Gomez and Cleary, 1998). It is estimated that only 1–5 % of infections are reported, but as a whole the incidence of *Salmonella* infections has steadily increased since World War II, which might be caused by an increasingly older population, changing agricultural and food distribution methods and increased consumption of raw or slightly undercooked foods (CDC, 1991). Additionally, an increasing number of immunocompromised or chronically ill people together with the deterioration of public health infrastructures may also be at fault (Altekruse *et al.*, 1997). The most common sources of *Salmonella* are beef, poultry, and eggs with eggs being increasingly important (Snoeyenbos *et al.*, 1969). Eggs can be contaminated through cracks in the shell or from an infected ovary (transovarally) and left at room temperature can quickly achieve concentrations of over 10^{11} cells per yolk (Snoeyenbos *et al.*, 1969).

Salmonella is capable of causing a variety of diseases ranging from enteric fever, bacteraemia, enterocolitis, and focal infections, of which enterocolitis is by far the most common. Enteric fever (typhoid fever) is caused primarily by *S. typhi* and *S. paratyphi*, while enterocolitis is typically caused by serovars Typhimurium, Enteritidis and Heidelberg (CDC, 1994).

It is generally believed that a large inoculum is required to overcome acidity in the stomach and compete against the normal flora of the intestinal tract. This becomes evident when considering that the infectious dose decreases significantly when *Salmonella* is consumed with food that either clears the stomach rapidly (fluids) or neutralises acidity such as in milk and milk products (Tauxe and Pavia, 1998). On the other hand, several *in vitro* and *in vivo* models have revealed that breastmilk can protect against infection (Wold and Adlerberth, 2000).

Salmonella infection occurs when cells invade the lumen of the small bowel, where they multiply. Afterwards, they penetrate the ileum and, to a lesser extent, the colon, where an inflammatory reaction occurs. Even though it is believed that the small intestine (Peyer's patches) is the primary site of involvement, some reports suggest that the large intestine is the primary location of infection (Boyd, 1985). Once an infection has established, severity ranges from slight to severe oedema with infiltration of polymorphonuclear leucocytes (PMNs) and monocytes, with a focal inflammation of the lamina propria, degeneration of the mucosa and extravasation of erythrocytes and PMNs. Lymphoid follicles become enlarged and may ulcerate. Overall, salmonellae seem to thrive particularly in increasingly anaerobic environments, which correlates with increased adherence and invasion of host tissues accompanied by increased intracellular survival within macrophages (Singh *et al.*, 2000). However, a typical case scenario and course of *Salmonella* infection is impossible to ascertain since most people infected are never hospitalised and the exact mechanisms of molecular pathogenesis are unclear (Darwin and Miller, 1999).

As with *E. coli* infection, α-lactalbumin and glycomacropeptide may be useful in treating or preventing *Salmonella* infection, especially in infants. A two-stage continuous culture system containing fresh faecal specimens of 1, 3 and 6 months old human infants showed that *Salmonella* counts decreased significantly in vessels containing α-lactalbumin and breast milk (Brück *et al.*, 2003a). However, since lactobacilli counts also increased, the inhibitory effects might be caused the by the formation of some antibacterial factors produced by lactobacilli which may facilitate the treatment of *Salmonella* with other prebiotics.

More traditional prebiotics, such as FOS, stimulate the protective gut microflora, resulting in an increased production of organic acids, which can result in increased luminal killing of acid-sensitive pathogens (Ten Bruggencate *et al.*, 2003). This was confirmed in a study where probiotics, prebiotics, vaccination and acidification of drinking water were used to assess the ability to reduce *Salmonella* transmission in pigs. Piglets weaned at 12 days of age and verified as *Salmonella* free were randomly assigned to one of 10 treatment groups consisting of formic acid in the water (0.02 %w/v), Ferlac-2 (a prebiotic), Flavomycin (0.5 g ton^{-1} feed), FOS in feed (1 %w/v), FOS in water (1 %w/v), egg yolk specific immunoglobulins (1 g pig^{-1} day^{-1}), endotoxin vaccine, a live attenuated *Salmonella* Cholerasuis vaccine, FOS + Ferlac-2, or a control. Each group was composed of 10 pigs, which received their treatments from 14 days before the *Salmonella* challenge. The research found that FOS, when supplemented in drinking water at 1 %, changed the pig faecal bacterial flora while reducing shedding of *Salmonella typhimurium*.

In comparison, in another study where rats were fed a 'humanised' purified diet containing 4 %(w/v) lactulose, FOS, resistant starch, wheat fibre, or cellulose, the opposite effect was observed (Bovee-Oudenhoven *et al.*, 2003). While the supplements did induce significant changes in faecal biochemical and microbiological parameters before an oral infection with *Salmonella enteritidis*, surprisingly, FOS and lactulose seemed to impair the resistance of rats to intestinal salmonella infection. In addition, despite the stimulation of intestinal lactobacilli and bifidobacteria, FOS and lactulose significantly enhanced *Salmonella* translocation. However, this does not indicate that any similar effect would occur in humans as the animals were fed a dose of calcium known to induce translocation, as well as a very high pathogen dose. This indicates that host defence against invasive pathogens, like salmonellae, not only depends on an increased production of organic acids but also on the natural barrier function of the intestinal mucosa. Since a rapid fermentation process of prebiotics leads to high concentrations of organic acids the barrier function may be impaired which in turn impairs the resistance to salmonella infection. Thus, FOS dose-dependently increased salmonella numbers in caecal contents and mucosa and caused a major increase in infection-induced diarrhoea.

In another study examining transgalactosylated oligosaccharides (TOS), the anti-infectious activity of bifidobacteria in combination with TOS against *Salmonella enterica* LT-2 was examined in mice with an antibiotic-induced murine infection (Asahara *et al.*, 2001). Intestinal colonisation by bifidobacteria given exogenously together with TOS during antibiotic treatment prevented antibiotic-induced disruption of colonisation resistance to oral infection with *S. enterica* serovar *typhimurium*, and the metabolic activity needed to produce organic acids and lower the intestinal pH is important in the anti-infectious activity of synbiotics against enteric infection with *Salmonella*.

The use of traditional prebiotics for the treatment of *Salmonella* infections seems to give conflicting results in various animal models but this largely depends upon the physicochemical conditions used. Nevertheless, oligosaccharides may be useful in building colonisation resistance and providing an acidic environment inhibitive to *Salmonella* colonisation. Dietary supplementation of anti-adhesive or antimicrobial prebiotics compounds such as α-lactalbumin or glycomacropeptide may provide yet another approach and a useful alternative to more traditional treatment even though further human volunteer trials are needed to assess activity *in vivo*.

7.6 *Clostridium difficile* – Pseudomembranous Colitis and Antibiotic-induced Colitis

C. difficle is a classic example of the opportunistic proliferation of an intestinal pathogen following breakdown of colonisation resistance. Although this organism is naturally present in the healthy gastrointestinal tract, it may become a major problem when the healthy indigenous microbiota is disrupted after the administration of antibiotics, in particular clindamycin, cephalosporins and penicillins (Reid *et al.*, 2003). After antibiotic intake, *C. difficile* releases two protein exotoxins, toxin A and toxin B, which mediate diarrhoea and colitis. The acquisition of toxin-producing *C. difficile* is the cause of approximately 20–40 % or 2700 cases per 100 000 cases of antibiotic-associated diarrhoea and is the major identifiable cause of nosocomial diarrhoea in the USA, infecting 15–25 % of adult hospitalised patients. This can have serious consequences such as pseudomembranous colitis, toxic megacolon, intestinal perforation and death, particularly in the elderly and debilitated population (Rolfe, 2000).

Standard treatment of *C. difficile* can be expensive and difficult, usually involving vancomycin or metronidazole. Additionally, about 25 % of patients relapse after treatment is discontinued and multiple relapses may occur with increasing severity from the original disease. This may be due to *C. difficile* spores surviving in the intestinal tract and then germinating and producing toxins after treatment is terminated. Thus, as an alternative to antibiotic treatment to restore homeostasis, probiotics or prebiotics may be used to re-establish a normal gut microflora and form an effective colonisation barrier.

To date, the most successful studies involve the use of *Lactobacillus* GG at a dose of 1×10^{10} viable organisms per day and *Saccharomyces boulardii* at a dose of $1\,g\,day^{-1}$. However, the mechanisms involved with these studies might simply be overwhelming the gut lumen with beneficial bacteria (Saavedra *et al.*, 1994). In a study with undernourished children, *Lactobacillus* GG decreased the incidence of diarrhoeal disease in non-breastfed children, but had little effect on those who were breastfed, suggesting that high risk groups may be the most benefited by the prophylactic use of probiotics for diarrhoeal disease (Oberhelman *et al.*, 1999).

Toxin A produced by *C. difficile* has been shown to bind to synthetic oligosaccharide sequences related to previously identified receptors for the toxin. Various inert supports (SYNSORB) containing a variety of oligosaccharides were examined for their potential to neutralise toxin A activity from toxin-containing solutions, as well as clinical stool samples from patients with either pseudomembranous colitis or antibiotic-associated diarrhoea. The results suggested that Synsorb supported oligosaccharides could effectively neutralise toxin A activity from stool samples and thus serve as a potential therapy for *C. difficile*-associated diarrhoea (Heerze *et al.*, 1994). A further study by Castagliuolo *et al.* (1996) examined intestinal inflammation by *C. difficile* and specific binding of toxin A to Galα 1-3Galβ 1-4GlcNAc trisaccharides on enterocyte receptors in rodents. Again, toxin A showed specific binding to Synsorb, bearing the specific trisaccharide. This suggests that an immobilised toxin A receptor sequesters toxin A in the intestinal lumen and inhibits its effects of ileal mucosa, which suggests a potential use for this agent in treating patients with *C. difficile* colitis.

The use of nondigestible oligosaccharides (NDO) may also present a practical approach for restoring or recovering colonisation resistance. The ability of NDO to stimulate bifidobacteria resident in the human colon, while simultaneously alleviating diarrhoeal symptoms has been documented in a variety of studies (e.g. Gibson *et al.*, 1995; Kleessen *et al.*, 1997; Kolida *et al.*, 2002). The ability of FOS to increase colonisation resistance against *C. difficile* infection in animal models was investigated by Gaskins *et al.* (1996). Generally, FOS increased the mean survival time of animals with antibiotic-induced colitis by reducing toxin A titres. Hopkins and Macfarlane (2003) examined the use of FOS, GOS and inulin preparations in normal and antibiotic-treated faecal microbiotas and their abilities to increase barrier resistance against colonisation by *C. difficile*. All NDO did stimulate bifidobacterial growth, while reducing *C. difficile* populations. GOS was the preferred substrate for *Bifidobacterium adolescentis*, *B. angulatum*, and *B. bifidum*, while FOS mainly stimulated *B. catenulatum*. Overall, in the absence of clindamycin, NDO enhanced colonisation resistance against *C. difficile*. However, in the presence of clindamycin, activity against bifidobacteria was augmented in the presence of NDO, resulting in a further loss of colonisation resistance. Similar results were observed by Ito *et al.* (1997) studying the inhibitory effect of lactulose on the growth of *C. difficile* in human faecal cultures. Lactobacilli predominated while bacteroides, and enterobacteria were also suppressed. Surprisingly, bifidobacteria also showed a significant decrease.

Treatment of *C. difficile* infections with traditional prebiotics largely seemed to hinder recovery from diarrhoea in combination with the use of antibiotics. Nonetheless, anti-adhesive oligosaccharides appear to be effective in various models. However, this effect requires further confirmation with human volunteer trials before specific recommendations can be given. Thus, it might be a more reasonable alternative to prevent recurring *C. difficile*-associated diarrhoea episodes with dietary supplementation of various NDO in order to improve colonisation resistance.

7.7 *Helicobacter pylori* Gastroenteritis

H. pylori is a gram-negative curved or spiral rod distinguished by multiple, sheathed flagellae and abundant urease activity. Its antigenic structures are not completely defined and no universal typing scheme has been developed. Thus, strains may be differentiated by genotypic methods such as restriction fragment length polymorphism (RFLP) or polymerase chain reaction (PCR). *H. pylori* is the bacterial pathogen generally responsible for peptic ulcers and type B gastritis (approximately 90 % of all cases), an inflammatory response in the gastric mucosa, ultimately leading to cellular hyperproliferation and malignant transformation. Where chronic infection with *H. pylori* is usually asymptomatic, acute infection may cause vomiting and upper gastrointestinal pain as hypochlorhydria and intense gastritis develop. The bacterium is sheltered from gastric acidity in the mucus layer with a small proportion of cells adhering to the gastric epithelium. While it does not appear to invade tissues, the production of urease, a vasculating cytotoxin (vacA), and the cagA-encoded protein which most likely seems to stimulate the production of chemotactic factors, is associated with injury to the gastric epithelium. About one-third of the world's population is infected with prevalence of infection increasing with age. Both PCR and RFLP diagnosis require endoscopy and biopsy specimens. However, less invasive techniques such as urea breath tests faecal antigen tests or serum antibody detection are more sensitive than diagnostic techniques involving biopsies.

Problems with current therapies for management of *H. pylori*-associated disease are patient compliance and increasing antibiotic resistance. Therefore, prevention of disease via nutritional intervention seems to a practical alternative approach (Campbell *et al.*, 2004). Most prebiotics and probiotics appear not to eradicate *H. pylori*, but may be able to reduce and suppress the bacterial load in infected patients (Felley *et al.*, 2001).

In a randomised study by Felley *et al.* (2001), 53 volunteers infected with *H. pylori* received either LC-1 (*Lactobacillus johnsonii*) or a placebo 180 ml twice a day for 3 weeks. During the last 2 weeks of acidified milk therapy, all subjects also received clarithromycin (500 mg). LC-1 ingestion reduced inflammation and gastritis while decreasing *H. pylori*. Clarithromycin eradicated the infection in 26 % of subjects with LC-1 not improving the antibiotic effect.

Michetti *et al.* (1999) reported that a whey-based culture supernatant of *Lactobacillus acidophilus* La1 has a partial, acid-independent, long-term suppressive effect on *H. pylori* in humans. The supernatant of LC-1 culture was shown to be bactericidal and to have a partial, acid-independent suppressive effect on *H. pylori*.

Breastfeeding may protect from infection by *H. pylori* during early life, in part because of antimicrobial activity of human milk resides in the caseins (Stromqvist *et al.*, 1995). The K-casein derivative, glycomacropeptide has previously been thought to inhibit the adhesion of *H. pylori* to the cell membrane by binding to the pathogen's receptor sites (Kawakami, 1997). Inhibition of attachment of *H. pylori* by sialylated glycoconjugates from bovine milk was studied in an experimental BALB/cA mouse model. Mice were given iron free lactoferrin, iron saturated lactoferrin or bovine milk fat globule membrane fractions once daily for 10 days and then killed to examine for bacterial colonisation and gastritis. Gastric colonisation and inflammation by *H. pylori* was decreased in all mice treated with bovine milk glycoconjugates than in infected control animals (Wang *et al.*, 2001). However, effects on the gastric microflora were not examined which makes it impossible to assess the supplements overall suitability as prebiotics. Overall, it is possible that the fucose containing carbohydrate (sialic acid) moieties of milk K-casein are the cause for this protection of *H. pylori* adhesion and infection.

On the whole, compounds produced by *L. johnsonii* may aid to suppress *H. pylori* infection while simultaneously reducing inflammation. However, dietary supplementation with glycomacropeptide or other conjugates with specific *H. pylori*-binding residues may provide another approach and a useful alternative to more traditional antibiotic and bismuth treatment.

7.8 Rotavirus Infection

Rotaviruses are important causes of infant morbidity and mortality worldwide. Rotavirus is the most common cause of severe diarrhoea among children, resulting in the hospitalisation of approximately 55 000 children each year in the USA and the death of over 600 000 children annually worldwide. The incubation period for rotavirus disease is approximately 2 days and the disease is characterised by vomiting and watery diarrhoea for 3–8 days. Fever and abdominal pain occur frequently. Immunity after infection appears to be incomplete. However, repeat infections tend to be less severe than the original infection. A rotavirus has a characteristic wheel-like appearance when viewed by electron microscopy (the name rotavirus is derived from the Latin *rota* meaning wheel). Rotaviruses are nonenveloped, double-shelled viruses. The genome is composed of 11 segments of double-stranded RNA, which code for six structural and five nonstructural proteins. The primary mode of transmission is faecal-oral, although some have reported low titres of virus in respiratory tract secretions and other body fluids. However, because the virus is stable in the environment, transmission may occur through ingestion of contaminated water or food and contact with contaminated surfaces. In the USA and other countries with a temperate climate, the disease has a winter seasonal pattern, with annual epidemics occurring from November to April. Highest rates of illness occur among infants and young children, and most children in the USA are infected by 2 years of age. Adults can also be infected, though disease tends to be mild. For persons with healthy immune systems, rotavirus gastroenteritis is a self-limited illness, lasting for only a few days. Treatment is nonspecific and consists of oral rehydration therapy to prevent dehydration.

Overall, the evidence suggests that pre- and probiotics can significantly reduce the duration of rotaviral diarrhoea and perhaps help to prevent it. Again, the role of probiotics in the treatment of rotavirus is established. One double-blind, placebo-controlled trial of 269 children (age 1 month to 3 years) with acute diarrhoea found that those treated with *Lactobacillus* GG recovered more quickly than those given placebo (Guandalini *et al.*, 2000). Another double-blind, placebo-controlled study of 81 hospitalised children found that treatment with *Lactobacillus* GG reduced the risk of developing diarrhoea, particularly rotavirus infection (Szajewska *et al.*, 2001). A further double-blind, placebo-controlled study found that *Lactobacillus* GG helped prevent diarrhoea in 204 undernourished children. Qiao *et al.* (2002) orally fed Balb/c pups *Bifidobacterium* species (*B. bifidum* and *B. infantis*), with or without prebiotic compounds (arabino-galactan, short-chain FOS, iso-malto-dextrins) to evaluate theirpotential synergistic effects on modulating the course of rhesus rotavirus (RRV) infection, as well as their ability to mediate the associated mucosal and humoral immune responses. Delayed onset and early resolution of diarrhoea were observed in bifidobacteria-treated, RRV-infected mice compared with RRV-infected control mice. However, the supplementation of prebiotics did not appear to have an effect.

Glycoprotein-based prebiotics might be an important factor for prevention of rotavirus infections and could provide an important alternative to treatment with rehydration therapies, especially in the immunocompromised and malnourished. Glycoproteins bind directly to rotaviruses and readily accessible sialic acid oligosaccharides are required for efficient rotavirus infection of MA-104 cells (Newburg, 1997). Thus, sialic acid oligosaccharides play an important role in the interactions of rotaviruses with both glycoproteins and cells that support rotaviral replication. The glycopeptide/proteins glycomacropeptide and lactadherin inhibit rotavirus binding and low levels of lactadherin in human milk are associated with a higher incidence of symptomatic rotavirus in breast-fed infants (Kawakami, 1997; Newburg, 1997). Altogether, although no human studies have yet examined the protective effect of glycomacropeptide on rotavirus infection, preliminary results seem encouraging and along with *Lactobacillus* GG might provide a genuine alternative to traditional treatment.

7.9 HIV/AIDS-associated Diarrhoea

Diarrhoea is a very serious consequence of human immunodeficiency virus (HIV) infection. The aetiology of this diarrhoea is frequently unknown and there are no effective treatments. In addition, HIV-infected individuals tend to routinely develop severe nutritional deficiencies while progressing towards AIDS. More recently, some AIDS wasting research has focused on defining the association between HIV and the gastrointestinal tract. Several reports have defined interrelationships between HIV, secretory IgA, tumor necrosis factor-alpha (TNFα) and the GI tract (Grunfeld and Feingold, 1992). HIV promotes TNFα production in an infected CD4 lymphocyte which in turn induces oxidative stress through increasing the metabolic rate. As cell death occurs, the associated oxidative stress results in local inflammation. This is most apparent in the intestine, where potential pathogens and toxins are normally present and mucosal inflammation may provoke mucosal adhesion of resident pathogens leading to opportunistic infections, leaky gut and diarrhoea (Lahdevirta *et al.*, 1988).

Prebiotics and probiotics may be able to correct dysbiosis of the gut while watery stools may be bulked up and bound by bismuth-containing agents (Gibson and Roberfroid, 1995). In addition, dietary isoflavones and other phytonutrients may be administered to decrease gut inflammation and prevent TNFα enhancement thus purging the downward spiral of oxidative stress and cell death. Since HIV/AIDS-related diarrhoea may have a variety of causative agents, specific recommendation for treatment is difficult and should be taken on a case-to-case basis. In general, prebiotic administration should be the logical preference over probiotic use for HIV/AIDS-diarrhoea management. However, available literature seems to emphasise the use of probiotics whereas prebiotics in the form of soluble and insoluble fibres have just been used as a bulking agent. Blehaut *et al.* (1992) reported a double blind study with 35 adult patients with AIDS-related diarrhoea. Aetiologies for the cases were *Cryptosporidium* (17 %), *Candida* (14 %), Karposi sarcoma (8 %), atypical *Mycobacterium* (8 %), cytomegalovirus (8 %), *Mycobacterium* (6 %), and unknown (39 %). *Saccharomyces boulardii* was used at a dose of 1.5 g twice daily for 1 week. At the end of the clinical study, 10 out of 18 patients receiving *Saccharomyces boulardii* had their symptoms resolved while this was the case for only 1 of 17 patients receiving placebo. Born *et al.* (1983) used *Saccharomyces boulardii* to treat 33 HIV patients with chronic diarrhoea. In these double-blind studies, 56 % of patients receiving *Saccharomyces boulardii* had resolution of diarrhoea compared with only 9 % of patients receiving placebo. The apparent preference of the use of probiotics over prebiotics in clinical trials examining HIV/AIDS-related diarrhoea might be clarified by bloating and other adverse effects sometimes observed with prebiotics. As such, symptoms might have a negative effect on HIV patients and in many cases worsen the overall outcome of the diarrhoea episode as observed in patients with inflammatory bowel disease.

7.10 Irritable and Inflammatory Bowel Conditions

Irritable bowel syndrome (IBS) is a common disorder of the intestines that affects 8–22 % of the general population (Parker *et al.*, 1995). It is characterised by bloating, abdominal pain, gas and changes in bowel habits. Some IBS sufferers have constipation, others have diarrhoea and some experience both. IBS is prevalent in all age groups, especially in women and is painful in most patients. It is therefore a very serious condition that occupies a significant proportion of GP time and constitutes over 50 % of patient referrals to outpatient clinics.

Several reports have suggested that there is an increased risk of IBS after having bacterial gastroenteritis caused by *Salmonella* sp. and *Campylobacter* sp. (Neal *et al.*, 1997; Rodríguez and Ruigómez, 1999). A recent study of 38 patients with *Salmonella enteritis* reported that 12 still had bowel dysfunction a year after the acute infection had subsided (McKendrick and Read, 1994). The onset of IBS after *Campylobacter enteritis* infection is associated with bacterial toxins that interact with toxin-sensitive epithelial cells (Hep-2 toxin-positive) resulting in the patient experiencing prolonged (at least 6 months) bouts of diarrhoea following infection (Thornley *et al.*, 2001). Similar results have been reported in patients whose upper gastrointestinal tract is infected with the stomach ulcer causing bacterium *H. pylori* (Gerards *et al.*, 2001).

Administration of prebiotics or synbiotic may be able to restore the predominance of a beneficial microflora in the gut. Even though the treatment most widely recommended for IBS patients is an increased intake of dietary fibre, current clinical trials do not support the use of prebiotics in the treatment of an inflammatory bowel disease (Sartor, 2004). Generally, meta-analysis of 17 studies suggested that soluble and insoluble fibres have diverse impacts on IBS, with soluble forms such as psyllium and isphaghula showing significant improvement in relief of constipation, whereas insoluble fibre (i.e. wheat bran) in many cases worsened the clinical outcome.

Probiotics, in comparison, used to naturally enhance gastrointestinal function may have a more beneficial effect in inflammatory bowel disease (IBD). Examining the impact of *E. coli* Nissle strain 1917 (Mutaflor), and a multiple organism product, VSL#3 (from VSL Pharmaceuticals, Fort Lauderdale, FL), initial reports have resulted in encouraging results (Floch, 2003). Boudeau *et al.* (2003) examined bacterial adhesion and invasion of intestinal epithelial cells (Intestine-407). The inhibitory effect of *E. coli* Nissle 1917 was determined after co-incubation with adherent-invasive *E. coli* strains or after pre-incubation of the intestinal epithelial cells with this probiotic strain prior to infection with adherent-invasive *E. coli* strains. *E. coli* strain Nissle 1917 exhibited dose- and time-dependent adherence to intestinal epithelial cells and inhibited the adhesion and invasion of various adherent-invasive *E. coli* strains. Pre-incubation of intestinal epithelial cells with strain Nissle 1917 reduced adherent-invasive *E. coli* adhesion by 97.2–99.9 %.

In a trial conducted at the Mayo Clinic, twenty-five patients with diarrhoea-predominant IBS (Rome II criteria) were randomly assigned to receive VSL#3 powder (450 billion lyophilised bacteria day^{-1}) or matching placebo twice daily for 8 weeks after a 2-week run-in period. While VSL#3 was tolerated well, no effect on individual symptoms with the exception of bloating was observed. However, this was unrelated to changes in gastrointestinal or colonic transit (Kim *et al.*, 2003). In a more recent, randomised study by Saggioro (2004), 50 patients (24 males, 26 females) with IBS (Rome II criteria) were assigned to receive either an active preparation containing *Lactobacillus plantarum* LP0 1 and *Bifidobacterium breve* BR0 at a concentration of 5×10 CFU ml^{-1}, or placebo powder containing starch identical to the study product, for 4 weeks. Treatment efficacy was evaluated using a pain score in different abdominal locations. Pain decreased in 38 % of the probiotics group versus 18 % in the placebo group after 14 days. After 28 days pain was decreased in 52 % versus 11 % of the groups. The severity score of characteristic IBD symptoms significantly decreased in probiotic group versus placebo group after 14 days 49.6 % versus 9.9 % which was confirmed after 28 days.

A somewhat encouraging effect for a future of FOS in IBS patients was observed by Olesen and Gudmand-Hoyer (2000), whose study examined the effect of regular consumption of the prebiotic in patients with IBS. It was found that the symptoms of IBS patients did not worsen significantly after daily ingestion of 20 g FOS for 12 weeks. Initially, seven FOS-treated patients reported abdominal pain compared with only one placebo-treated patient. At the end of the study, however, no difference in symptoms between the placebo and the FOS group were apparent. This colonic adaptation effect is also seen with the prolonged ingestion of lactulose or lactose in lactose-intolerant persons which may suggest that other oligosaccharide-based prebiotics may have a similar effect in IBS patients. Thus, it may be possible that the combinatory treatment of prebiotics and probiotics (synbiotics) may provide some future direction for IBS/IBD management. However, the significant lack of positive studies examining the role of prebiotics and the obvious ethical conflicts of exposing patients to additional suffering may indicate that other directions will have to be taken.

7.11 Atopic Eczema

The increase of atopic diseases such as atopic dermatitis (eczema), allergic rhinitis and asthma has been reaching epidemic proportions (Isolauri, 2004). Reasons for the steep rise of these conditions are unapparent but factors such as family size, breast-feeding, frequency of infectious diseases, passive smoking and diet may be associated. On the molecular level, an imbalance between the T-helper 1 (TH1) and T-helper 2 (TH2) cell immune response has been proposed as a possible cause and new strategies target the persistence of atopic TH2-skewed immune responders (Martinez, 1994). Other studies have shown that interaction of microbes with the intestinal mucosa-associated lymphoid tissue (MALT) plays an important role (Isolauri, 2001). This was reported by Matricardi *et al.* (2000), who showed that frequent intestinal microbial exposure, rather than respiratory exposure prevented atopy in young men with and without atopy. Significantly, lactobacilli and bifidobacteria were more common in nonallergic children, whereas staphylococci and coliforms were more numerous in allergic children and a reduced ratio of bifidobacteria and clostridia preceded the onset of atopy (Björksten *et al.*, 1999; Kalliomaki *et al.*, 2001). Thus, it may seem that hypersensitivity is largely determined in early childhood, when changes to environmental influences are most affecting the immature gut microflora. However, since the disturbance of the gastrointestinal microflora continues throughout life, different lifestyle factors, along with dietary choices approximating a reduced or unnatural microbial spectrum may also be of importance. Thus, exposure to pets and nonpasteurised milk may prevent the development of atopy, instead of inducing it (Schmidt, 2004).

Outcompeting the unfavourable flora with beneficial gut bacteria may also be possible. This has been demonstrated with a diet rich in nondigestible food items such as oligosaccharides abundant in legumes, leek, asparagus, chicory, and unprocessed grains amongst others which favour the growth of endemic bifidobacteria while compromising the growth of clostridia through lowering gut pH (Schmidt, 2004). This effect may further be enhanced by the addition of prebiotics which have been refined and concentrated from such plant material in order to reduce overall intake of such food items. On the other hand, probiotic organisms such as *Lactobacillus* and *Bifidobacterium* may achieve a similar effect by enhancing gut-specific IgA responses and promote gut barrier function through the direct restoration of beneficial gut microbes. However, a role in the TH1 enhancement and TH2 reduction remains to be proven. In a double-blind, randomised pilot study, *Lactobacillus* GG was given both to prenatal mothers and postnatally to their infants. The frequency of atopic eczema in the probiotic group was half that of the placebo group (Kalliomaki *et al.*, 2001). Further clinical trials with *Lactobacillus* GG and

Bifidobacterium lactis Bb-12 also appears to have been useful in infants allergic to cow's milk and a perhaps a combination of prebiotics with probiotics may provide the most useful treatment (Isolauri *et al.*, 2000; Majamaa and Isolauri, 1996).

7.12 Concluding Remarks

The gut microflora and the mucosa itself can act as a barrier against invasion by potential pathogens (Isolauri *et al.*, 1999a,b). Specifically, the lactic microflora of the human gastrointestinal tract is thought to play a significant role in improved colonisation resistance (Gibson *et al.*, 1997). Indigenous bifidobacteria and lactobacilli may inhibit pathogens like *E. coli* and *Salmonella* sp., while combating disorders such as atopic eczema or IBD (Gibson and Wang, 1994). A number of possible mechanisms are in operation:

(1) metabolic end products such as acids excreted by these microorganisms may lower the gut pH, in a microniche, to levels below those at which pathogens are able to effectively compete;
(2) competitive effects from occupation of normal colonisation sites;
(3) direct antagonism through natural antimicrobial excretion;
(4) stimulation of the immune response such that pathogens are inhibited;
(5) balancing of skewed immune responses;
(6) competition for nutrients.

As a whole, the healthy human microflora is well adapted to occupy potential colonisation sites on the gut mucosa while using endogenous substances such as mucin. Attempts have been made to try and identify possible differences between the bacteriological profiles of the mucosally associated populations in diseased and healthy individuals. This is of importance since the ability of bacteria to attach to the host has an enormous impact on the microbial ecology of the large intestine, which subsequently confers both advantageous and detrimental effects to the individual. In the case of IBD, the mucosal microflora has been implicated in being at least partly responsible for the detrimental effects of the disease. This is in part due to the high recovery rate of L-forms, otherwise known as cell-wall deficient organisms or spheroplasts. These L-forms bacteria are resistant to the host defences since they lack cell wall antigens, which facilitates intracellular survival (Chadwick and Anderson, 1995). This is an important consideration for the fermentation of prebiotics since too often, the lumenal flora is only considered when developing prebiotics, which is a poor reflection of overall gut microecology.

The idea of combining prebiotic properties with anti-adhesive activities is currently under investigation and seems to be a promising new area for the development of prebiotics. Binding of pathogens to these receptors is the first step in the colonisation process and many intestinal pathogens utilise monosaccharides or short oligosaccharide sequences as receptors. Knowledge of these receptor sites has relevance for engineering biologically enhanced prebiotics (Karlsson, 1989; Finlay and Falkow, 1989). Anti-adhesive properties would thus add major functionality to the approach of altering gut pathogenesis. These agents are multivalent derivatives of sugars and act as

Table 7.2 *Known receptor saccharides for gastrointestinal pathogens and toxins*

Galα4Gal	*E.* coli (P-piliated), Vero cytotoxin
GalNAcβ4Gal	*P. aeruginosa, H. influenzae, S. aureus, K. pneumoniae*
Galβ4GlcNAcβ3Gal	*S. pneumoniae*
Galα3Galβ4GlcNAc	*C. difficile* toxin A
GlcNAc	*E. coli, V. cholerae*
Sialic acids	*E. coli* (S-fimbriated)
Fucose	*V. cholerae*
Mannose	*E. coli, K. aerogenes, Salmonella* sp., (Type 1-fimbriated)

'blocking factors', dislodging the adherent pathogen (Heerze *et al.*, 1994). There is much potential for developing novel prebiotics. Incorporating a receptor monosaccharide or oligosaccharide sequence and peptides with sugar moieties, such as the glycomacropeptide, might be a good alternative for traditional prebiotics (Table 7.2). Since such compounds would be present before a pathogen and bind to its receptor sites before infecting, such compounds could be thought of as 'decoy prebiotics' (Figure 7.1). Alternatively, prebiotic proteins might also yield as of yet undiscovered potentials and α-lactalbumin might just be the beginning of a variety of novel 'antimicrobial prebiotics'. Whilst both options at this time are more a preventative measure than a therapeutic option, such multifunctional prebiotics could be administered as a preventative measure by increasing host resistance to infection, reducing the likelihood of pathogen establishment and subsequent elaboration of virulence.

Altogether, the best evidence for the success of prebiotics lies in their ability to improve resistance to pathogens by increasing, for example, bifidobacteria and lactobacilli, which are known for their inhibitory properties by lowering the gut pH to levels below those at which pathogens are able to effectively compete. However, when designing novel prebiotics, it is also important to keep in mind possible adverse effects of prebiotics observed *in vitro* and in experimental animals since the potential health implications for introducing such food substrates to a diet may be severe.

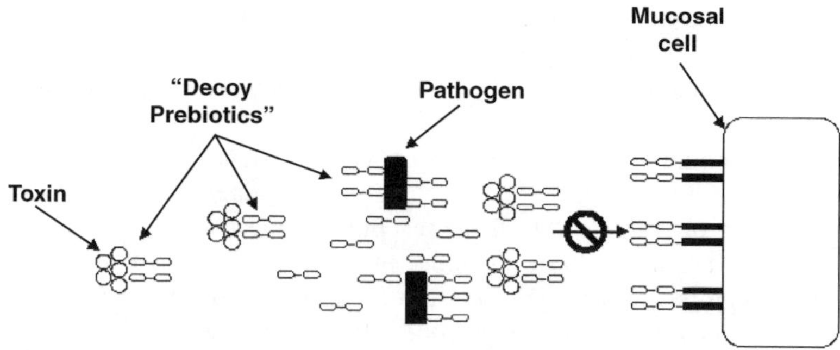

Figure 7.1 *Inhibitory action of decoy prebiotics*

References

Altekruse, S. F., Cohen, M. L. and Swerdlow, D. L. (1997) Emerging foodborne diseases. *Emerging Infect.* **3**, 285–293.

Asahara, T., Nomoto, K., Shimizu, K., Watanuki, M. and Tanka, R. (2001) Increased resistance of mice to *Salmonella enterica* serovar *Typhimurium* infection by synbiotic admmistration of bifidobacteria and tansgalactosylated oligosaccharides. *J. Appl. Microbiol.* **91** (6), 985–996.

Ballongue, J., Schumann, C. and Quignon, P. (1997) Effects of lactulose and lactitol on colonic microflora and enzymatic activity. *Scand. J. Gastroenterol.* **32**, 41–44.

Björksten, B., Naaber, P., Sepp, E. and Mikelsaar, M. (1999) The intestinal microflora in allergic Estonian and Swedish 2-year-old children. *Clin. Exp. Allergy* **29** (3); 342–346.

Blehaut, H., Saint-Marc, T. and Touroine, J. L. (1992) Double blind trial of *Saccharomyces boulardii* in AIDS related diarrhea. International Conference on AIDS/Third STD World Congress, Abstract 2120.

Born, P., Lersch, C., Zimmerhackl, B. and Classen, M. (1993) The *Saccharomyces boulardii* therapy of HIV-associated diarrhea. *Dtsch Med. Wochenschr.* **118** (20), 765.

Boudeau, J., Glasser, A. L., Julien, S., Colomberl, J. F. and Darefeuille-Michaud, A. (2003) Inhibitory effect of probiotic *Escherichia coli* strain Nissle 1917 on adhesion to and invasion of intestinal epithelial cells by adherent-invasive *E. coli* strains isolated from patients with Crohn's disease. *Aliment. Pharmacol Ther.* **18** (1), 45–56.

Bovee-Oudenhoven, I. M., ten Bruggencate, S. J., Lettink-Wissink, M. L. and van der Meer, R. (2003) Dietary fructo-oligosaccharides and lactulose inhibut intestinal colonisation but stimulate translocation of salmonella in rats. *Gut* **52** (11), 1572–1578.

Boyd, J. F. (1985) Pathology of the alimentary tract in *Salmonella typhimurium* food poisoning. *Gut* **26** (9), 935–944.

Brück, W. M., Graverholt, G. and Gibson, G. R. (2002) Use of batch culture and a two-stage continuous culture system to study the effect of supplemental alpha-lactalbumin and glycomacropeptide on mixed populations of human gut bacteria. *FEMS Microbiol. Ecol.* **41**, 231–237.

Brück, W. M., Graverholt, G. and Gibson, G. R. (2003a) A two-stage continuous culture system to study the effect of supplemental alpha-lactalbumin and glycomacropeptide on mixed cultures of human gut bacteria challenged with enteropathogenic *Escherichia coli* and *Salmonella typhimurium. J. App. Microbiol.* **95**, 33–53.

Brück, W. M., Kelleher, S. L., Gibson, G. R., Nielsen, K. E., Chatterton, D. E. W. and Lönnerdal, B. (2003b) rRNA probes used to quantify the effects of glycomacropeptide and alpha-lactalbumin supplementation of the predominant groups of intestinal bacteria of infant Rhesus monkeys challenged with enteropathogenic *Escherichia coli. J. Pediatr. Gastroenterol. Nutr.* **37**, 273–280.

Camara, L. M., Carbonare, S. B., Silva, M. L. and Carneiro-Sampaio, M. M. (1994) Inhibition of enteropathogenic *Escherichia coli* (EPEC) adhesion to HeLa cells by human colostrum: detection of specific sIgA related to EPEC outer-membrane proteins. *Int. Arch. Allergy Immunol.* **103** (3), 307–310.

Campbell, M. A., Kolev, Y., Stahl, B., Boehm, G., Butler, R. N. and Stevenson, L. M. (2004) The role of oligosaccharides and *Helicobacter pylori*-specific antibodies in disease prevention. *Asia Pac. J. Clin. Nutr.* **13** (Suppl.), S92.

Castagliuolo, I., LaMont, J. T., Qiu, B., Nikulasson, S. T. and Pothoulakis, C. (1996) A receptor decoy inhibits the enterotoxic effects of *Clostridium difficile* toxin A in rat ileum. *Gastroenterology* **111** (2), 433–438.

CDC. (1991) Multistate outbreak of *Salmonella poona* infections: United States and Canada. *Morbid Mortal Weekly Rep.* **40**, 549.

CDC. (1994) *Salmonella Surveillance – Annual Summary – 1992*. Centers for Disease Control and Prevention, Atlanta.

Chadwick, V. S. and Anderson, R. P. (1995) The role of intestinal bacteria in etiology and maintenance of inflammatory bowel disease. In *Human Colonic Bacteria: Role in Nutrition, Physiology and Pathology*, G. R. Gibson, G. T. Macfarlane, Eds. CRC Press, Boca Raton, pp. 227–256.

Collins, M. D., Gibson, G. R. (1999) Probiotics, prebiotics, and synbiotics: approaches for modulating the microbial ecology of the gut. *Am. J. Clin. Nutr.* **69** (Suppl.), 1052S–1057S.

Craviota, A., Tello, R. E. and Ortega, R. (1988) Prospective study of diarrhoeal disease in a cohort of rural Mexican children: incidence and isolated pathogens during the first two years. *Epidemiol. Rev.* **101**, 123.

Darwin, K. H. and Miller, V. L. (1999) Molecular basis of the interaction of *Salmonella* with the intestinal mucosa. *Clin. Microbiol. Rev.* **12** (3), 405–428.

Donnenberg, M. S. (1995) Enteropathogenic *Escherichia coli*. In *Infections of the Gastrointestinal Tract*, M. J. Blaser, P. D. Smith, J. I. Ravdin, H. B. Greenberg, R. L. Guerrant, Eds. Raven Press, New York, NY. pp. 709–726.

Felley, C. P., Corthesy-Theulaz, I., Rivero, J. L., Sipponen, P., Kaufmann, M., Bauerfeind, P., Wiesel, P. H., Brassart, D., Pfeifer, A., Blum, A. L. and Michetti, P. (2001) Favourable effect of an acidified milk (LC-1) on *Helicobacter pylori* gastritis in man. *Eur. J. Gastroenterol. Hepatol.* **13** (1), 25–29.

Figueroa-Quintanilla, E., Salazar-Lindo, E., Sack, R. B., Leon-Barua, S., Sarabia-Arce, S., Campos-Sanchez, M. and Eyzaguirre-Maccan, E. (1993) A controlled trial of bismuth subsalicylate in infants with watery diarrheal disease. *N. Engl. J. Med.* **328**, 1653–1658.

Finlay, B. B. and Falkon, S. (1989) Common themes in microbial pathogenicity. *Microbiol. Rev.* **53**, 210–230.

Floch, M. H. (2003) Probiotics, irritable bowel syndrome, and inflammatory bowel disease. Curr. *Treat. Options Gastroenterol.* **6** (4), 283–288.

Fuller, R. and Gibson, G. R. (1997) Modification of the intestinal microflora using probiotics and prebiotics. *Scand. J. Gastroenterol.* **222** (Suppl.), 28–31.

Gerards, C., Leodolter, A., Glasbrenner, B. and Malfertheiner, P. (2001) *H. pylori* infection and visceral hypersensitivity in patients with irritable bowel syndrome. *Dig. Dis.* **19** (2), 170–173.

Gibson, G. R. and Wang, X. (1994) Regulatory effects of bifidobacteria in the growth of other colonic bacteria. *J. Appl. Bacteriol.* **77**, 412–420.

Gibson, G. R. and Roberfroid, M. B. (1995) Dietary modulation of the human colonic microbiota: introducing the concept of prebiotics. *J. Nutr.* **125** (6), 1401–1412.

Gibson, G. R., Beatty, E. R., Wang, X. and Cummings, J. H. (1995) Selective stimulation of bifidobacteria in the human colon by oligofructose and inulin. *Gastroenterology*, **108** (4), 975–982.

Gibson, G. R., Saavedra, J. M., Macfarlane, S. and Macfarlane, G. T. (1997) Gastrointestinal Microbial Disease. In *Probiotics 2: Application and Practical Aspects*, R. Fuller, Ed. Chapman and Hall, Andover, pp. 10–39.

Gomez, H. F. and Cleary, G. G. (1998) *Salmonella*, 4th edn, Vol. 1. The EB Saunders Co., Philadelphia.

Grunfeld, C. and Feingold, K. (1992) Metabolic disturbances and wasting in the acquired immunodeficiency syndrome. *N. Engl. J. Med.* **327**, 329–337.

Guandalini, S., Pensabene, L., Zikri, M. A., Dias, J. A., Casali, L. G., Hoekstra, H., Kolacek, S., Massar, K., Micetic-Turk, D., Papadopoulou, A., de Sousa, J. S., Sandhu, B., Szajewska, H. and Weizman, Z. (2000) *Lactobacillus* GG administered in oral rehydration solution to children with acute diarrhea: a multicenter European trail. *J. Pediatr. Gastroenterol. Nutr.* **30** (1), 54–60.

Heerze, L. D., Kelm, M. A., Talbot, J. A. and Armstrong, G. D. (1994) Oligosaccharide sequences attached to an inert support (SYNSORB) as potential therapy for antibiotic-associated diarrhea and pseudomembranous colitis. *J. Infect. Dis.* **169** (6), 1291–1296.

Hill, S. M., Philips, A. D. and Walker-Smith, J. A. (1991) Enteropathogenic *Escherichia coli* and life-threatening chronic diarrhoea. *Gut* **32**, 154–158.

Hopkins, M. J. and Macfarlane, G. T. (2003) Nondigestible oligosaccharides enhance bacterial colonization resistance against *Clostridium difficile in vitro Appl. Environ. Microbiol.* **69** (4), 1920–1927.

Ito, Y., Moriwaki, H., Muto, Y., Kato, N., Watanabe, K. and Ueno, K. (1997) Effect of lactulose on short-chain fatty acids and lactate production and on the growth of faecal flora, with special reference to *Clostridium difficile.J. Med. Microbiol.* **46** (1), 80–84.

Isolauri, E. (2001) Probiotics in the prevention and treatment of allergic disease. *Pediatr. Allergy Immunol.* **12** (Suppl. 14), 56–59.

Isolauri, E. (2004) Dietary modification of atopic disease: use of probiotics in the prevention of atopic dermatitis. *Curr. Allergy Asthma Rep.* **4** (4), 270–275.

Isolauri, E., Salminen, S. and Mattila-Sandholm, T. (1999a) New functional foods in the treatment of food allergy. *Ann Med.* **31** (4), 299–302.

Isolauri, E., Arvilommi, H. and Salminen, S. (1999b) Gastrointestinal infections. In *Colonic Microbiota, Nutrition and Health*, G. R. Gibson, M. B. Roberfroid, Eds. Kluwer Academic Publishers, Dordrecht, pp. 267–279.

Isolauri, E., Arvola, T., Sutas, Y., Moilanen, E. and Salminen, S. (2000) Probiotics in the management of atopic eczema. *Clin. Exp. Allergy* **30**, 1604–1610.

Kalliomaki, M., Kirjavainen, P., Eerola, E., Kero, P., Salminen, S. and Isolauri, E. (2001) Distinct patterns of neonatal gut microflora in infants in whom atopy was and was not developing. *J. Allergy Clin. Immunol.* **107** (1), 129–134.

Karlsson, K.-A. (1989) Animal glycosphingolipids as membrane attachment sites for bacteria. *Ann. Rev. Biochem.* **58**, 309–350.

Kawakami, H. (1997) Biological significance of sialic acid-containing substances in milk and their application. *Recent Res. Dev. Agric. Biol. Chem.* **1**, 193–208.

Kawasaki, Y., Isoda H., Tanimoto, M., Dosako, S., Idota, T. and Ahiko, K. (1992) Inhibition by lactoferrin and k-casein glycomacropeptide of binding of *Cholera* toxin to its receptor. *Biosci. Biotech. Biochem.* **56**, 195–198.

Kawasaki, Y., Isoda, H., Shinmoto, H., Tanimoto, M., Dosako, S., Idota, T. and Nakajima, I. (1993) Inhibition by k-casein glycomacropeptide and lactoferrin of influenza virus hemaglutination. *Biosci. Biotech. Biochem.* **57**, 1214–1215.

Kelleher, S. L. and Lönnerdal, B. (2001) Immunological activities associated with milk. *Adv. Nutr. Res.* **10**, 39–65.

Kim, H. J., Camilleri, M., McKinzie, S., Lempke, M. B., Burton, D. D., Thomforde, G. M. and Zinsmeister, A. R. (2003) A randomized controlled trail of a probiotic, VSL#3, on gut transit and symptoms in diarrhoea-predominant irritable bowel syndrome. *Aliment. Pharmacol. Ther.* **17** (7), 895–904.

Kleessen, B., Bunke, H., Tovar, K., Noak, J. and Sawatzki, G. (1995) Influence of two infant formulas and human milk on the development of the faecal flora in newborn infants. *Acta Paediatr.* **84**, 1347–1356.

Kleessen, B., Stoof, G., Prol, J., Schmiedl, D., Noack, J. and Blaut, M. (1997) Feeding resistant starch affects fecal and cecal microflora and short-chain fatty acids in rats. *J anim. Sci.* **75** (9), 2453–2462.

Kolida, S., Tuohy, K. and Gibson, G. R. (2002) Prebiotic effects of inulin and oligofructose. *Br. J. Nutr.* **87** (suppl. 2), S193–S197.

Lahdevirta, J., Maury, C. P. J., Teppo, A. M. and Repo, H. (1988) Elevated levels of circulating cachectin/tumor necrosis factor in patients with acquired immunodeficiency syndrome. *Am. J. Med.* **85**, 289–291.

Levine, M. M. and Edelman, R. (1984) Enteropathogenic *Escherichia coli* of classic serotypes associated with infant diarrhoea: epidemiology and pathogenesis. *Epideniol. Rev.* **6**, 31–51.

Levine, M. M., Bergquist, E. J., Nalin, D. R., Waterman, D. H., Hornick, R. B., Young, C. R., Sotman, S. and Rowe, B. (1978) *Escherichia coli* that cause diarrhoea but do not produce heat-labile or heat-stable enterotoxins and are non-invasive. *Lancet* **1**, 1119–1122.

Liehr, H., Englisch, G. and Rasenack, U. (1980) Lactulose – a drug with antiendotoxin effect. *Hepatogastroenterology* **27** (5), 356–360.

Majamaa, H. and Isolauri, E. (1996) Evaluation of the gut mucosal barrier: evidence for increased antigen transfer in children with atopic eczema. *J. Allergy Clin. Immunol.* **97**, 985–990.

Martinez, F. D. (1994) Role of viral infections in the inception of asthma and allergies during childhood: could they be protective? *Thorax* **49** (12), 1189–1191.

Mathew, M., Mathan, M. M., Mani, K., George, R., Jebakumar, K., Dharamsi, R., Kirubakaran, C., Pereira, S. and Mathan, V. I. (1991) The relationship of microbial pathogens to acute infectious diarrhoea of childhood. *J. Trop. Med. Hyg.* **94** (4), 253–260.

Matricardi, P. M., Rosmini, F., Riondino, S., Fortini, M., Ferrigno, L., Rapicetta, M. and Bonini, S. (2000) Exposure to foodborne and orofecal microbes versus airborne viruses in relation to atopy and allergic asthma: epidemiological study. *Br. Med. J.* **320** (7232), 412–417.

McKendrick, M. W. and Read, N. W. (1994) Irritable bowel syndrome–post salmonella infection. *J. Infect.* **29** (1), 1–3.

Michetti, P., Dorta, G., Wiesel, P. H., Brassart, D., Verdu, E., Herranz, M., Felley, C., Porta, N., Rouvet, M., Blum, A. L. and Corthesy-Theulaz, I. (1999) Effect of whey-based culture supernatant of *Lactobacillus acidophilus* (*johnsonii*) La1 on *Helicobacter pylori* infection in humans. *Digestion* **60** (3), 203–209.

Nataro, J. P. and Kaper, J. B. (1998) Diarrheagic *Escherichia coli*. *Clin. Microbiol. Rev.* **11**, 142–201.

Neal, K. R., Hebden, J. and Spiller, R. (1997) Prevalence of gastrointestinal symptoms six months after bacterial gastroenteritis and risk factors for development of the irritable bowel syndrome: postal survey of patients *Br. Med. J.* **314**, 779–782.

Newburg, D. S. (1997) Do the binding properties of oligosaccharides in milk protect human infants from gastrointestinal bacteria? *J Nutr.* **127** (5 Suppl.), 980S–984S.

Nzegwu, H. C. and Levin, R. C. (1994) Neutrally maintained hypersecretion in undernourished rat intestine activated by *E.coli* STa enterotoxin and cyclic nucleotides *in vitro*. *J. Physiol. (London)* **479**, 159–169.

Oberhelman, R. A., Gilman, R. H., Sheen, P., Taylor, D. N., Black, R. E., Cabrera, L., Lescano, A. G., Meza, R. and Madico, G. (1999) A placebo-controlled trail of *Lactobacillus* GG to prevent diarrhea in undernourished Peruvian children *J. Pediatr.* **134** (1), 15–20.

Olesen, M., Gudmand-Hoyer, E. (2000) Efficacy, safety, and tolerability of fructooligosaccharides in the treatment of irritable bowel syndrome. *Am. J. Clin. Nutr.* **72** (6), 1570–1575.

Parker, N., Tsai, H. H., Ryder, S. D., Raouf, A. H. and Rhodes, J. M. (1995) Increased rate of sialylation of colonic mucin by cultured ulcerative colitis mucosal explants. *Digestion* **56** (1), 52–56.

Pelligrini, A., Thomas, U., Bramaz, N., Hunziker, P. and von Fellenberg, R. (1999) Isolation and identification of three bactericidal domains in the bovine alpha-lactalbumin molecule. *Biochim. Biophys. Acta* **1426** (3), 439–448.

Pihlanto-Leppälä, A., Marnila, P., Hubert, L., Rokka, T., Korhonen, H. J. T. and Karp, M. (1999) The effect of α-lactalbumin and β-lactoglobulin hydrolysates on the metabolic activity of *Escherichia coli* JM103. *J. Appl. Microbiol.* **87**: 540–545.

Qiao, H., Duffy, L. C., Griffiths, E., Dryja, D., Leavens, A., Rossman, J., Rich, G., Riepenhoff-Talty, M. and Locniskar, M. (2002) Immune responses in rhesus rotavirus-challenged BALB/c mice treated with bifidobacteria and prebiotic supplements. *Pediatr. Res.* **51** (6), 750–755.

Reid, G., Sanders, M. E., Gaskins, H. R., Gibson, G. R., Mercenier, A., Rastall, R., Roberfroid, M., Rowland, I., Cherbut, C. and Klaenhammer, T. R. (2003) New Scientific paradigms for probiotics and prebiotics. *J. Clin. Gastroenterol.* **37** (2), 105–118.

Ribeiro Jr, H. (2000) Diarrheal disease in a developing nation. *Am. J. Gastroenterol.* **95** (Suppl.), S14–S15.

Roberfroid, M. B. (2001) Prebiotics: preferential substrates for specific germs? *Am. J. Clin. Nutr.* **73** (Suppl.), 406S–409S.

Robins-Browne, R. M., Still, C. S., Miliotis, M. D., Richardson, N. J., Koornhof, H. J., Freiman, I., Schoub, B. D., Lecatsas, G. and Hartman, E. (1980) Summer diarrhoea in African infants and children. *Arch. Dis. Child.* **55** (12), 923.

Rodríguez, L. A. and Ruigómez, A. (1999) Increased risk of irritable bowel syndrome after bacterial gastroenteritis: cohort study. *Br. Med. J.* **318**, 565–566.

Rolfe, R. D. (2000) The role of probiotic cultures in the control of gastrointestinal health. *J. Nutr.* **130** (2S Suppl.), 396S–402S.

Saavedra, J. M., Bauman, N. A., Oung, I., Perman, J. A. and Yolken, R. H. (1994) Feeding of *Bifidobacterium bifidum* and *Streptococcus thermophilus* to infants in hospital for prevention of diarrhoea and shedding of rotavirus. *Lancet* **344**, 1046–1049.

Saggioro, A. (2004) Probiotics in the treatment of irritable bowel syndrome. *J. Clin. Gastroenterol.* **38** (6 Suppl.), S104–S106.

Saito, T., Miyakawa, H., Tamura, Y., Shimamura, S. and Tomita, M. (1991) Potent bactericidal activity of bovine lactoferrin hydrolysates produced by heat treatment at acidic pH. *J. Dairy Sci.* **74**, 3724–3730.

Salminen, S., Bouley, C., Boutron-Ruault, M.-C., Cummings, J. H., Franck, A., Gibson, G. R., Isolauri, E., Moreau, M.-C., Roberfroid, M. and Rowland, I. (1998) Functional food science and gastrointestinal physiology and function. *Br. J. Nutr.* **80** (Suppl. 1), S147–S171.

Sartor, R. B. (2004) Therapeutic manipulation of the enteric microflora in inflammatory bowel diseases: antibiotics, probiotics, and prebiotics. *Gastroenterology* **126** (6), 1620–1633.

Schmidt, W.-P. (2004) Model of the epidemic of childhood atopy. *Med. Sci. Monit.* **10** (2), HY5–HY9.

Singh, R. D., Khullar, M., Ganguly, N. K. (2000) Role of anaerobiosis in virulence of *Salmonella typhimurium*. *Mol. Cell. Biochem.* **215**, 39–46.

Snoeyenbos, G. H., Smyser, C. F. and Van Roekel, H. (1969) *Salmonella* infections of the ovary and peritoneum of chickens. *Avian Dis.* **13**, 668–670.

Stalheim-Smith, A. and Fitch, G. K. (1993) *Understanding Human Anatomy and Physiology*. West Publishing Company, St. Paul, pp. 783–844.

Steer, T., Carpenter, H., Tuohy, K. and Gibson, G. R. (2000) Perspectives on the role of the human gut microbiota and its modulation by pro- and prebiotics. *Nutr. Res. Rev.* **13**, 229–254.

Stromqvist, M., Falk, P., Bergstrom, S., Hansson, L., Lonnerdal, B., Normark, S. and Hernell, O. (1995) Human milk kappa-casein and inhibition of *Helicobacter pylori* adhesion to human gastric mucosa. *J. Pediatr. Gastroenterol. Nutr.* **21** (3), 288–296.

Szajewska, H. and Mrukowicz, J. Z. (2001) Probiotics in the treatment and prevention of acute infections diarrhea in infants and children: a systematic review of published randomized, double-blind, placebo-controlled trails. *J. Pediatr. Gastroenterol. Nutr.* **33** (Suppl.), S17–S25.

Tannock, G. W. (1998) Studies of the intestinal microflora: a prerequisite for the development of probiotics. *Int Dairy J.* **8**, 527–533.

Tauxe, R. V. and Pavia, A. T. (1998) Salmonellosis: nontyphoidal. In *Bacterial Infections of Humans, Epidemiology and Control*, A. S. Evans, P. S. Brachman, Eds. Plenum Medical Book Co., New York, pp. 223–242.

Ten Bruggencate, S. J., Bovee-Oudenhoven, I. M., Lettink-wissink, M. L. and Van der Meer, R. (2003) Dietary fructo-oligosaccharides dose-dependently increase translocation of salmonella in rats. *J. Nutr.* **133** (7), 2313–2318.

Thornley, J. P., Jenkins, D., Neal, K., Wright, T., Brough, J. and Spiller, R. C. (2001) Relationship of *Campylobacter* toxigenicity in vitro to the development of postinfectious irritable bowel syndrome. *J. Infect. Dis.* **184** (5), 606–609.

Tuohy, K. M., Kolida, S., Lustenberger, A. M. and Gibson, G. R. (2001) The prebiotic effects of biscuits containing partially hydrolysed guar gum and fructo-oligosaccharides – a human volunteer study. *Br. J. Nutr.* **86** (3), 341–348.

Van Loo, J. A. E., Coussement, P., De Leenheer, L., Hoebregs, H. and Smits, G. (1995) On the presence of inulin and oligofructose as natural ingredients in the Western diet. *CRC Cri. Rev. Food Sci. Nutr.* **35**: 525–552.

Wang, X., Hirmo, S., Willen, R. and Wadstrom, T. (2001) Inhibition of *Helicobacter pylori* infection by bovine milk glycoconjugates in a BAlb/cA mouse model. *J. Med. Microbiol.* **50** (5), 430–435.

Willis, C. L. and Gibson, G. R. (1999) The natural microflora of humans. In *Encyclopedia of Food Microbiology*, R. Robinson, C. Batt, P. Patel, Eds. Academic Press, London, pp. 1351–1355.

Wilson, R., Feldman, R. A., Davis, J. and LaVenture, M. (1982) Salmonellosis in infants: the importance of intrafamilial transmission. *Pediatrics* **69**, 436–438.

Wold, A. E. and Adlerberth, I. (2000) Breast-feeding and the intestinal microflora of the infant – implications for protection against infectious diseases. In *Short and Long Term Effects of Breast Feeding on Child Health*, B. Keletzo *et al.*, Eds. Kluwer Academic/Plenum Publishers, Dordrecht, pp. 77–93.

Xue, Y., Lui, J. N., Sun, Z., Ma, Z., Wu, C. and Zhu, D. (2001) α-Lactalbumin mutant acting as lysozyme. *Proteins: Struct. Funct. Genet.* **42**, 17–22.

Ziemer, C. J. and Gibson, G. R. (1998) An overview of probiotics, prebiotics and synbiotics in the functional food concept: perspectives and future strategies. *Int. Dairy J.* **8**, 473–479.

8

Dietary Intervention for Improving Human Health: Chronic Disorders

Natalie R. Bullock and Mark R. Jones

8.1 Introduction

There is little doubt today that food and health are inextricably linked. Epidemiological, pre-clinical and clinical studies provide irrefutable evidence that foods are much more than simply a source of energy. On a global scale, more countries are faced with an expanding population and increased individual life expectancy, both of which resultantly increase the cost of healthcare. Thus, there is a growing awareness of the need for diet to combat the effects of an ageing society through reducing the incidence of life-style related diseases such as allergies, diabetes, osteoporosis, cardiovascular disorders and cancer. In line with this need, the concept of 'functional foods' has arisen, defined as 'foods that by virtue of physiologically active food components, provide health benefits beyond basic nutrition' (Milner, 2002).

Many foods are already associated with health promotion and disease prevention. The mechanisms by which these foods exert their effect are varied and complex. Nutrients serve as antioxidants, promote the activity of detoxification enzymes, block carcinogen metabolism or formation or produce shifts in hormonal balance (Green, 1999; Bingham, 2000). Others directly influence cell homeostasis, controlling functions such as cellular metabolism, differentiation and apoptosis (Reddy, 1994; Chang *et al.*, 1998; Chapkin *et al.*, 2000). Whilst the perceived health benefit of a food may be a metabolic response that lowers risk for a particular disease, the actual target for the food or food component may be the gastrointestinal tract (GIT). Serving as an interface between the body and the external environment, the GIT is the largest endocrine organ of the human anatomy and

Prebiotics: Development and Application Edited by G.R. Gibson and R.A. Rastall

releases an array of peptides in response to nutrient intake. These may interact with other peptides, as well as with the nervous system, to produce reactions that directly influence health. Functional foods may thereby delay or accelerate gastric emptying, alter mixing within the intestinal contents and/or alter the availability of digestive enzymes.

Approximately 10^{12} microorganisms reside in the GIT, the majority of which are found in the large intestine. Whilst some of these are considered harmful, causing a variety of diseases, others have proved beneficial and often essential to human wellbeing, synthesising vitamins, facilitating mineral absorption and stimulating the immune system. Nondigestible carbohydrates are a major energy source for the colonic microflora and much attention has now been directed to the potential impact of the metabolism of gut microbes in reducing the risk of chronic diseases. Fermentation of carbohydrates in the large intestine increases microbial cell mass, contributing towards stool bulk and aiding laxation (Chen *et al.*, 1998). Products of the microbial activity include ammonia, gases (principally hydrogen, methane and carbon dioxide) and short-chain fatty acids (SCFA) (Table 8.1). Some SCFA, such as propionic acid, appear in the general circulation where they are used by the body for energy and have beneficial effects on carbohydrate and lipid metabolism. Other SCFA are utilised directly by colonocytes. Butyric acid appears to be essential in the maintenance of a healthy colonic mucosa (Roediger, 1982).

Recently, interest has focused on prebiotics as functional foods, for use both as prophylactic measures and pharmacological agents against gastrointestinal disorders. Prebiotics are nondigestible carbohydrates that are selectively fermented by certain bacterial groups resident in the colon, such as bifidobacteria, lactobacilli and eubacteria, considered to be beneficial for the human host. Prebiotic supplements, particularly inulin and fructooligosaccharides (FOS), have been added to food products to influence the colonic flora. Prebiotics have a low energy value (<9 kJ g^{-1}) and increase stool volume. However, excessive use has been shown to produce dose-related fermentation and osmotic side effects, including excessive flatus, bloating, abdominal cramps and diarrhoea (Stone-Dorshow and Levitt, 1987; Ito *et al.*, 1990). Increased production of certain SCFA by the colonic microflora may protect against development of colorectal cancer, and the use of prebiotics to influence this activity is therefore of growing interest (Hughes and Rowland, 2001; Augenlicht *et al.*, 2002). Additional areas in which prebiotic therapy may be useful include prevention of intestinal infections and modulation of the intestinal immune response in inflammatory bowel disease (IBD).

The use of live microbial feed supplements (probiotics) in the treatment of certain diseases [ulcerative colitis, irritable bowel syndrome (IBS), etc.] has already been documented and some are considered in this chapter. The use of prebiotics confers certain advantages over the probiotic approach in terms of survivability through the GIT and long-term stability in food products. As such, prebiotics offer exciting potential in the prevention and treatment of major gastrointestinal disorders.

8.2 Gastritis and Peptic Ulcer Disease

Gastritis includes a myriad of disorders that involve inflammatory changes in the gastric mucosa, including erosive gastritis caused by a noxious irritant [stress, alcohol, bile, and nonsteroidal anti-inflammatory drugs (NSAIDs)], reflux gastritis from exposure to bile

Table 8.1 *Predominant products of carbohydrate metabolism in the human colon*

End product	Bacterial group involved	Metabolic fate
Acetate	Bacteroides, bifidobacteria, eubacteria, lactobacilli, clostridia, ruminococci, peptococci, veillonella, peptostreptococci, propionibacteria, fusobacteria, butyrivibrio	Metabolised in muscle, kidney, heart and brain
Propionate	Bacteroides, propionibacteria, veillonella	Cleared by the liver, possible gluceogenic precusor, suppresses cholesterol synthesis
Butyrate	Clostridia, fusobacteria, butyrivibrio, eubacteria, peptostreptococci	Metabolised by the colonic epithelium, regulator of cell growth and differentiation
Ethanol, succinate, lactate, pyruvate	Bacteroides, bifidobacteria, lactobacilli, eubacteria, peptostreptococci, clostridia, ruminococci, actinomycetes, enterococci, fusobacteria	Absorbed, electron sink products, further fermented to SCFA
Hydrogen	Clostridia, ruminococci, fusobacteria	Partially excreted in breath, metabolised by hydrogenotrophic bacteria

Data from: Gibson, 1999.

and pancreatic fluids, hemorrhagic gastritis, infectious gastritis, and gastric mucosal atrophy. Included in this is peptic ulcer disease (PUD) – a discrete mucosal defect in the portions of the GIT (gastric or duodenal) exposed to acid and pepsin secretion. Infection with *Helicobacter pylori* is the leading cause of PUD and is associated with all ulcers not induced by NSAIDs. *H. pylori* colonises the deep layers of the mucosal gel that coats the gastric mucosa and disrupts its protective properties (Appelmelk and Negrini, 1997).

H. pylori is thought to infect virtually all patients with chronic active gastritis. Viral or fungal gastritis may develop in people with a prolonged illness or an impaired immune system, such as those with AIDS or cancer, or those taking immunosuppressant drugs. Ménétrier's disease, whose cause is unknown, is a type of gastritis in which the stomach wall develops thick, large folds; enlarged glands; and fluid-filled cysts. The disease may be due to an abnormal immune reaction and has also been associated with *H. pylori* infection. Furthermore, patients diagnosed with duodenal ulcer caused by *H. pylori* are twice as likely to go on to develop gastric cancer when compared with people with a normal stomach (Sipponen *et al.*, 1994).

Despite a plethora of clinical trials and reported case studies, no single therapeutic regimen for the treatment of *H. pylori* infection has been established. A 7-day course combining twice-daily proton pump inhibitors (e.g. omeprazole) and two antibiotics has been used to eradicate the organism. However, there is increasing resistance emerging amongst *Helicobacter* species to antibiotics such as clarithromycin and metronidazole. In addition, current therapy does not address the issue of faecal-oral reinfection due to colonic carriage of *H. pylori*.

H. pylori is a target for prebiotic eradication therapy. The antagonistic actions of some lactic acid bacteria against this organism have already been demonstrated *in vitro*. These effects were found to be independent of acid production. *Bifidobacterium breve, B. catenulatum, B. magnum* and *B. bifidum* all inhibited the growth of *H. pylori* through secretion of heat-labile bacteriocins (Eun-Ah *et al.*, 2000). Oral *Lactobacillus acidophilus* and *L. rhamnosus* suppressed colonisation of the gastric mucosa by *H. pylori* and also reduced inflammation of the gastric antrum in C57BL/6 mice (Johnson-Henry *et al.*, 2004). A significant reduction of the urease activity was reported in 20 patients treated with a supernatant of *L. johnsonii* combined with omeprazole (Michetti *et al.*, 1999). Two randomised controlled trials reported that the ingestion of a fermented dairy product containing *L. johnsonii* or a heat killed *L. acidophilus* could help to decrease the gastric colonisation by *H. pylori* (Canducci *et al.*, 2000; Felley *et al.*, 2001). More recently, anecdotal evidence has been presented for the efficacy of garlic in the eradication of *H. pylori* during active PUD. A number of theories exist, including stimulation of inherent probiotic species by the prebiotic inulin contained in the garlic and/or a direct antibiotic effect of thiosulphinate (Jonkers *et al.*, 1999). Thus, prebiotics may be employed in the treatment of gastritis through their ability to stimulate the growth of bifidobacteria and lactobacilli, subsequently inhibiting the survival of *H. pylori* in the gastric mucosa. They may also prevent faecal-oral transmission by reducing the colonic carriage of *H. pylori*, mediated via these proteinaceous components.

8.3 Irritable Bowel Syndrome

IBS is a common complaint of the intestine that affects almost one quarter of the UK general population. Female sufferers outnumber males by 1.4 to 1.0. It is of major socio-economic importance, as it probably occupies more GP time than any other gastrointestinal disorder (Jones and Lydeard, 1992). IBS is characterised by abdominal pain, diarrhoea and/or constipation, flatulence, variability in the bowel habit and general malaise. Despite this, there are no detectable structural or biochemical abnormalities. Although not life-threatening, sufferers have a reduced quality of life. The causes of IBS are still unclear, although onset has been attributed to a variety of stimuli, including the use of antibiotics, ovarian hormones, abdominal operations, dietary fibre deficiency, gastrointestinal infections, alcohol consumption, stress, high fat intake and food intolerance. Often, patients report a single definite event as the trigger for symptoms (Hunter and Alun-Jones, 1985).

It is thought that the microbiota of the large intestine plays a key role in both IBS onset and maintenance. There are good reasons to believe that there may be changes in the nature of the intestinal flora in IBS. When compared with healthy subjects, IBS patients have lower numbers of colonic bifidobacteria, lactobacilli and coliforms (Balsari *et al.*, 1982; Bayliss *et al.*, 1984). They may also have abnormal fermentation, manifest as excess gas production (King *et al.*, 1998), which resolves when symptoms are relieved.

Several reports have highlighted an increased risk of developing IBS following bacterial gastroenteritis caused by *Salmonella*, *Shigella* or *Campylobacter*. Studies following individuals with infectious gastroenteritis involving these species have reported that one-quarter to one-third of patients then go on to develop IBS for the first time (McKendrick and Read, 1994; Gwee *et al.*, 1999). In fact, patients are 10 times more likely than healthy people to develop IBS for the first time following an acute attack of gastroenteritis (Rodriguez and Ruigomez, 1999). This may be due to the intestinal infection producing a local inflammatory response via mast cell activation, resulting in gut hypersensitivity (Gui, 1998).

A relationship between carriage of yeasts such as *Candida* species and symptoms associated with the disorder has also been noted, where it is thought that toxins produced by the fungi may trigger onset. Often the antifungals (e.g. nystatin) used to treat yeast infections, e.g. candida-induced recurrent thrush, may trigger IBS. It is also known that immunocompromised individuals, such as those who have recently received antibiotics, are more likely to be colonised by *Candida albicans* and suffer the symptoms of IBS (Scheurlen, 1996).

Thus the aetiology of IBS remains unknown and there seems to be no satisfactory consensus amongst clinicians as to how the disorder should be managed. Symptomatic treatment usually involves supplementing fibre (up to a total intake of 30 g per day for individuals with constipation), loperamide or other opioids for diarrhoea, and antispasmodics for pain (Camilleri, 2001). Until relatively recently, alternative biological therapies were largely disregarded, possibly as they were considered to lie outside the realms of conventional medicine. However, due to the growing body of evidence for the involvement of intestinal microflora in the pathogenesis of IBS, there was a logical conclusion that manipulation of this population, either by replacing missing organisms or encouraging a more favourable intestinal milieu, might have beneficial effects.

Studies have emerged whereby probiotics have been successfully employed in the treatment of IBS. The therapeutic effect of *Lactobacillus plantarum* 299V was compared with two drugs used for the treatment of IBS. It was found that administration of *L. plantarum* with or without the drugs produced a greater improvement of symptoms than administration of the drugs alone (Niedzielin *et al.*, 1998). In follow-up studies, the group also demonstrated that administration of *L. plantarum* 299 V could decrease abdominal pain and normalise stool frequency in constipated patients. The efficacy of other lactic acid bacteria (including separate and mixed cultures of *Streptococcus thermophilus, S. faecium, Bifidobacterium longum, B. breve, B. infantis, L. acidophilus, L. plantarum, L. casei, L. delbrueckii* subspecies *bulgaricus, L. helveticus, Escherichia coli* and *Enterococcus faecalis*) has also been confirmed in several controlled studies, with patients reporting marked improvements in abdominal pain, flatus and aerophagia (Nobaek *et al.*, 2000; De Simone *et al.*, 2001).

Although probiotics have demonstrated clinical efficacy in the treatment of IBS, it can be argued that their use is of limited value because colonisation resistance prevents them from becoming established permanently in the GIT and patients' symptoms return shortly after withdrawing treatment. Prebiotics have the potential to overcome this by stimulating the growth of existing probiotic species present in the colon. Further rationale is provided by the fact that FOS increase counts of bifidobacteria – a species known to be deficient in colon of IBS patients (Gibson *et al.*, 1995). Hunter *et al.* (1999) conducted a double-blind, placebo-controlled study with 21 IBS patients given 2 g oligofructose (Raftilose P95) three times daily versus sucrose (1 g) three times daily. It was concluded that this dose of prebiotic was too low to show an appreciable effect, as previous studies in constipation have suggested that 20 g is effective in increasing stool weight and frequency (Kleesen *et al.*, 1997). Olesen and Gudmand-Hoyer (2000) carried out a 12-week placebo-controlled study where 50 IBS patients were randomised to receive 20 g FOS per day. Administration of the prebiotic improved symptom score (including abdominal distension, flatulence and pain) after 4–6 weeks and alleviated general worsening of symptoms over the entire study period.

An increase in inflammatory cells (such as neutrophils) has been observed in the gut of some IBS patients and emerging literature has demonstrated the immunomodulation of the motor system of the gut during this complaint (Kristjansson *et al.*, 2004). Prebiotics can alter various properties of the immune system, including those of the gut-associated lymphoid tissues (GALT). Changes in the intestinal microflora that occur with the consumption of prebiotics may potentially mediate immune changes via the direct contact of lactic acid bacteria or bacterial products (cell wall or cytoplasmic components) with immune cells in the intestine, the production of SCFA from fibre fermentation, or by changes in mucin production (Chen *et al.*, 1999; Hove et al., 1999; Sakata *et al.*, 1999). There is convincing preliminary data to suggest that fermentable carbohydrates can be exploited in the treatment of IBS through their ability to modulate immune parameters in GALT, secondary lymphoid tissues and peripheral circulation.

8.4 Bowel Cancer

Bowel cancer is defined as cancer of any part of the colon or rectum and in western countries is a prolific disease (it is predicted that 35 000 will be diagnosed with the disease this year in the UK alone; 940 000 worldwide) with a 50 % mortality rate. Although colorectal cancer is found predominantly in more mature people, it may affect virtually any age group. Addressing this disease is considered to be crucial; the UK

National Health Service (NHS) has committed to spending £37.5 million over 2 years on a national screening programme.

As with many cancers, if caught early colorectal cancer is curable. The pre-cancerous state first appears as a polyp which, if untreated, may develop either into a cancer protruding into the bowel's lumen, often causing an obstruction or ulcerating, or it may cross the gut wall and spread to nearby organs.

The symptoms indicative of carcinogenesis are diverse and include persistent stomach ache, increased emergency faecal excretion, rectal bleeding, and anaemia. Although these symptoms do not necessarily indicate colorectal cancer the only way to be sure is to undergo proven clinical tests. Various treatments may be adopted depending on the area of the tumour, its cause and progression. These range from gut resectioning (rare as the initial treatment) to chemotherapy and radiotherapy. The former is invasive surgery, with its attendant risks, while the latter may cause collateral damage. Over recent years, attention has turned to the use of pro- and prebiotics to prevent colorectal cancer and treat the pre-cancerous state.

There appear to be various reasons for developing bowel cancer; having a family history of colorectal cancer increases risk as does having ulcerative colitis. An adverse lifestyle (i.e. high alcohol intake and no exercise) and obesity also increase risk. Diet seems to play a key role both in the development of, and treatment at, the pre-cancerous stage and there are links between floral content of the gut and development of the disease; some bacteria conferring a degree of protection against colorectal cancers while others encourage it.

Butyrate, a SCFA produced in the intestine of people with high fibre diets, is directly implicated in reduced incidence of colorectal cancer (Yuan *et al*., 2004). IL-6 and its receptor act by phosphorylating the transcription factor STAT1; the active STAT1 then up-regulates the expression of the Bcl-xl gene, a gene responsible for cell survival. Parallel to this, IL-6 inhibits Fas-ligation induced apoptosis. Butyrate reduces the expression of the IL-6 Rα chain of the receptor, in this way effectively blocking these two pathways. Interestingly though, it may be a by-product of SCFA β-oxidative metabolism carried out in the mitochondria that provides the control mechanism (via the mitochondrial membrane potential) for at least one of these pathways in the large intestine (Augenlicht *et al*., 2004).

Butyrate helps regulate the β-catenin transcription factor (BCT) dependent Wnt pathway which facilitates modulate proliferation, differentiation and tumour progression genes of colonic carcinoma cell lines SW620 and HCT116 (Bordonaro *et al*., 2002). However, these two cell types differ in their responses to butyrate; the former undergoing down-regulation of the TK-tcf, MMP7 and cyclin-D1 promoters and subsequently differentiation and apoptosis, the latter, up-regulation of the same promoters then reversible growth arrest. The MMP family is involved not only in extracellular matrix breakdown but cell adhesion, cytokine and growth factor production, while MMP7 localises to the leading edge in lamellipodia (Wroblewski *et al*., 2003). MMP7 is lacking in normal intestinal cells but is up-regulated in both gastric cancer and bacterial contact (for example with *H. pylori*) with the epithelium of the colon. This induction of MMP7 is considered a subversion of the protective host response to epithelial damage.

Thus, butyrate appears to be a major influence over several potentially tumourigenic pathways although the method of control may vary between tissue types. Clearly then, bacterial strains that metabolise prebiotics to produce butyrate should aid in the treatment of colorectal cancer. Using an *in vitro* human gut model and the prebiotic inulin, Van de Weile *et al.* (2004) demonstrated that the floral content of the colon could be shifted towards lactobacilli and bifidobacteria with a concomitant increase in butyrate and propionic acid. Another study using endoscopic biopsies during colonoscopy of 15 colorectal patients showed a similar floral response to the prebiotics oligofructose and inulin (Langlands *et al.*, 2004). Similarly, these two chicory derived fructans increased apoptosis in the crypt cells of the rat colon and showed a tendency to decrease ammonia concentration in that area (Hughes and Rowland, 2001).

Measurement of faecal mutagens has been used to study the potential effect of prebiotic wheat bran on colon cancer risk. Reddy *et al.* (1989) found that increased intake of fermentable fibres lowered levels of mutagens in the stools of 19 healthy human subjects. A study of dietary risk factors (Howe *et al.*, 1992) has provided unequivocal evidence that colonic fermentation of carbohydrates decreases the incidence of colorectal cancer. This was an analysis of data from 13 case-controlled studies, comparing 5287 subjects with colorectal cancer with 10 470 disease free individuals. Increased intake of fermentable carbohydrate was directly correlated with reduced risk of both left- and right-sided colon and rectal cancers in men and women.

In conclusion, the study of prebiotics in the treatment of colorectal cancers is a relatively new field. There have been many mechanistic approaches that are beginning to clarify the road to cancer initiation but more study is required to evaluate the potential of various prebiotics. In particular, recent reviews have highlighted lack of data from human clinical studies in cancer patients. It is hoped that a holistic, non-invasive approach can eventually be adopted to help combat the development of colorectal cancers.

8.5 Ulcerative Colitis

Ulcerative colitis (UC) and Crohn's disease (CD) are grouped in the category of idiopathic IBD. UC is a chronic inflammatory condition of the large intestine. Patients may present at any age, although it is primarily a disease of young people with a peak incidence between the ages of 10 and 40 years. Men and women are equally affected. Current incidence in the West is around 10–20 cases per 100 000 population (Probert *et al.*, 1992), with a reported prevalence of 100–200 per 100 000. The incidence remains stable, but the prevalence is likely to be an underestimate, because this implies an average disease duration (prevalence/incidence) of 10 years for a condition that is known to last for life. There is a genetic predisposition to IBD, with a positive familial history being an important risk factor in 6–10 % of UC cases (Orholm *et al.*, 1991). There are also marked differences between ethnic groups with some (such as Ashkenazi Jews) having a particularly high incidence. In total, up to 240 000 people are affected by IBD in the UK.

UC is characterised by acute noninfectious inflammation of the colonic mucosa, manifesting physically as rectal bleeding and/or bloody diarrhoea and often accompanied by left-sided abdominal pain (Ghosh *et al.*, 2000). It always affects the rectum and often involves variable contiguous segments of the colon. Active lesions consist of oedema, erythema, lack of normal vascular pattern, bleeding, exudation of mucous and ulceration. The disease cycles between active and remissive phases, which may be anywhere from a few days to several months' duration. The goal of the current medical therapy is generally focused on attenuation of local inflammation in the colon. Corticosteroids and 5-aminosalicylate (5-ASA) compounds are widely given as a first line drug therapy. However, these drugs are often associated with adverse side effects (gastrointestinal disturbances, headaches and arthralgia), which limits their therapeutic value. Although more potent immunosuppressive agents like cyclosporin A, 6-mercaptopurine and azathiopurine have proven efficacious in the treatment of UC, again their side effects are a major concern.

The aetiology of UC is still largely unknown. Even though a genetic predisposition is well established, several studies have shown that the presence of a luminal microflora is absolutely essential for the disease to develop in both induced and spontaneous animal models of colitis (Rath *et al.*, 1996; Hans *et al.*, 2000). Specific species of bacteria have been implicated in the pathogenesis of UC. *Bacteroides vulgatus* induces gastrointestinal inflammation in rodent models of colitis and was also found in greater numbers in colonic biopsies taken from UC patients when compared with healthy individuals (Matsuda *et al.*, 2000; Rath *et al.*, 2001; Setoyama *et al.*, 2003). Serum antibody responses against *B. vulgatus* and *B. fragilis* were also higher in these patients. There is an indication that *Bacteroides* may express a specific outer membrane protein that triggers an inflammatory response (Bamba *et al.*, 1995). Onderdonk *et al.* (1998) reported that high concentrations of *E. coli* and *Enterococcus* species were associated with severe colitis in B27 TG rats. *Clostridium difficile* toxin has also been found to exacerbate inflammation in patients with chronic colitis (Sartor *et al.*, 1996).

There is even stronger evidence to support the involvement of sulphate-reducing bacteria (SRB) in UC, as their metabolic end product, hydrogen sulphide, is a highly cytotoxic compound. Acting through inhibition of butyrate oxidation in colonocytes, it leads to chronic inflammation and cell death (Roediger *et al.*, 1997). The prevalence of SRB in the faeces of patients with UC is significantly higher than that of healthy individuals (100 % versus 50 %) (Gibson *et al.*, 1988; Pitcher *et al.*, 1995). Specific species such as *Desulfovibrio piger* have been detected in the colon of a high number of UC patients (Loubinoux *et al.*, 2002). Zinkevich and Beech (2000) also demonstrated the ubiquitous presence of SRB in the colitic human colonic mucosa.

Amelioration of the symptoms of UC in both experimental models and human patients has been produced by administration of probiotic bacterial species. Germ-free interleukin (IL)-10 gene-deficient mice pre-treated with *L. plantarum* developed significantly less severe colitis when later exposed to a specific pathogen free flora (Schultz *et al.*, 2002). A reduction in mucosal inflammatory activity was also observed with *L. salivarius* subspecies *salivarius* (O'Mahony *et al.*, 2001). Bifidobacteria-fermented milk was shown to maintain remission when given as a dietary adjunct to UC patients, possibly reducing the relative proportion of *B. vulgatus* through local alterations in organic acid concentration (Ishikawa *et al.*, 2003). A mixed probiotic preparation containing *L. casei*, *L. plantarum*, *L. acidophilus*, *L. delbruekii* subspecies *bulgaricus*, *B. longum*, *B. breve*, *B. infantis* and *Streptococcus salivarius* subspecies *thermophilus* (VSL#3) was effective in maintaining remission in UC patients intolerant to conventional 5-ASA treatment (Venturi *et al.*, 1999). The resulting increase in colonic concentrations of lactobacilli and bifidobacteria was subsequently found to normalise colonic physiological function and barrier integrity (Madsen *et al.*, 2001). Mechanistically, probiotic bacteria may reduce colonic inflammation by enhancing natural and acquired immunity. *L. rhamnosus*, *L. acidophilus* and *B. lactis* fed to mice increased the phagocytic activity of peripheral blood leucocytes and peritoneal macrophages, and significantly elevated splenic interferon-γ production (Gill *et al.*, 2000). Orally administered *L. casei* and *L. bulgaricus* were able to activate macrophages in mice, inferring that probiotic bacteria could enhance the host immune response. Furthermore, different lactic acid bacteria strains produce distinct mucosal cytokine profiles in BALB/c mice – increased IL-10 and IL-4 was observed mainly in mice fed with *L. casei* and *L. delbrueckii* subspecies *bulgaricus*, whilst a significant induction of IL-2 and IL-12 was observed with *L. acidophilus* (Perdigon *et al.*, 2002). If lactic acid bacteria are able to suppress populations of pathogenic species, such as *Bacteroides* or SRB, through stimulation of the host immune response and production of SCFA, it might be expected that prebiotics would aid in the treatment of UC by enhancing natural populations of lactobacilli and bifidobacteria in the colon.

Numerous studies have demonstrated the efficacy of prebiotics in maintaining remission and reducing colonic damage in UC patients. In a 4-month placebo-controlled study, Hallert *et al.* (1991) reported that ispaghula husk significantly relieved gastrointestinal symptoms in 69 % of patients with quiescent UC. Comparing 102 UC patients who were in remission, randomised into groups to receive oral treatment with *Plantago ovata* seeds (10 g b.i.d.), mesalamine (500 mg t.i.d.), or *Plantago ovata* seeds plus mesalamine at the same doses, Fernandez-Banares *et al.* (1999) concluded that the prebiotic was as effective as conventional treatment in maintaining remission over a 12-month period. Faecal butyrate levels were significantly increased during *Plantago ovata* seed administration, which was thought to mediate the beneficial effects observed. More recently, trials have been conducted using a prebiotic mixture derived from germinated barley (GBF) that contains low-lignified hemicellulose. Therapeutic efficacy was first demonstrated in transgenic rodent models of colitis, where GBF fed to HLA-B27 rats reduced the presence of faecal occult blood and colonic mucosal hyperplasia. These effects were brought about by increased microbial butyrate production, which suppressed colonic mucosal NF-κB-DNA binding activity (an important regulatory factor of pro-inflammatory cytokine production) and the manufacture of IL-8 (Kanauchi *et al.*, 1999a). GBF was then shown to stimulate the growth of *Bifidobacterium* and *Eubacterium limosum* in healthy human volunteers, which resulted in an increase in colonic butyrate concentration (Kanauchi *et al.*, 1999b). The same group went on to study the efficacy of GBF in UC patients. Clinical activity index scoring of the disease improved in 21 sufferers with mild to moderate, active UC when their normal drug regime was supplemented with 20–30 g day^{-1} GBF for 24 weeks. When compared with drugs alone, a similar dose for 12 months, combined with conventional therapy (5-ASA and/or tapering steroids), significantly prolonged remission in 22 patients (Kanauchi *et al.*, 2003; Hanai *et al.*, 2004).

In conclusion, prebiotics provide clinical benefits to UC patients both by prolonging remission and improving symptoms during active disease (especially occult faecal blood and diarrhoea). Furthermore, unlike conventional drug therapy, there are virtually no side effects associated with administration. By reducing colonic inflammation, prebiotics may improve the immediate quality of life for sufferers and minimise the risk of developing colon cancer in the long term.

8.6 Crohn's Disease

Crohn's disease (CD) is a subacute or chronic relapsing inflammatory condition of the GIT, found anywhere between the mouth and anus. The incidence of CD is around 5–10 per 100 000 per year with a prevalence of 50–100 per 100 000, but as with UC this may be an underestimate due to the fact that the disease is present throughout life. In contrast to UC, the incidence of Crohn's is increasing (Rubin *et al.*, 2000). There is a north/south distinction, with a lower incidence of CD in southern European countries.

Symptoms of CD are similar to those seen in UC and it is characterised by longitudinal ulcers and noncaseating epithelioid granulomas. The aetiology remains unknown, although studies have shown that the disease has both environmental (e.g. the detrimental effect of smoking) and genetic components. Multiple epidemiologic investigations have documented a significant familial predisposition to CD. The greatest risk is seen in first-degree relatives, with the sibling of a patient with CD having a 30-fold increased risk of developing the disease when compared with the normal population (Satsangi *et al.*, 1994). This concordance rises to 44 % in monozygotic twins (Tysk *et al.*, 1988). Genetic factors appear to be more important in CD than UC and inheritance does not follow simple Mendelian lines. The search for candidate genes has linked CD to a locus on chromosome 16. Independent groups have reported that mutations in the *NOD2* gene, in a region known as IBD1 on this chromosome, are associated with Crohn's disease (Hugot *et al.*, 2001; Ogura *et al.*, 2001). *NOD2* is expressed in monocytes in response to bacterial antigens, where it is considered to function as a cytosolic receptor for pathogenic components. The *NOD2* leucine-rich repeat region senses lipopolysaccharide and activates an NF-κB signalling pathway which results in the inflammatory response through expression of tumour necrosis factor and other inflammatory cytokines (Figure 8.1). Administration of corticosteroids to treat CD blocks NF-κB. Thus, identification of this gene provides a clear link between immune response to enteric bacteria and development of the disease.

Evidence supporting the role of luminal bacteria in CD exists in both animal and human disease studies. As with UC, the presence of a colonic microflora is necessary for inflammation to develop in animal models. When the faecal stream is diverted by an ileostomy proximal to the ileocolonic anastomosis in CD patients, inflammation ceases. However, it promptly recurs within 6 months after restoration of ileal continuity (Rutgeerts *et al.*, 1991). Furthermore, infusion of a CD patient's ileostomy effluent into the excluded normal ileal loop of the same patient rapidly induces local immune activation (infiltration of mononuclear cells, eosinophils, and polymorphonuclear cells) and *de novo* inflammation (D'Haens *et al.*, 1998).

Much work has been dedicated to identifying causative agents within the microflora that trigger the disease. The symptoms bear a marked resemblance to intestinal *Mycobacterium avium* subspecies *paratuberculosis* (*M. paratuberculosis*) infection, which usually targets the ileocecal region, and to Johne's disease, which is a naturally occurring granulomatous enterocolitis in cattle. DNA from, and serum antibodies to, *M. paratuberculosis* have been detected in patients with CD more frequently than in healthy individuals (Hermon-Taylor *et al.*, 2000). There is currently much public debate as to whether CD can be induced by consumption of water or milk contaminated with this

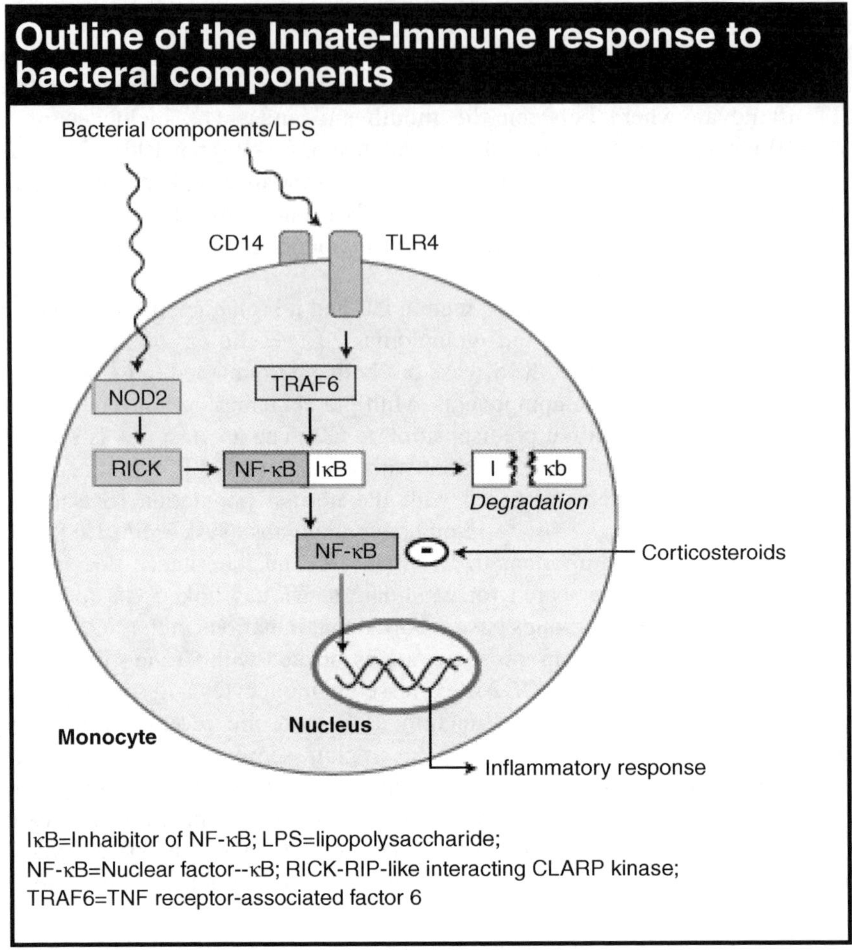

Figure 8.1 *Pathway of inflammatory response to bacterial antigens. Reprinted from The Lancet, Vol. 357, David A. Van Heel et al., Crohn's disease: genetic susceptibility, bacteria, and innate immunity, pp. 1902–1904. Copyright 2001, with permission from Elsevier*

organism. Subsequent studies using quantitative molecular methods have failed to consistently isolate *M. paratuberculosis* genomes directly from the site of inflammation, which has led some researchers to conclude that other species of bacteria may be involved. A small number of investigations have reported the presence of *Helicobacter* species in biopsies from CD patients. *H. pylori* is capable of causing chronic, spontaneously relapsing PUD, associated with nonimmunologically mediated tissue injury. However, its relationship to the development of CD has never been firmly established. Virulent strains of *E. coli* may play a pathogenic role (Elson *et al.*, 1995), as might mucosa-associated *B. vulgatus*. Both species have been detected in tissue samples taken from patients with IBD (Fujita *et al.*, 2002).

The concept that bacteria in the gut lumen are involved in the pathogenesis of CD has led to several controlled studies using antibiotics (e.g. metronidazole and ciprofloxacin). However, remission rates have not been significantly improved beyond that of more conventional steroid treatment (Prantera *et al.*, 1996). Both therapies are subject to side effects, which has led to the application of alternative treatment regimens for CD (and IBD in general). Helminths were found to diminish immune responsiveness in naturally colonised humans and reduce inflammation in experimental colitis. A recent study fed 2500 live *Trichuris suis* ova to 29 individuals with active CD every 3 weeks for 24 weeks; 80 % of patients remitted, with no adverse side effects. Colonisation with helminthes was postulated to improve disease score by inhibiting production of Th1 cytokines and thereby reducing inflammation severity (Summers *et al.*, 2005). The use of probiotics to alleviate inflammation in IBD has been discussed elsewhere (see section on Ulcerative Colitis) and similar species have been trialled in the treatment of CD.

Following successful viability studies in healthy volunteers, studies are currently being conducted to investigate the application of a synbiotic, combining *B. bifidum* and *B. lactis* with an inulin-based prebiotic (Synergy1), for the treatment of CD (Bartosch *et al.*, 2005). Prebiotics exhibiting a bifidogenic effect may improve the prognosis for CD patients by encouraging the growth of lactic acid bacteria, reducing the establishment of causative organisms through local effects on luminal pH (SCFA production) and colonisation inhibition. The effectiveness of GBF has already been demonstrated in IBD sufferers (Kanauchi *et al.*, 2003) and prebiotics remain a promising potential approach for CD.

8.7 Pseudomembranous Colitis

Pseudomembranous colitis is an acute, exudative colitis most often caused by *Clostridium difficile*, a gram positive toxin releasing bacterium. It commonly occurs as a complication of antibiotic therapy (most usually clindamycin, ampicillin and the cephalosporins). *C. difficile* is only responsible for 10–25 % of antibiotic induced diarrhoea, but virtually all cases of pseudomembranous colitis. The organism was first recognised as the cause of the condition in 1977 (George *et al.*, 1978). Unfortunately, antibiotic treatment facilitates the development of the colitis by reducing the patient's resistance to colonisation with *C. difficile*. Owing to the fact that antibiotic treatment and colonisation may occur at different times, antibiotic-susceptible strains of *C. difficile* are able to cause disease. Pseudomembranous colitis can thus be considered an opportunistic infection. Symptoms vary from mild, self-limiting diarrhoea to incapacitating and prolonged diarrhoea. Patients may also develop sudden chills, fever and abdominal pain. The diagnosis is made on the basis of clinical findings in association with the presence of *C. difficile* toxin in stool specimens.

Treatment of pseudomembranous colitis includes barrier nursing to prevent spread of the infection, stopping the causative antibiotic and giving oral metronidazole or vancomycin to eradicate the organism. This can be expensive and difficult. In addition, whilst antibiotics may be effective in alleviating the symptoms of infection, they do not prevent relapse. Reoccurrence rates may be as high as 25 % in patients with *C. difficile* diarrhoea (Bartlett *et al.*, 1980; Walters *et al.*, 1983).

Prebiotics have been examined for their value in restoring or improving colonisation resistance in compromised patients. This approach is considered superior to probiotics, as it utilises the indigenous microflora as opposed to introducing additional foreign organisms. Studies revealed that FOS could prevent the growth and toxin elaboration by *C. difficile* in C57BL/6NHsd mice treated orally with a broad-spectrum antibiotic (Gaskins *et al.*, 1996). Addition of FOS enhanced the growth of *Bifidobacterium* and *Lactobacillus*, but inhibited *Clostridium* and *E. coli* in the small intestinal and proximal colonic contents of pigs. Lactulose was subsequently found to suppress the growth of both clostridia and bacteroides species when added to an *in vitro* human facecal incubation system (Ito *et al.*, 1997). Using a similar model, FOS, GOS and inulin increased bifidobacterial counts following reduction by clindamycin treatment. *C. difficile* was out competed by the normal microbiota in the presence of each prebiotic indicating that they might be used prophylactically against pseudomembranous colitis (Hopkins and Macfarlane, 2003).

8.8 Conclusions

Our understanding of digestive physiology and the role of the gastrointestinal microflora in health and disease has increased dramatically over the last few decades. In a society of abundance, optimised nutrition aims to maximise physiological function, ensuring both well-being and health, but also minimising the risk of disease. Prebiotics offer an exciting and challenging concept in digestive function. By selectively stimulating the growth of beneficial bacteria in the large bowel, they have been shown unequivocally to benefit health. In addition to their role as fibre in the diet, nondigestible oligosaccharides such as inulin and oligofructose have been shown to improve calcium bioavailability, reduce the risk of developing precancerous lesions in the colon, ameliorate mucosal inflammation in numerous gastrointestinal disorders, and induce hypotriglyceridemia and hypoinsulinemia. Furthermore, prebiotics are free from the side effects associated with many traditional drug therapies. Evidence gained through animal models of disease and anecdotal case reports is currently being substantiated with controlled, appropriately powered, clinical trials. This may lead to the advocation of carefully selected prebiotics in the prevention and first-line treatment of gastrointestinal, and other, disorders.

References

Appelmelk, B. J. and Negrini, R. (1997). *Helicobacter pylori* and gastric autoimmunity. *Current Opinion in Gastroenterology* **13**, 31–34.

Augenlicht, L. H., Mariadason, J. M., Wilson, A., Arango, D., Yang, W.-C., Heerdt, B. G. and Velcich, A. (2002). Short chain fatty acids and colon cancer. *Journal of Nutrition* **132**, 3804S–3808S.

Balsari, A., Ceccarelli, A., Dubini, F., Fesce, E. and Poli, G. (1982). The fecal microbial population in the irritable bowel syndrome. *Microbiologica* **5**, 185–194.

Bamba, T., Matsuda, H., Endo, M. and Fujiyama, Y. (1995). The pathogenic role of *Bacteroides vulgatus* in patients with ulcerative colitis. *Journal of Gastroenterology* **30**, 45–47.

Bartlett, J. G., Tedesco, F. J., Shull, S., Lowe, B. and Chang, T. (1980). Symptomatic relapse after oral vancomycin therapy of antibiotic-associated pseudomembranous colitis. *Gastroenterology* **78**, 431–434.

Bartosch, S., Woodmansey, E. J., Paterson, J. C. M., McMurdo, M. E. T. and Macfarlane, G. T. (2005). Microbiological effects of consuming a synbiotic containing *Bifidobacterium bifidum*, *Bifidobacterium lactis*, and oligofructose in elderly persons, determined by real-time polymerase chain reaction and counting of viable bacteria. *Clinical Infectious Diseases* **40**, 28–37.

Bayliss, C. E., Houston, A. P., Alun-Jones, V., Hishon, S. and Hunter, J. O. (1984). Microbiological studies on food intolerance. *Proceedings of the Nutrition Society* **43**, 16A.

Bingham, S. A. (2000). Diet and colorectal cancer prevention. *Biochemical Society Transactions* **28**, 12–16.

Bordonaro, M., Lazarova, D. L., Augenlicht, L. H. and Sartorelli, A. C. (2002). Cell type- and promoter-dependent modulation of the Wnt signaling pathway by sodium butyrate. *International Journal of Cancer.* **97**, 42–51.

Camillieri, M. (2001). Management of the irritable bowel syndrome. *Gastroenterology* **121**, 1527–1528.

Canducci, F., Armuzzi, A., Cremonini, F., Cammarota, G., Bartolozzi, F., Pola, P., Gasbarrini, G. and Gasbarrini, A. (2000). A lyophilized and inactivated culture of *Lactobacillus acidophilus* increases *Helicobacter pylori* eradication rates. *Alimentary Pharmacology and Therapeutics* **14**, 1625–1629.

Chang, W. C., Chapkin, R. S. and Lupton, J. R. (1998). Fish oil blocks azoxymethane-induced tumorigenesis by increased cell differentiation and apoptosis rather than decreased cell proliferation. *Journal of Nutrition* **128**, 491–497.

Chapkin, R. S., Fan, Y.-Y. and Lupton, J. R. (2000). Effect of diet on colonic-programmed cell death: molecular mechanism of action. *Toxicology Letters* **112–113**, 411–414.

Chen, H.-L., Haack, V. S., Janecky, C. W., Vollendorf, N. W. and Marlett, J. A. (1998). Mechanisms by which wheat bran and oat bran increase stool weight in humans. *American Journal of Clinical Nutrition* **68**, 711–719.

Chen, T., Isomaki, P., Rimpilainen, M. and Toivanen, P. (1999). Human cytokine responses induced by Gram-positive cell walls of normal intestinal microbiota. *Clinical and Experimental Immunology* **118**, 261–267.

De Simone, C., Famularo, G., Salvadori, B. B., Moretti, S., Marcellini, S., Trinchieri, V. and Santini, G. (2001). Treatment of irritable bowel syndrome (IBS) with the newer probiotic VSL#3: a multicenter trial. *American Journal of Clinical Nutrition* **73** (Suppl. 2), 491S–491S.

D'Haens, G. R., Geboes, K., Peeters, M., Baert, F., Penninckx, F. and Rutgeerts, P. (1998). Early lesions of recurrent Crohn's disease caused by infusion of intestinal contents in excluded ileum. *Gastroenterology* **114**, 262–267.

Elson, C. O., Sartor, R. B., Tennyson, G. S. and Riddell, R. H. (1995). Experimental models of inflammatory bowel disease. *Gastroenterology* **109**, 1344–1367.

Eun-Ah, B., Kim, D.-H. and Han, M. J. (2000). Anti-*Helicobacter pylori* activity of *Bifidobacterium* spp. *Journal of Microbiology and Biotechnology* **10**, 532–534.

Felley, C. P., Corthesy-Theulaz, I., Rivero, J. L., Sipponen, P., Kaufmann, M., Bauerfeind, P., Wiesel, P. H., Brassart, D., Pfeifer, A., Blum, A. L. and Michetti, P. (2001). Favourable effect of an acidified milk (LC-1) on *Helicobacter pylori* gastritis in man. *European Journal of Gastroenterology and Hepatology* **13**, 25–29.

Fernandez-Banares, F., Hinojosa, J., Sanchez-Lombrana, J. L., Navarro, E., Martinez-Salmeron, J. F., Garcia-Puges, A., Gonzalez-Huix, F., Riera, J., Gonzalez-Lara, V., Dominguez-Abascal, F., Gine, J. J., Moles, J., Gomollon, F. and Gassull, M. A. (1999). Randomized clinical trial of *Plantago ovata* seeds (dietary fiber) as compared with mesalamine in maintaining remission in ulcerative colitis. *American Journal of Gastroenterology* **94**, 427–433.

Fujita, H., Eishi, Y., Ishige, I., Saitoh, K., Takizawa, T., Arima, T. and Koike, M. (2002). Quantitative analysis of bacterial DNA from *Mycobacteria* spp., *Bacteroides vulgatus*, and *Escherichia coli* in tissue samples from patients with inflammatory bowel diseases. *Journal of Gastroenterology* **37**, 509–516.

Gaskins, H. R., Mackie, R. I., May, T. and Garleb, K. A. (1996). Dietary fructo-oligosaccharide modulates large intestinal inflammatory responses to *Clostridium difficile* in antibiotic-compromised mice. *Microbial Ecology in Health and Disease* **9**, 157–166.

George, R. H., Symonds, J. M., Dimock, F., Brown, J. D., Arubi, Y. and Shinagawa, N. (1978). Identification of *Clostridium difficile* as a cause of pseudomembranous colitis. *British Medical Journal* **1**, 695.

Ghosh, S., Shand, A. and Ferguson, A. (2000). Regular review – ulcerative colitis. *British Medical Journal* **320**, 1119–1123.

Gibson, G. R. (1999). Dietary modulation of the human gut microflora using the prebiotics oligofructose and inulin. *Journal of Nutrition* **129**, 1438S–1441S.

Gibson, G. R., Macfarlane, G. T. and Cummings, J. H. (1988). Occurrence of sulphate-reducing bacteria in human feces and the relationship of dissimilatory sulphate reduction to methanogenesis in the large gut. *Journal of Applied Bacteriology* **65**, 103–111.

Gibson, G. R., Beatty, E. B., Wang, X. and Cummings, J. H. (1995). Selective stimulation of bifidobacter in the human colon by oligofructose and inulin. *Gastroenterology* **108**, 975–982.

Gill, H. S., Rutherford, K. J., Prasad, J. and Gopal, P. K. (2000). Enhancement of natural and acquired immunity by *Lactobacillus rhamnosus* (HN001), *Lactobacillus acidophillus* (HN017) and *Bifidobacterium lactis* (HN019). *British Journal of Nutrition* **83**, 167–176.

Green, L. S. (1999). Asthma, antioxidant stress and diet. *Nutrition* **15**, 899–907.

Gui, X.-Y. (1998). Mast cells: a possible link between psychological stress, enteric infection, food allergy and gut hypersensitivity in the irritable bowel syndrome. *Journal of Gastroenterology and Hepatology* **13**, 980–989.

Gwee, K.-A., Leong, Y. L., Graham, C., McKendrick, M. W., Collins, S. M., Walters, S. J., Underwood, J. E. and Read, N. W. (1999). The role of psychological and biological factors in postinfective gut dysfunction. *Gut* **44**, 400–406.

Hallert, C., Kaldma, M. and Petersson, B. G. (1991). Ispaghula husk may relieve gastrointestinal symptoms in ulcerative colitis in remission. *Scandinavian Journal of Gastroenterology* **26**, 747–750.

Hanai, H., Kanauchi, O., Mitsuyama, K., Andoh, A., Takeuchi, K., Takayuki, I., Araki, Y., Fujiyama, Y., Toyonaga, A., Sata, M., Kojima, A., Fukuda, M. and Bamba, T. (2004). Germinated barley foodstuff prolongs remission in patients with ulcerative colitis. *International Journal of Molecular Medicine* **13**, 643–647.

Hans, W., Scholmerich J., Gross, V. and Falk, W. (2000). The role of the resident intestinal flora in acute and chronic dextran sulfate sodium-induced colitis in mice. *European Journal of Gastroenterology and Hepatology* **12**, 267–273.

Hermon-Taylor, J., Bull, T. J., Sheridan, J. M., Cheng, J., Stellakis, M. L. and Sumar, N. (2000). Causation of Crohn's disease by *Mycobacterium avium* subspecies *paratuberculosis*. *Canadian Journal of Gastroenterology* **14**, 521–539.

Hopkins, M. J. and Macfarlane, G. T. (2003). Nondigestible oligosaccharides enhance bacterial colonization resistance against *Clostridium difficile in vitro*. *Applied and Environmental Microbiology* **69**, 1920–1927.

Hove, H., Norgaard, H. and Mortensen, P. B. (1999). Lactic acid bacteria and the human gastrointestinal tract. *European Journal of Clinical Nutrition* **53**, 339–350.

Howe, G. R., Bentino, E. and Castelleto, R. (1992). Dietary intake of fiber and decreased risk of cancers of the colon and rectum: evidence from the combined analysis of 13 case-control studies. *Journal of the National Cancer Institute* **84**, 1887–1896.

Hughes, R. and Rowland, I. R. (2001). Stimulation of apoptosis by two prebiotic chicory fructans in the rat colon. *Carcinogenesis* **22**, 43–47.

Hugot, J. P., Chamaillard, M., Zouali, H., Lesage, S., Cezard, J. P., Belaiche, J., Almer, S., Tysk, C., O'Morain, C. A., Gassull, M., Binder, V., Finkel, Y., Cortot, A., Modigliani, R., Laurent-Puig, P., Gower-Rousseau, C., Macry, J., Colombel, J. F., Sahbatou, M. and Thomas, G. (2001). Association of NOD2 leucine-rich repeat variants with susceptibility to Crohn's disease. *Nature* **411**, 599–603.

Hunter, J. O. and Alun-Jones, V. (1985). Studies on the pathogenesis of irritable bowel syndrome produced by food intolerance. In: *Irritable Bowel Syndrome*, pp. 185-190 (N. W. Read, editor). London: Grune & Stratton Ltd.

Hunter, J. O., Tuffnell, Q. and Lee, A. J. (1999). Controlled trial of oligofructose in the management of unitable bowel syndrome. *Journal of Nutrition* **129**, 1451S–1453S.

Ishikawa, H., Akedo, I., Umesaki, Y., Tanaka, R., Imaoka, A. and Otani, T. (2003). Randomised controlled trial of the effect of bifidobacteria-fermented milk on ulcerative colitis. *Journal of the American College of Nutrition* **22**, 56–63.

Ito, M., Deguchi, Y. and Miyamori, A. (1990). Effects of administration of galactooligosaccharides on the human faecal microflora, stool weight and abdominal sensation. *Microbial Ecology in Health and Disease* **3**, 285–292.

Ito, Y., Moriwaki, H., Muto, Y., Kato, N., Watanabe, K. and Ueno, K. (1997). Effect of lactulose on short-chain fatty acids and lactate production and on the growth of faecal flora, with special reference to *Clostridium difficile*. *Journal of Medical Microbiology* **46**, 80–84.

Johnson-Henry, K. C., Mitchell, D. J., Avitzur, Y., Galindo-Mata, E., Jones, N. L. and Sherman, P. M. (2004). Probiotics reduce bacterial colonization and gastric inflammation in *H. pylori*-infected mice. *Digestive Diseases and Sciences* **49**, 7–8.

Jones, R. and Lydeard, S. (1992). Irritable bowel syndrome in the general population. *British Medical Journal* **304**, 87–90.

Jonkers, D., van den Brock, E., van Dooren, I., Thijs, C., Dorant, E., Hageman, G. and Stobberingh, E. (1999). Antibacterial effect of garlic and omeprazole on *Helicobacter pylori*. *Journal of Antimicrobial Chemotherapy* **43**, 837–839.

Kanauchi, O., Andoh, A., Iwanaga, T., Fujiyama, Y., Mitsuyama, K., Toyonaga, A. and Bamba, T. (1999a). Germinated barley foodstuffs attenuate mucosal damage and mucosal nuclear factor kappa B activity in a spontaneous colitis model. *Journal of Gastroenterology and Hepatology* **14**, 1173–1179.

Kanauchi, O., Fujiyama, Y., Mitsuyama, K., Araki, Y., Ishii, T., Nakamura, T., Hitomi, Y., Agata, K., Saiki, T. Andoh, A. Toyonaga, A. and Bamba, T. (1999b). Increased growth of *Bifidobacterium* and *Eubacterium* in germinated barley foodstuff, accompanied by enhanced butyrate production in healthy volunteers. *International Journal of Molecular Medicine* **3**, 175–179.

Kanauchi, O., Mitsuyama, K., Homma, T., Takahama, K., Fujiyama, Y., Andoh, A., Araki, Y., Suga, T., Hibi, T., Naganuma, M., Asakura, H., Nakano, H., Shimoyama, T., Hida, N., Haruma, K., Koga, H., Sata, M., Tomiyasu, N., Toyonaga, A., Fukuda, M., Kojima, A. and Bamba, T. (2003). Treatment of ulcerative colitis patients by long-term administration of germinated barley foodstuff: multi-center open trial. *International Journal of Molecular Medicine* **12**, 701–704.

King, T. S., Elias, M. and Hunter, J. O. (1998). Abnormal colonic fermentation in irritable bowel syndrome. *Lancet* **352**, 1187–1189.

Kleessen, B., Sykura, B., Zunft, H. J. and Blaut, M. (1997). Effects of inulin and lactose on fecal microflora, microbial activity, and bowel habit in elderly constipated persons. *American Journal of Clinical Nutrition* **65**, 1397–1402.

Kristjansson, G., Venge, P., Wanders, A. Loof, L. and Hallgren, R. (2004). Clinical and subclinical intestinal inflammation assessed by the mucosal patch technique: studies of mucosal neutrophil and eosinophil activation in inflammatory bowel diseases and irritable bowel syndrome. *Gut* **53**, 1806–1812.

Langlands, S. J., Hopkins, M. J., Coleman, N. and Cummings, J. H. (2004). Prebiotic carbohydrates modify the mucosa associated microflora of the human large bowel. *Gut* **53**, 1610–1616.

Loubinoux, J., Bronowicki, J. P., Pereira, A. C., Mougenel, J. L. and Le Faou, A. E. (2002). Sulfate-reducing bacteria in human feces and their association with inflammatory bowel diseases. *FEMS Microbiology and Ecology* **40**, 107–112.

Madsen, K. L., Cornish, A., Soper, P., McKaigney, C., Jijon, H., Yachimec, C., Doyle, J., Jewell, L. and De Simone, C. (2001). Probiotic bacteria enhance murine and human intestinal epithelial barrier function. *Gastroenterology* **121**, 580–591.

Matsuda, H., Fujiyama, Y., Andoh, A., Ushijima, T., Kajinami, T. and Bamba, T. (2000). Characterization of antibody responses against rectal mucosa-associated bacterial flora in patients with ulcerative colitis. *Journal of Gastroenterology and Hepatology* **15**, 61–68.

McKendrick, M. W. and Read, N. W. (1994). Irritable bowel syndrome – post salmonella infection. *Journal of Infection* **29**, 1–3.

Michetti, P., Dorta, G., Wiesel, P. H., Brassart, D., Verdu, E., Herranz, M., Felley, C., Porta, N., Rouvet, M., Blum, A. L. and Corthesy-Theulaz, I. (1999). Effect of whey-based culture supernatant of *Lactobacillus acidophilus (johnsonii)* La1 on *Helicobacter pylori* infection in humans. *Digestion* **60**, 203–209.

Milner, J. A. (2002). Functional foods and health: a US perspective. *British Journal of Nutrition* **88** (Suppl. 2), S151–S158.

Niedzielin, K., Kordecki, H. and Kosik, R. (1998). New possibility in the treatment of irritable bowel syndrome: probiotics as a modification of the microflora of the colon. *Gastroenterology* **114** (Suppl. 2), G1640.

Nobaek, S., Johansson, M. L., Molin, G., Ahrne, S. and Jeppsson, B. (2000). Alteration of intestinal microflora is associated with reduction in abdominal bloating and pain in patients with irritable bowel syndrome. *American Journal of Gastroenterology* **95**, 1231–1238.

Ogura, Y., Bonen, D. K., Inohara, N., Nicolae, D. L., Chen, F. F., Ramos, R., Britton, H., Moran, T., Karaliuskas, R., Duerr, R. H., Achkar, J. P., Brant, S. R., Bayless, T. M., Kirschner, B. S., Hanauer, S. B., Nunez, G. and Cho, J. H. (2001). A frameshift mutation in NOD2 associated with susceptibility to Crohn's disease. *Nature* **114**, 603–606.

Olesen, M. and Gudmand-Hoyer, E. (2000). Efficacy, safety, and tolerability of fructooligosaccharides in the treatment of irritable bowel syndrome. *American Journal of Clinical Nutrition* **72**, 1570–1575.

O'Mahony, L., Feeney, M., O'Halloran, S., Murphy, L., Kiely, B., Fitzgibbon, J., Lee, G., O'Sullivan, G., Shanahan, F. and Collins, J. K. (2001). Probiotic impact on microbial flora, inflammation and tumour development in IL-10 knockout mice. *Alimentary Pharmacology and Therapeutics* **15**, 1219–1225.

Onderdonk, A. B., Richardson, J. A., Hammer, R. E. and Taurog, J. D. (1998). Correlation of cecal microflora of HLA-B27 transgenic rats with inflammatory bowel disease. *Infection and Immunity* **66**, 6022–6023.

Orholm, M., Munkholm, P., Langholz, E., Nielsen, O. H., Sorensen, I. A. and Binder, V. (1991). Familial occurrence of inflammatory bowel disease. *New England Journal of Medicine* **324**, 84–88.

Perdigon, G., Maldonado-Galdeano, C., Valdez, J. C. and Medici, M. (2002). Interaction of lactic acid bacteria with the gut immune system. *European Journal of Clinical Nutrition* **56**, S21–S26.

Pitcher, M. C. L., Beatty, E. R., Gibson, G. R. and Cummings J. H. (1995). Sulphate-reducing bacteria – prevalence in active and inactive ulcerative colitis. *Gastroenterology* **108**, A894.

Prantera, C., Zannoni, F., Scribano, M. L., Berto, E., Andreoli, A., Kohn, A. and Luzi, C. (1996). Antibiotic regimen for the treatment of active Crohn's disease: a randomized, controlled clinical trial of metronidazole plus ciprofloxacin. *American Journal of Gastroenterology* **91**, 328–332.

Probert, C. S., Jayanthi, V., Pinder, D., Wicks, A. C. and Mayberry, J. F. (1992). Epidemiologic-study of ulcerative proctocolitis in Indian migrants and the indigenous population of Leicestershire. *Gut* **33**, 687–693.

Rath, H. C., Herfarth, H. H., Ikeda, J. S., Grenther, W. B., Hamm, T. E., Balish, E., Taurog, J. D., Hammer, R. E., Wilson, K. H. and Sartor, R. B. (1996). Normal luminal bacteria, especially bacteroides species, mediate chronic colitis, gastritis, and arthritis in HLA-B27/human beta (2) microglobulin transgenic rats. *Journal of Clinical Investigation* **98**, 945–953.

Rath, H. C., Schultz, M., Freitag, R., Dieleman, L. A., Li, F. L., Linde, H. J., Scholmerich, J. and Sartor, R. B. (2001). Different subsets of enteric bacteria induce and perpetuate experimental colitis in rats and mice. *Infection and Immunity* **69**, 2277–2285.

Reddy, B. S. (1994). Chemoprevention of colon cancer by dietary fatty acids. *Cancer Metastasis Reviews* **13**, 285–302.

Reddy, B., Engle, A. and Katsifis, S. (1989). Biochemical epidemiology of colon cancer: effect of types of dietary fiber on fecal mutagens, acid, and neutral sterols in healthy subjects. *Cancer Research* **49**, 4629–4635.

Rodriguez, L. A. G. and Ruigomez, A. (1999). Increased risk of irritable bowel syndrome after bacterial gastroenteritis. *British Medical Journal* **318**, 565–566.

Roediger, W. E. (1982). Utilization of nutrients by isolated epithelial cells of the rat colon. *Gastroenterology* **83**, 424–429.

Roediger, W. E., Moore, J. and Babidge, W. (1997). Colonic sulfide in pathogenesis and treatment of ulcerative colitis. *Digestive Diseases and Sciences* **42**, 1571–1579.

Rubin, G. P., Hungin, A. P. S., Kelly, P. J. and Ling, J. (2000). Inflammatory bowel disease: epidemiology and management in an English general practice population. *Alimentary Pharmacology and Therapeutics* **14**, 1553–1559.

Rutgeerts, P., Goboes, K., Peeters, M., Hiele, M., Penninckx, F., Aerts, R., Kerremans, R. and Vantrappen, G. (1991). Effect of faecal stream diversion on recurrence of Crohn's disease in the neoterminal ileum. *Lancet* **338**, 771–774.

Sakata, T., Ichikawa, H. and Inagaki, A. (1999). Influences of lactic acid, succinic acid and ammonia on epithelial cell proliferation and motility of the large bowel. *Asia Pacific Journal of Clinical Nutrition* **8**, S9–S13.

Sartor, R. B., Rath, H. C. and Sellon, R. K. (1996). Microbial factors in chronic intestinal inflammation. *Current Opinions in Gastroenterology* **12**, 327–333.

Satsangi, J., Rosenberg, W. C. M. and Jewell, D. P. (1994). The prevalence of inflammatory bowel disease in relatives of patients with Crohn's disease. *European Journal of Gastroenterology and Hepatology* **6**, 413–416.

Scheurlen, M. (1996). Pathogenicity of fungi in the intestines–current status of the discussion. *Forshung Medica* **114**, 319–321.

Setoyama, H., Imaoka, A., Ishikawa, H. and Umesaki, Y. (2003). Prevention of gut inflammation by *Bifidobacterium* in dextran sulfate-treated gnotobiotic mice associated with *Bacteroides* strains isolated from ulcerative colitis patients. *Microbes and Infection* **5**, 115–122.

Shultz, M., Veltkamp, C., Dieleman, L. A., Grenther W. B., Wyrick, P. B., Tonkonogy, S. L. and Sartor, R. B. (2002). *Lactobacillus plantarum* 299V in the treatment and prevention of spontaneous colitis in interleukin-10-deficient mice. *Inflammatory Bowel Disease* **8**, 71–80.

Sipponen, P., Riihela, M., Hyvarinen, H. and Seppala, K. (1994) Chronic nonatropic 'superficial' gastritis increases risk of gastric carcinoma. *Scandinavian Journal of Gastroenterology* **29**, 336–340.

Stone-Dorshow, T. and Levitt, M. D. (1987). Gaseous response to ingestion of a poorly absorbed fructooligosaccharide sweetener. *American Journal of Clinical Nutrition* **46**, 61–65.

Summers, R. W., Elliott, D. E., Urban Jr., J. F., Thompson, R. and Weinstock, J. V. (2005). *Trichuris suis* therapy in Crohn's disease. *Gut* **54**, 6–8.

Tysk, C., Lyndberg, E., Jarnerot, G. and Floderus-Myrhed, B. (1988). Ulcerative colitis and Crohn's disease in an unselected population of monozygotic and dizygotic twins. A study of heritability and the influence of smoking. *Gut* **29**, 990–996.

Van de Wiele, T., Boon, N., Possemiers, S., Jacobs, H. and Verstraete, W. (2004). Prebiotic effects of chicory inulin in the simulator of the human intestinal microbial ecosystem. *FEMS Microbiology and Ecology* **51**, 143–153.

Van Heel, D., McGovern, D. P. B. and Jewell, D. P. (2001). Crohn's disease: genetic susceptibility, bacteria, and innate immunity. *Lancet* **357**, 1902–1904.

Venturi, A., Gionchetti, P., Rizzello, F., Johansson, R., Zucconi, E., Bigidi, P., Matteuzzi, D. and Campieri, M. (1999). Impact on the composition of the faecal flora by a new probiotic preparation: preliminary data on maintenance treatment of patients with ulcerative colitis. *Alimentary Pharmacology and Therapeutics* **13**, 1103–1108.

Walters, B. A., Roberts, R., Stafford, R. and Seneviratne, E. (1983). Relapse of antibiotic associated colitis: endogenous persistence of *Clostridium difficile* during vancomycin therapy. *Gut* **24**, 206–212.

Wroblewski, L. E., Noble, P. J., Pagliocca, A., Pritchard, D. M., Hart, C. A., Campbell, F., Dodson, A. R., Dockray, G. J. and Varro, A. (2003). Stimulation of MMP-7 (matrilysin) by *Helicobacter pylori* in human gastric epithelial cells: role in epithelial cell migration. *American Journal of Cell Science* **116**, 3017–3026.

Yuan, H., Liddle, F. J., Mahajan, S. and Frank, D. A. (2004). IL-6-induced survival of colorectal carcinoma cells is inhibited by butyrate through down-regulation of the IL-6 receptor. *Carcinogenesis* **25**, 2247–2255.

Zinkevich, V. and Beech, I. B. (2002). Screening of sulfate-reducing bacteria in colonoscopy samples from healthy and colitic human gut mucosa. *FEMS Microbiology and Ecology* **34**, 147–155.

9

Extra Intestinal Effects of Prebiotics and Probiotics

Gregor Reid

9.1 Introduction

To date virtually all prebiotic applications have been targeted at the gastrointestinal tract. In principle, however, the concept of modulating a complex microbial ecosystem through the means of a selectively fermented carbohydrate could be applied much more widely. This process is already happening with probiotics. The evolution of the definition of probiotics from one focused on the intestine to the current one 'Live microorganisms which when administered in adequate amounts confer a health benefit on the host' (FAO/WHO, 2001), reflects the broad applicability of microbes to the host. In this chapter, examples of extra intestinal applications of probiotics will be provided with particular emphasis on the urogenital tract and the potential for extending these applications to prebiotics considered where appropriate.

The rationale for probiotics is relatively simple, yet the present society's dependence on antibiotics and the domination of medical practice by the pharmaceutical industry has pushed aside the necessary understanding of this concept. So, let us start by going back in time.

A billion years before the continents were divided and humans emerged, this planet was colonised solely by bacteria. Amongst them were species perfectly capable of killing humans and animals. However, when homosapiens emerged, they did not do this. Instead, they allowed, and arguably took part in, the emergence of humans, and evolved with us in such a way as to dominate the human body. Given the amount of time that they have been

Prebiotics: Development and Application Edited by G.R. Gibson and R.A. Rastall

involved in this evolutionary process, one might think that they play a major role in much of how humans function. After all, without them we simply die.

Of the ones we live with, what could they possibly do? A recent study suggests no less than them being able to confer longer life (Brummel *et al.*, 2004). That being the case, mankind's greatest wish could be reachable, to live forever! To examine such things, a study is being undertaken on the indigenous microbiota of a women in France who died at the age of 125 years. Hardly living forever, but still 50 years better than average. The molecular and microbiological tools are now available to see which organisms are present, although harnessing all of them is not feasible given the current inability to grow many species.

The probiotic concept is simply a means to give to the host living organisms that we believe can boost the present occupation force and fight off any organisms which shorten the longevity of the host's life. This assumes that bacteria do not want to leave the host and prefer its warm and nutritious environment. Yet daily, leave they must through swallowing, defacation, and cleaning activities of us humans. The dispatched bacteria may take a while to find another host, although humans and animals have a pretty reliable way to helping them out, as salmonella outbreaks can attest.

Of the species that do make the body their home, it is somewhat of a first-come first-gets-to-stay event. The host has some say in which stays, for example the mouth has adapted not to harbour *Escherichia coli* on a regular basis and the skin and urethra are generally not the preferred site for *Streptococcus mutans*. Thus, even if we wanted to apply a recombinant *S. mutans* with all the right adhesins and anti-infective properties to prevent urethritis, the chances of it being able to establish itself as a natural coloniser is close to nil. Such is the competitive pressure of organisms that have been indigenous in the urethra since infancy.

And so to infancy, and how this all started. There is no question that most of our ingrained microbes joined us in the first 6–12 months of life. Yet, how did we select which ones we wanted and how did the ones that stayed beat out the ones that didn't? Clearly, humans have done nothing to purposely become colonised by specific species, although breast feeding does influence the selection process, with lactic acid bacteria thriving on the home produce. The exciting future of probiotics comes from deciphering which species we should be getting, how much and when, and to what end. This will only come with fully understanding the colonisation process in conjunction with the nature of each person's genomic code.

Society and the medical community (and litigation lawyers) in particular are terrified of dabbling with early childhood. This caution is useful to a point, but often it is based on fear rather than common sense. The hygiene hypothesis (Romagnani, 2004) has drawn our attention to the need to consider the extent to which we protect our newborns from the microbes that they need. To truly progress this issue, we need to find the right balance and the right means of administering reliable, clinically tested, properly manufactured probiotics.

9.2. Probiotics for the Nose

No applications of prebiotics or probiotics to the nose have been published, however the potential exists for these to occur. The nose's filtration system helps prevent some pathogens, dust and other undesirable elements from entering the respiratory system. In doing, so it can then harbour these pathogens and potentially pass them on to other people who are more susceptible to their infectious traits. Thus, it is not uncommon to find potentially pathogenic bacteria (PPB) such as *Staphylococcus aureus, Streptococcus pneumoniae*, beta-hemolytic streptococci, and *Haemophilus influenzae* in the nose (Gluck and Gebbers, 2003).

Rather than implanting probiotic competitors into the nose itself, an open, prospective trial, of 209 volunteers was undertaken to use the link between enhancing the mucosal immunity of the gut with that of other mucosal surfaces. The subjects were randomly assigned to consume either a probiotic, fermented milk drink with *Lactobacillus rhamnosus* GG or standard yogurt daily for 3 weeks (Gluck and Gebbers, 2003). Nasal microbiota were analysed on days 1, 21 and 28. The outcome was a significant reduction (19 %; $P < 0.001$) in the occurrence of nasal PPB in the group who consumed the probiotic drink but not in the group who consumed yogurt. This led the authors to conclude that regular intake of a *L. rhamnosus* probiotic stimulated intestinal B lymphocytes of the GALT, which then migrated to the upper respiratory tract and induced production of SIgA2 which reduced the PPB numbers. If this is indeed confirmed by others, the specificity is quite remarkable and the applicability enormous, especially in hospital settings where transfer of PBB can be lethal to susceptible patients. A question that immediately arises is what did these immune cells do to pathogens at other mucosal sites like the mouth, throat and urogenital tract? Also, what is the long term effect? It seems hard to imagine that daily ingestion of a living food (which did not function on its own and needed an additional lactobacilli added) could be so effective, and that the PBB cannot adapt to the altered immune environment, especially as these organisms form biofilm structures well known for resistance to immune attack (Meluleni *et al.*, 1995; Donlan and Costerton, 2002).

9.3 Probiotics to Prevent Dental Caries

Lactobacilli have long been known to produce acid that plays a role in dental caries, a disease common to humans and associated primarily with organisms like *S. mutans*. Yet, two studies, once again using *L. rhamnosus* GG, have shown benefits in fighting caries. Thirty-five healthy volunteers were randomised into three groups to receive lactobacilli and/or placebo for 45 days: Group A ($n = 14$) received probiotics in capsules and placebo in liquid form; group B ($n = 16$) took liquid probiotics and placebo in capsules, and group C ($n = 5$) used placebo in both liquid and capsule form (Montalto *et al.*, 2004). The salivary counts of lactobacilli and *S. mutans* were measured semi-quantitatively. Compared with placebo, the oral administration of probiotics, both in capsules and in liquid form, significantly increased salivary counts of lactobacilli ($P = 0.005$ and $P = 0.02$, respectively). In a double-blind, placebo controlled experiment, 594 children aged 1–6 years old were given either normal milk or milk containing *L. rhamnosus* GG for 7 months (Nase *et al.*, 2001). There were significantly fewer cavities in the children receiving the probiotic. The mechanisms of action are unknown but the concept of the lactobacilli displacing or inhibiting the growth of *S. mutans* or influencing intestinal mucosal immunity and the body's immune system per se, have been postulated.

One other study of 40 female students who consumed *L. reuteri* SD2112 (ATCC 55730) yogurt for 2 weeks, reported that salivary carriage of *S. mutans* decreased. However, a close examination of the findings showed that the decrease compared with placebo only occured in one group, and in a second group placebo faired better (Nikawa *et al.*, 2004). Still, the authors concluded that both groups benefited and gave an explanation that only *L. reuteri* inhibited *S. mutans* yogurt *in vitro*. This seems far fetched and speaks more to bad experimental design than any specific strain effect. Validation of this study by others is needed.

Of interest in all these studies, is the potential for benefits to occur in areas of the body that are not the primary site of probiotic application.

9.4 Probiotics to Prevent Tonsillitis and Nasopharyngeal Infections

Over 10 years ago, a double blind, randomised, placebo-controlled study was undertaken to attempt to restore the normal alpha-streptococcal flora by reimplantation of 'probiotic' alpha-streptococci (Roos, 1993). Thirty-six patients with recurrent streptococcal group A tonsillitis were treated with antibiotics followed by either placebo (19 patients) or a pool of four selected alpha-streptococcal strains (17 patients) that inhibited growth of clinical isolates of beta-streptococci. The results showed that no probiotic-treated patients had recurrent tonsilitis over a 2-month follow-up, but 7/19 arose in those treated with antibiotics and placebo. After 3 months, 1 in the probiotic group and 11 in the placebo-treated group had recurrences. Surprisingly, these encouraging results have neither led to further studies nor to a commercially available product.

The concept of using alpha-streptococci dates back to the 1960s with work by Sprunt and others (Sprunt and Redman, 1968; Sprunt and Leidy, 1988). The studies used alpha-hemolytic streptococcus strain 215 inoculated into the nasopharynx of neonates in the intensive care unit. A single inoculation changed abnormal colonisation of the pharynx to 'normal' (alpha-strep predominant) in 48–72 h in most neonates. The researchers found that *Streptococcus* with similar characteristics to strain 215 occurred naturally in 17–6 % of neonates in that particular intensive care unit, emphasising the replenishment concept of probiotics.

9.5 Probiotics for Prevention of Respiratory Infections and Atopic Diseases

In a similar vein to the examples cited above, studies have examined the potential for probiotics to influence the health of the respiratory tract. Unlike the nose and throat, the effects have not been restricted to interfering with bacteria, but have appeared to include anti-viral activity. According to the World Health Organisation's World Health Report, the top five respiratory diseases account for 17.4 % of all deaths. The agency states that due to many changes in the population of today's society, changes in health care systems, schooling, income, and tobacco use, the threat of chronic respiratory diseases including asthma, will increase.

Asthma is a lasting lung disease that affects millions of people daily and is the most common chronic disease in children. Symptoms of asthma include episodes of wheezing, coughing, shortness of breath, and increased mucus production. Asthma and other allergic reactions are multifactorial processes with both complex genetic and environmental components. To reduce the risk of asthma, breastfeeding has been recommended during the first 4–6 months of life and consumption of dairy products delayed until 1 year (Stanaland, 2004). However, probiotic consumption by the mother is recommended. Findings from a Finnish study are somewhat contrary to the recommendation that no probiotics be given to newborns. In addition to lactating mothers ingesting a milk-based probiotic *Lactobacillus* GG product, the newborns also received this treatment for 6 months with significantly reduced cases of severe atopic dermatitis (Kalliomaki *et al.*, 2004). The mechanisms of action are not defined, but modulation of a Th1/Th2 balance, enhancement of IgA responses and reducing intestinal permeability have been considered.

Based upon the increased traffic of circulating CD34 + hemopoietic precursors cells (HPC) being an important feature of systemic allergic inflammation, a recent study gave a mixture of *L. acidophilus, L. delbrueckii* and *S. thermophilus* for 30 days to 14 people with clinical symptoms of asthma and/or conjunctivitis, rhinitis, urticaria, atopic dermatitis, food allergy and irritable bowel syndrome (Mastrandrea *et al.*, 2004). It was theorised that bacterial stimulation of transcription of maturational cytokines IL12 and interferons through the activation of toll-like-receptor and the subsequent nuclear translocation of the NF-kappaB factor, would lead to improved treatment of asthma. Circulating CD34 + cell values were significantly ($P < 0.001$) reduced with this therapy, providing some encouragement for other studies using probiotics. In another treatment study with a double-blind, placebo-controlled, crossover design, 6 weeks of therapy with *L. rhamnosus* 19070-2 and *L. reuteri* DSM 122460 resulted in 56 % versus 15 % improvement of eczema in children ($P = 0.001$)(Rosenfeldt *et al.*, 2003). However, the total eczema index did not change significantly, nor did production of cytokines IL-2, IL-4, IL-10 or IFN-gamma.

In the longest follow-up studies reported to date, a Czech group used a questionnaire to determine if 150 full term and 77 premature infants who had received an *E.coli* strain at birth, showed any reduction in allergies (Lodinova-Zadnikova *et al.*, 2003). They found statistically higher rates of allergy 10 and 20 years later in patients given the *E. coli* versus those who had not received it ($P < 0.01$). Specific serum IgE antibodies confirmed the presence of allergies in 100 % of 10-year-old and 91 % of 20-year-old patients with clinical symptoms of allergy in the controls.

Soaring rates of atopic diseases throughout the last decade makes this line of investigation timely and important. Atomic dermatitis manifests itself by skin breakouts, itchiness, swelling and scaling on the face, hands and arms. Eczema accounts for 10–20 % of all referrals to dermatologists, affecting up to 7 % of the total population. Thus, the encouraging Finnish and Czech studies need to be expanded to determine if probiotics truly can prevent allergy and help with its treatment.

9.6 Urogenital Benefits of Probiotics and Prebiotics

The rationale for the administration of lactobacilli to the vagina, and subsequent beneficial outcomes have been summarised previously (Reid and Bruce, 2001, 2003; Reid and

Burton, 2002; Reid *et al.*, 2003a). Rather than repeat these points, a brief summary will be given, followed by an updated perspective on how this area is evolving.

The concept was first considered in a paper by Bruce in 1973 (Bruce *et al.*, 1973). He noted that women who did not suffer from bladder infections had a lactobacilli dominated vaginal microbiota. It was theorised that this acted as a barrier to the ascension of uropathogens from the rectum to the bladder. Thus, in the early 1980s, strains of *Lactobacillus* were selected for their ability to potentially replenish the vaginal microbiota as a means of reducing the recurrence of infections. The criteria at that time were the ability to adhere to and colonise the vagina and prevent pathogens from attaching (Reid *et al.*, 1987). Over 22 years later, three strains, *L. rhamnosus* GR-1, *L. reuteri* (formerly *fermentum*) B-54 and *L. reuteri* (formerly *acidophilus* and *fermentum*) RC-14 have been shown to colonise the vagina and significantly reduce pathogen colonisation and infection in the bladder and vagina (Reid *et al.*, 1995, 2001a,b, 2003b, 2004; Cadieux *et al.*, 2002).

Several interesting aspects of this research have emerged in recent times. Unlike the *L. rhamnosus* GG strain, the GR-1 organism, originally isolated from the urethra, appears better adapted to colonisation of the vagina (Cadieux *et al.*, 2002) and competition against urogenital pathogens (Reid *et al.*, 2001b; Colodner *et al.*, 2003). This illustrates that the same species do not necessarily act the same way *in vivo*, and although the reasons are not clear, the findings emphasise that companies should not be selling generic strains and claiming that they will therefore confer benefits. Likewise, *L. reuteri* RC-14, chosen for its production of anti-pathogenic biosurfactant (Velraeds *et al.*, 1996), colonises the vagina unlike the commercial *L. reuteri* SD2112 strain, never shown to colonise the vagina, and chosen for production of an antibiotic reuterin that may or may not be produced in the gut, and may or may not be relevant to preventing infection. Indeed, reuterin has been proposed as a blood substitute (Chen *et al.*, 2004) and tissue fixative (Sung *et al.*, 2003), and the weak antibacterial effects of it and reutericin 6 (Arques *et al.*, 2004; Kawai *et al.*, 2004) except against gut pathogens *E. coli* 0157:H7, *Salmonella choleraesuis* and *Yersinia enterocolitica* (Arques *et al.*, 2004), raises questions as to its importance. But, such inhibition is not restricted to *L. reuteri*, and strains such as *L. rhamnosus* GR-1 have been shown to likewise inhibit enteropathogens (Reid *et al.*, 2002).

There are many reports on bacteriocins and antibiotics produced by lactic acid bacteria. These may well play a role in the reduction of pathogen growth *in vivo*, and in creating a niche in which lactobacilli can prosper. But such ideas, mostly from the 1980s and 1990s, are not being borne out to be either commercially viable (except for nisin's limited applications) or ecologically crucial. For example, lactobacilli and urogenital pathogens can be found together on the same vaginal surface (Devillard *et al.*, 2004), and in any given month many women develop bacterial vaginosis, a condition in which lactobacilli are displaced or eradicated by mostly anaerobic Gram negative rods (Reid and Bocking, 2003).

In terms of more likely mechanisms of action of lactobacilli in the vagina and urethra, three are being studied at present. The ability of lactobacilli to modulate and down-regulate inflammation appears realistic as the absence of lactobacilli correlates with a 19-fold increase in IL-8 (Cauci *et al.*, 2003). An animal study suggests that

L. reuteri can reduce inflammation in a colitis gut (Holma *et al.*, 2001), while another suggests *L. reuteri* strain ML1 can induce inflammatory cytokines in the gut (Maassen *et al.*, 2000). In our studies of humans with inflammatory bowel disease, one-third had significantly downregulated inflammation following daily ingestion of probiotic yogurt containing *L. reuteri* RC-14 (unpublished). This follows a study in which RC-14 and its biosurfactant by-products were shown to reduce inflammation in an animal wound model (Gan *et al.*, 2002).

The wound study led to another avenue of investigation, namely the production of signalling molecules that down regulate virulence of pathogens. To date, *L. reuteri* RC-14 has been found to produce signals that clearly block exotoxin and a larger pathogenicity island in *S. aureus* (McCormick *et al.*, 2004; Laughton *et al.*, 2006), as well as the hemolytic toxin associated with *E. coli* 0157: H7 (Anand *et al.*, 2006). The precise identity of these protein/peptide signals remains to be determined. Cell–cell signalling between lactobacilli and the host has been described with respect to mucin induction (Mack *et al.*, 2003) and physiological regulation (Johnsborg *et al.*, 2003), but these are the first reports of lactobacilli adversely affecting pathogens.

The ability of *Lactobacillus* strains GR-1 and RC-14 to kill viruses such as HIV (Cadieux *et al.*, 2002) (Table 9.1) is likely an acid-based feature, but it illustrates how women with lactobacilli may have a lower risk of HIV infection (Sewankambo *et al.*, 1997). The production of organic acids and hydrogen peroxide has long been regarded as attributes that help lactobacilli create and maintain an ecological niche (Kawai *et al.*, 2004). It is not clear to what extent these allow penetration and eradication of pathogenic bacterial biofilms, nor is it clear to what extent hormone levels override or alter the environmental pH and glycogen levels thereby allowing pathogens an opportunity to thrive. Some microbes like yeast and enterococci can tolerate lower pH and both are commonly found in the vagina. Indeed, some enterococci can produce bacteriocins lethal to lactobacilli, thereby providing one potential way for BV to arise (Kelly *et al.*, 2003).

Table 9.1 *In vitro results of experiments carried out in which 10^9 viable viral cells were incubated with culture supernatants from Lactobacillus rhamnosus GR-1 or Lactobacillus reuteri RC-14. The percentage reduction values are shown as compared with controls with culture medium alone. The data show the anti-viral activity of the organisms. No effect was found due to hydrogen peroxide but organic acids cannot be ruled out*

Virus exposed to lactobacilli supernatants for 10 min	Percentage reduction in viable viral counts	
	Lactobacillus RC-14	*Lactobacillus* GR-1
HIV-1	99.99	99.98
Poliovirus Type 1	0	0
Influenza virus	100	100
Rotavirus	43.8	82.2
Coxsackievirus B3	0	0
Adenovirus type 2	0	43.8
Adenovirus type 5	100	100
Vesicular stomatitis virus	100	100

The concept of stimulating the indigenous lactobacilli in the vagina using a prebiotic has been explored to some extent. While skimmed milk is not a prebiotic per se, its instillation into the vagina was found to significantly increase the lactobacilli counts in two separate studies (Reid *et al.*, 1994, 1995). This was believed to aid in reduction of urinary tract infections through indigenous lactobacilli interference of pathogen ascension from the vagina into the bladder. However, in many patients with recurrent urogenital infections, the lactobacilli count is low or absent, and therefore a prebiotic application would be less likely to function. Further *in vitro* studies have been performed to develop a new prebiotic formula that would allow lactobacilli to grow while not supporting growth of pathogens and not having any toxicity effects against the host. This task proved more difficult than was foreseen, especially as *Candida albicans* was able to grow in most media combinations. Nevertheless, the data suggested that some vitamins, mineral and other compounds had the potential to be prebiotics (Reid *et al.*, 1988). The ultimate test of this approach will be to test if such prebiotics override the local mucosal factors (nutrients, receptor sites) used by pathogens to colonise and infect the host.

9.7 Summary

In summary, probiotic applications to areas of the body other than the intestine have a long history, likely starting with treatment of wounds and throat infections in paleolithic times. The resurgence of interest in this area is multidimensional, in part due to antibiotic failures and side effects, increasing infection rates despite current therapies, and a desire for more 'natural' therapies. For the most part, probiotics have been sold as foods and dietary supplements where health claims are limited by bureaucratic governmental agencies. Thus, some probiotics proven to be better than or equivalent to so-called standard drug therapy, (such as antibiotics which receive approval with known side effects) and better than placebo, will have to be introduced as medicinal drugs. As pharmaceutical companies are now entering this area, we can look out for many new product formulations, some using dead bacteria (not probiotic but derivatives of), others comprising recombinant bacteria and others bases upon by-products from probiotic cells. The potential of this technology is enormous, as is the onus on scientists and companies to maintain high standards in bringing products to market, and making appropriate and proven claims.

There is scope for developing novel prebiotic applications for extra intestinal sites. However, several sites are not ideal environments for prebiotic application due to rapid transit times (for instance in the oral cavity) and the resultant difficulties in achieving persistence of the bioactive agent. The use of prebiotics to complement probiotic activity (the 'synbiotic' concept) (Gibson and Roberfroid, 1995) is worthy of consideration, as shown from at least one study on recurrent urinary tract infection (Reid *et al.*, 2001a). The careful selection of the carbohydrate component of a synbiotic can influence many aspects of the anti-disease process, for example changing the antimicrobial spectrum of the probiotic (Tzortzis *et al.*, 2004). Synbiotic combinations can also enhance survival of the chosen probiotic or indigenous beneficial organisms present (Rastall and Maitin, 2002). Without question, as the field of prebiotic and probiotic research and development continues to flourish, extra intestinal applications will be a focus of increased activity in the future.

Acknowledgements

Thanks to Ms Lindsay Boam for her research into the topic and Dr Duane Charbonneau for assistance in obtaining the anti-viral data. Funding from OMAF is appreciated.

References

Anand, S. K., Medellin-Pena, M., Wang, H., Griffiths, M. W. Probiotic bacteria control expressions of virulence genes in *E.coli* 0157: H7 through a signaling pathway. Manuscript submitted 2006.

Arques, J. L., Fernandez, J., Gaya, P., Nunez, M., Rodriguez, E., Medina, M. Antimicrobial activity of reuterin in combination with nisin against food-borne pathogens. *Int J Food Microbiol* 2004; **95** (2): 225–9.

Bruce, A. W., Chadwick, P., Hassan, A., VanCott, G. F. Recurrent urethritis in women. *Can Med Assoc J* 1973; **108** (8): 973–6.

Brummel, T., Ching, A., Seroude, L., Simon, A. F., Benzer, S. Drosophila lifespan enhancement by exogenous bacteria. *Proc Natl Acad Sci USA* 2004; **101** (35): 12974–9.

Cadieux, P., Burton, J., Kang, C. Y., Gardiner, G., Braunstein, I., Bruce, A. W., Reid, G. *Lactobacillus* strains and vaginal ecology. *JAMA* 2002; **287**: 1940–1.

Cauci, S., Guaschino, S., De Aloysio, D., Driussi, S., De Santo, D., Penacchioni, P., Quadrifoglio, F. Interrelationships of interleukin-8 with interleukin-1beta and neutrophils in vaginal fluid of healthy and bacterial vaginosis positive women. *Mol Hum Reprod* 2003; **9** (1): 53–8.

Chen, Y. C., Chang, W. H., Chang, Y., Huang, C. M., Sung, H. W. A natural compound (reuterin) produced by *Lactobacillus reuteri* for hemoglobin polymerization as a blood substitute. *Biotechnol Bioeng* 2004; **87** (1): 34–42.

Colodner, R., Edelstein, H., Chazan, B., Raz, R. Vaginal colonization by orally administered *Lactobacillus rhamnosus* GG. *Isr Med Assoc J* 2003; **5** (11): 767–9.

Devillard, E., Burton, J. P., Hammond, J.-A., Lam, D., Reid, G. Novel insight into the vaginal microflora in postmenopausal women under hormone replacement therapy as analyzed by PCR-Denaturing Gradient Gel Electrophoresis. *Eur J Obstet Gynecol* 2004; **117** (1): 76–81.

Donlan, R. M., Costerton, J. W. Biofilms: survival mechanisms of clinically relevant microorganisms. *Clin Microbiol Rev* 2002; **15** (2): 167–93.

FAO/WHO. Evaluation of health and nutritional properties of powder milk and live lactic acid bacteria. Food and Agriculture Organization of the United Nations and World Health Organization Expert Consultation Report. http://www.fao.org/es/ESN/Probio/probio.htm posted 2001.

Gan, B. S., Kim, J., Reid, G., Cadieux, P., Howard, J. C. *Lactobacillus fermentum* RC-14 inhibits *Staphylococcus aureus* infection of surgical implants in rats. *J Infect Dis* 2002; **185**: 1369–72.

Gibson, G. R., Roberfroid, M. B. Dietary modulation of the human colonic microbiota: introducing the concept of prebiotics. *J Nutr* 1995; **125**: 1401–12.

Gluck, U., Gebbers, J. O. Ingested probiotics reduce nasal colonization with pathogenic bacteria (*Staphylococcus aureus, Streptococcus pneumoniae*, and beta-hemolytic streptococci). *Am J Clin Nutr* 2003; **77** (2): 517–20.

Holma, R., Salmenpera, P., Lohi, J., Vapaatalo, H., Korpela, R. Effects of *Lactobacillus rhamnosus* GG and *Lactobacillus reuteri* R2LC on acetic acid-induced colitis in rats. *Scand J Gastroenterol* 2001; **36** (6): 630–5.

Johnsborg, O., Diep, D. B., Nes, I. F. Structural analysis of the peptide pheromone receptor PlnB, a histidine protein kinase from *Lactobacillus plantarum. J Bacteriol* 2003; **185** (23): 6913–20.

Kalliomaki, M., Salminen, S., Arvilommi, H., Kero, P., Koskinen, P., Isolauri, E. Probiotics in primary prevention of atopic disease: a randomised placebo-controlled trial. *Lancet* 2001; **357** (9262): 1076–9.

Kawai, Y., Ishii, Y., Arakawa, K., Uemura, K., Saitoh, B., Nishimura, J., Kitazawa, H., Yamazaki, Y., Tateno, Y., Itoh, T., Saito, T. Structural and functional differences in two cyclic bacteriocins with the same sequences produced by lactobacilli. *Appl Environ Microbiol* 2004; **70** (5): 2906–11.

Kelly, M. C., Mequio, M. J., Pybus, V. Inhibition of vaginal lactobacilli by a bacteriocin-like inhibitor produced by *Enterococcus faecium* 62-6: potential significance for bacterial vaginosis. *Infect Dis Obstet Gynecol* 2003; **11** (3): 147–56.

Laughton, J. M., Devillard, E., Heinrichs, D. E., Reid, G., McCormick, J. K. Inhibition of expression of a staphylococcal superantigen-like protein by a secreted signaling factor from *Lactobacillus reuteri* R.C. 14. *Microbiology* 2006. In press.

Lodinova-Zadnikova, R., Cukrowska, B., Tlaskalova-Hogenova, H. Oral administration of probiotic *Escherichia coli* after birth reduces frequency of allergies and repeated infections later in life (after 10 and 20 years). *Int Arch Allergy Immunol* 2003; **131** (3): 209–11.

Maassen, C. B., van Holten-Neelen, C., Balk, F., den Bak-Glashouwer, M. J., Leer, R. J., Laman, J. D., Boersma, W. J., Claassen, E. Strain-dependent induction of cytokine profiles in the gut by orally administered *Lactobacillus* strains. *Vaccine* 2000; **18** (23): 2613–23.

Mack, D. R., Ahrne, S., Hyde, L., Wei, S., Hollingsworth, M. A. Extracellular MUC3 mucin secretion follows adherence of *Lactobacillus* strains to intestinal epithelial cells *in vitro*. *Gut* 2003; **52** (6): 827–33.

Mastrandrea, F., Coradduzza, G., Serio, G., Minardi, A., Manelli, M., Ardito, S., Muratore, L. Probiotics reduce the CD34 + hemopoietic precursor cell increased traffic in allergic subjects. *Allerg Immunol (Paris)* 2004; **36** (4): 118–22.

McCormick, J., Laughton, J., Heinrichs, D., Reid, G. Anti-infective signaling in *Lactobacillus reuteri* RC-14. International Scientific Association for Probiotics and Prebiotics, Copper Mountain, Colorado, 30 August 2004.

Meluleni, G. J., Grout, M., Evans, D. J., Pier, G. B. Mucoid *Pseudomonas aeruginosa* growing in a biofilm *in vitro* are killed by opsonic antibodies to the mucoid exopolysaccharide capsule but not by antibodies produced during chronic lung infection in cystic fibrosis patients. *J Immunol* 1995; **155** (4): 2029–38.

Montalto, M., Vastola, M., Marigo, L., Covino, M., Graziosetto, R., Curigliano, V., Santoro, L., Cuoco, L., Manna, R., Gasbarrini, G. Probiotic treatment increases salivary counts of lactobacilli: a double-blind, randomized, controlled study. *Digestion* 2004; **69** (1): 53–6.

Nase, L., Hatakka, K., Savilahti, E., Saxelin, M., Ponka, A., Poussa, T., Korpela, R., Meurman, J. H. Effect of long-term consumption of a probiotic bacterium, *Lactobacillus rhamnosus* GG, in milk on dental caries and caries risk in children. *Caries Res* 2001; **35**: 412–20.

Nikawa, H., Makihira, S., Fukushima, H., Nishimura, H., Ozaki, Y., Ishida, K., Darmawan, S., Hamada, T., Hara, K., Matsumoto, A., Takemoto, T., Aimi, R. *Lactobacillus reuteri* in bovine milk fermented decreases the oral carriage of mutans streptococci. *Int J Food Microbiol* 2004; **95** (2): 219–23.

Rastall, R. A., Maitin, V. Prebiotics and synbiotics: towards the next generation. *Curr Opin Biotechnol* 2002; **13** (5): 490–8.

Reid, G., Bocking, A. The potential for probiotics to prevent bacterial vaginosis and preterm labor. *Am J Obstet Gynecol* 2003; **189** (4): 1202–8.

Reid, G., Bruce, A. W. Selection of *Lactobacillus* strains for urogenital probiotic applications. *J Infect Dis* 2001; **183** (S1): S77–80.

Reid, G., Bruce, A. W. Urogenital infections in women – Can probiotics help? *Postgrad Med J* 2003; **79**: 429–32.

Reid, G., Burton, J. Use of *Lactobacillus* to prevent infection by pathogenic bacteria. *Microbes Infect* 2002; **4**: 319–24.

Reid, G., Bruce, A. W., Taylor, M. Instillation of *Lactobacillus* and stimulation of indigenous organisms to prevent recurrence of urinary tract infections. *Microecol Ther* 1995; **23**: 32–45.

Reid, G., Bruce, A. W., Tomeczek, L. The role of probiotics in the urogenital tract. In *Human Health: The Contribution of Microorganisms* S. A. W. Gibson (ed.). Springer Series in Applied Biology. Springer Verlag IFAB Communications, 1994, Ch. 7, pp. 111–8.

Reid, G., Cook, R. L., Bruce, A. W. Examination of strains of lactobacilli for properties which may influence bacterial interference in the urinary tract. *J Urol* 1987; **138**: 330–5.

Reid, G., Bruce, A. W., Soboh, F., Mittelman, M. Effect of nutrient composition on the *in vitro* growth of urogenital *Lactobacillus* and uropathogens. *Can J Microbiol* 1988; **44**: 1–6.

Reid, G., Bruce, A. W., Fraser, N., Heinemann, C., Owen, J., Henning, B. Oral probiotics can resolve urogenital infections. *FEMS Microbiol Immunol* 2001a; **30**: 49–52.

Reid, G., Beuerman, D., Heinemann, C., Bruce, A. W. Probiotic *Lactobacillus* dose required to restore and maintain a normal vaginal flora. *FEMS Immunol Med Microbiol* 2001b; **32**: 37–41.

Reid, G., Charbonneau, D., Gonzalez, S., Gardiner, G., Erb, J., Bruce, A. W. Ability of *Lactobacillus* GR-1 and RC-14 to stimulate host defences and reduce gut translocation and infectivity of *Salmonella typhimurium. Nutraceut Food* 2002; **7**: 168–73.

Reid, G., Jass, J., Sebulsky, T., McCormick, J. Potential uses of probiotics in clinical practice. *Clin Microbiol Rev* 2003a; **16** (4): 658–72.

Reid, G., Charbonneau, D., Erb, J., Kochanowski, B., Beuerman, D., Poehner, R., Bruce, A. W. Oral use of *Lactobacillus rhamnosus* GR-1 and *L. fermentum* RC-14 significantly alters vaginal flora: randomized, placebo-controlled trial in 64 healthy women. *FEMS Immunol Med Microbiol* 2003b; **35**: 131–4.

Reid, G., Burton, J., Hammond, J.-A., Bruce, A. W. Nucleic acid based diagnosis of bacterial vaginosis and improved management using probiotic lactobacilli. *J Med Food* 2004; **7** (2): 223–8.

Romagnani, S. The increased prevalence of allergy and the hygiene hypothesis: missing immune deviation, reduced immune suppression, or both? *Immunology* 2004; **112** (3): 352–63.

Roos, K., Holm, S. E., Grahn, E., Lind, L. Alpha-streptococci as supplementary treatment of recurrent streptococcal tonsillitis: a randomized placebo-controlled study. *Scand J Infect Dis* 1993; **25** (1): 31–5.

Rosenfeldt, V., Benfeldt, E., Nielsen, S. D., Michaelsen, K. F., Jeppesen, D. L., Valerius, N. H., Paerregaard, A. Effect of probiotic *Lactobacillus* strains in children with atopic dermatitis. *J Allergy Clin Immunol* 2003; **111** (2): 389–95.

Sewankambo, N., Gray, R. H., Wawer, M. J., Paxton, L., McNaim, D., Wabwire-Mangen, F., Serwadda, D., Li, C., Kiwanuka, N., Hillier, S. L., Rabe, L., Gaydos, C. A., Quinn, T. C., Konde-Lule, J. HIV-1 infection associated with abnormal vaginal flora morphology and bacterial vaginosis. *Lancet* 1997; **350** (9077): 546–50.

Sprunt, K., Leidy, G. The use of bacterial interference to prevent infection. *Can J Microbiol* 1988; **34** (3): 332–8.

Sprunt, K., Redman, W. Evidence suggesting importance of role of interbacterial inhibition in maintaining balance of normal flora. *Ann Intern Med* 1968; **68** (3):579–90.

Stanaland, B. E. Therapeutic measures for prevention of allergic rhinitis/asthma development. *Allergy Asthma Proc* 2004; **25** (1): 11–5.

Sung, H. W., Chen, C. N., Liang, H. F., Hong, M. H. A natural compound (reuterin) produced by *Lactobacillus reuteri* for biological-tissue fixation. *Biomaterials* 2003; **24** (8): 1335–47.

Tzortzis, G., Baillon, M. L. A., Gibson, G. R., Rastall, R. A. Modulation of anti-pathogenic activity in canine derived *Lactobacillus* species by carbohydrate growth substrate. *J Appl Microbiol* 2004; **96**: 552–9.

Velraeds, M. C., van der Mei, H. C., Reid, G., Busscher, H. J. Inhibition of initial adhesion of uropathogenic *Enterococcus faecalis* by biosurfactants from *Lactobacillus* isolates. *Appl Environ Microbiol* 1996; **62**: 1958–63.

10

Prebiotic Impacts on Companion Animals

Kelly S. Swanson and George C. Fahey Jr

10.1 Introduction

As is the case for other animals, the gastrointestinal microbiota of the dog and cat influence overall health of the host animal. In addition to their role in gastrointestinal immunity, commensal microbial populations are critically important in pathogenic resistance and production of short-chain fatty acids (SCFA), one of which (butyrate) is the major energy source of colonocytes (Roediger, 1980). Also, intestinal microbes are capable of modulating the expression of genes associated with several important intestinal functions such as nutrient absorption, mucosal barrier fortification, xenobiotic metabolism, angiogenesis, and postnatal intestinal maturation (Hooper *et al.*, 2001). As developments in molecular biology and nanotechnology are increasingly applied to the gastrointestinal tract, researchers will continue to make discoveries important to understanding microbe–host relationships. Although these relationships are not well understood at present, appropriate gut microbial balance is crucial for long-term health.

Dietary ingredient composition, in addition to host genotype and environmental factors, is key in establishing the commensal microbiota population and maintaining balance over time. The type of endogenous and exogenous substrates that avoid enzymatic/hydrolytic digestion in the stomach and small intestine, and thus reach the colon, greatly affect the compounds produced by microbial fermentation. Microbial fermentation of carbohydrates is important for intestinal health, as it leads to the production of SCFA. SCFA limit growth of pathogenic microbes by decreasing luminal pH, and by influencing intestinal gene expression. In contrast, microbial degradation of

Prebiotics: Development and Application Edited by G. R. Gibson and R. A. Rastall

nitrogenous compounds results in the production of numerous putrefactive catabolites (e.g., ammonia, phenols, sulfate-containing compounds) that are implicated as the major odour components of faeces (Spoelstra, 1980; O'Neill and Phillips, 1992). More importantly, many of these protein catabolites (e.g., phenols, sulfate-containing compounds) have been associated with intestinal disease.

The dietary inclusion of prebiotics, nondigestible ingredients that selectively stimulate the growth and/or activity of a limited number of bacteria in the colon, has been a popular approach used to attain and maintain the microbial balance in humans (Gibson and Roberfroid, 1995). Although the canine and feline prebiotic literature is not as robust as that of humans, this approach has gained steady support over the past decade. This chapter was written to provide readers with a summary of canine and feline prebiotic experiments published to date. In addition, Tables 10.1 (canine experiments) and 10.2 (feline experiments) were constructed to provide quantitative information on each experiment individually.

10.2 *In Vitro* Prebiotic Experiments Using Canine and Feline Fecal Inoculum

Although numerous *in vitro* fermentation experiments using canine and feline inoculum have been performed, few have used this approach to evaluate prebiotics. Vickers *et al.* (2001) were the first to use an *in vitro* fermentation to test prebiotic fermentability quantified as SCFA production. In this experiment, substrates were fermented at 39 °C for 6, 12 and 24 h. Lactate and SCFA concentrations were determined and used as an indicator of fermentability. In addition to beet pulp, cellulose, and soy fibre, common fibre sources in pet diets, substrates included yeast cell wall (YCW) containing mannanoligosaccharides (MOS), short-chain fructooligosaccharides (scFOS), and four inulin sources. Inulin sources varied as regards degrees of polymerisation (dp) and solubility: (1) dp = 2–8; (2) dp = 9, but insoluble in water; (3) dp = 9, but soluble in water; and (4) dp > 12. While YCW fermentation resulted in moderate concentrations of total SCFA, all inulin sources and scFOS were highly fermentable. Very low concentrations of lactate were produced as a result of YCW fermentation. In contrast, fermentation of scFOS and all inulin sources resulted in greater ($P < 0.05$) lactate concentrations than YCW and control fibre sources. Fructan chain length (dp) also affected SCFA production over time, with scFOS resulting in greater ($P < 0.05$) lactate concentrations than three of the four inulin sources after 12 and 24 h of fermentation. Results from this experiment demonstrate the prebiotic nature of fructans, with rapid production of SCFA and lactate. YCW, however, was only moderately fermentable and lacked the ability to promote growth of lactic acid bacteria.

Tzortzis *et al.* (2004a) evaluated the effect of various carbohydrate sources, including prebiotics, on the production of extracellular antagonistic compounds against two *Escherichia coli* strains and *Salmonella enterica* serotype Typhimurium by three canine-derived lactobacilli strains. *Lactobacillus* strains tested included *L. mucosae*, *L. acidophilus* and *L. reuteri*. Carbohydrates screened included: (1) Biotose®, a starch sweetener consisting of 41 % glucose, 20 % isomaltose, 9 % isomaltotriose, 9 % panose, 7 % maltose, and 14 % other sugars; (2) cellobiose; (3) gentiobiose; (4) glucose;

Table 10.1 *In vivo experiments, listed in chronological order, evaluating the effects of prebiotics in dogs*

Reference	Outcome variables quantified	Animals/ treatment	Dietary information	Daily prebiotic dose; source	Time on treatment	Major findings
Terada *et al.*, 1992	(a) Fecal characteristics (b) Fecal metabolites (c) Fecal microbial ecology	8 adults (4 M; 4 F)	(a) Diet type: dry (b) Primary ingredients: not provided (c) Chemical composition: 92 % DM, 26 % CP, 11 % fat, 48 % NFE, 22 % CF and 6 % ash	(a) Control (no prebiotic) (b) 1.5 g lactosucrose	2 weeks	Lactosucrose: (a) ↑ bifidobacteria** (b) ↓ lecithinase-positive clostridia*** (c) ↓ fecal DM %** (d) ↓ fecal ammonia, phenol, ethylphenol, indole and skatole concentrations** (e) ↓ fecal butyrate concentrations**
Willard *et al.*, 1994	(a) Duodenal fluid microbial ecology (b) Duodenal mucosal microbial ecology	16 M and F IgA- deficient adults	(a) Diet type: dry (b) Primary ingredients: poultry meat, PBM, corn, poultry fat, rice, and beet pulp (c) Chemical composition: 31 % CP, 22 % fat and 6 % ash	(a) Control (no prebiotic) (b) 1 % scFOS	46–51 days	scFOS: (a) ↓ aerobic and anaerobic bacteria in intestinal tissue** (b) ↓ aerobic bacteria in the duodenal/ jejunal fluid**
Diez *et al.*, 1997	(a) Fecal characteristics (b) Blood biochemistry (c) Total tract nutrient digestibility	8 M adults	(a) Diet type: dry (b) Primary ingredients: beef, corn, and corn oil (c) Chemical composition: not provided	(a) Control (no prebiotic) (b) 4 % scFOS + 1 % beet pulp (c) 8 % scFOS + 2 % beet pulp	6 weeks	scFOS/beet pulp blend: (a) ↑ wet fecal volume** (b) ↓ fecal DM %** (c) ↓ total tract CP digestibility** (d) ↓ preprandial urea, cholesterol and triglyceride concentrations** (e) ↓ postprandial glucose, urea and triglyceride concentrations

Table 10.1 *(Continued)*

Reference	Outcome variables quantified	Animals/ treatment	Dietary information	Daily prebiotic dose; source	Time on treatment	Major findings
Diez *et al.*, 1998	(a) Fecal characteristics (b) Blood biochemistry (c) Total tract nutrient digestibility	8 adults 2 M; 6 F)	(a) Diet type: dry (b) Primary ingredients: beef, corn, corn oil (c) Chemical composition: 23–25 % CP, 13–14 % fat, and 10–16 % TDF	(a) Control (no prebiotic) (b) 7 % beet fiber (c) 7 % guar gum (d) 7 % inulin	4 weeks	Inulin: (a) ↑ wet fecal volume** (b) ↓ fecal DM %** (c) ↓ total tract OM, CP and fat digestibility**
Russell, 1998	Study 1 (a) Fecal characteristics (b) Fecal metabolites (c) Fecal microbial ecology (d) Total tract nutrient digestibility	Study 1 8 adults (gender not defined)	Study 1 (a) Diet type: dry (b) Primary ingredients: soy only ingredient mentioned (c) Chemical composition: not provided	Study 1 (a) 12 % soy (control) (b) 12 % soy + 1 % scFOS (c) 12 % soy + 3 % chicory	Study 1 4 weeks	Study 1 scFOS: (a) ↓ fecal clostridia** (b) ↑ fecal bifidobacteria** Chicory: (a) ↓ fecal clostridia* (b) ↑ fecal bifidobacteria** (c) ↑ fecal volume**
	Study 2 (a) Fecal characteristics (b) Fecal microbial ecology (d) Total tract nutrient digestibility	Study 2 8 adults (gender not defined)	Study 2 (a) Diet type: dry (b) Primary ingredients: soy-free (c) Chemical composition: not provided	Study 2 (a) Control (no prebiotic) (b) 1.5 % chicory (c) 3 % chicory (d) 5 % chicory	Study 2 4 weeks	Study 2 All chicory concentrations: (a) ↑ fecal bifidobacteria**
	Study 3 (a) Fecal characteristics (b) Fecal microbial ecology (c) Total tract nutrient digestibility	Study 3 8 adults (gender not defined)	Study 3 (a) Diet type: dry (b) Primary ingredients: not provided (c) Chemical composition: not provided	Study 3 (a) No prebiotic; no soy (b) 0.5 % chicory; no soy (c) 2 % chicory; no soy (d) 4 % chicory; no soy (e) 0 % chicory; 6 % soy (f) 0 % chicory; 12 % soy (g) 0.5 % chicory; 6 % soy (h) 2 % chicory; 6 % soy (i) 4 % chicory; 6 % soy	Study 3 4 weeks	Study 3 All chicory concentrations: (a) ↑ fecal bifidobacteria** 2 % and 4 % chicory: (b) ↑ fecal volume** (c) ↓ total tract nutrient digestibility**

			(j) 0.5 % chicory; 12 % soy (k) 2 % chicory; 12 % soy (l) 4 % chicory; 12 % soy			
Buddington *et al.*, 1999	(a) Blood biochemistry (b) Gut morphology	5 adults (M and F)	(a) Diet type: dry (b) Primary ingredients: poultry meat, PBM, rice, poultry fat and fish meal (c) Chemical composition: 94 % DM, 26 % CP, 12 % fat, and 6 % ash	(a) Control (3.6 % cellulose) (b) 1 % OF + 4.2 % beet pulp	6 weeks	OF/beet pulp blend: (a) ↑ intestinal length, surface area, wet weight and dry weight** (b) ↑ carrier-mediated glucose uptake by small intestine**
Howard *et al.*, 1999	(a) Colonic blood flow (b) Gut morphology	7 F adults	(a) Diet type: dry (b) Primary ingredients: pregelatinized cornstarch, PBM, poultry fat (c) Chemical composition: not provided	(a) 6 % cellulose (b) 6 % beet pulp (c) 1.5 % scFOS (d) Fiber blend (6 % beet pulp, 2 % gum talha, and 1.5 % scFOS)	35 days	scFOS: (a) ↑ colonic blood flow** (b) ↓ proximal colon proliferation***
Flickinger *et al.*, 2000	(a) Fecal characteristics (b) Fecal microbial ecology (c) Ileal nutrient digestibility (d) Total tract nutrient digestibility	6 F adults	(a) Diet type: enteral formula (b) Primary ingredients: sodium caseinate, soy protein isolate, corn oil, corn syrup and sucrose (c) Chemical composition: 97 % OM, 15–16 % CP and 15 % fat	(a) Control (no prebiotic) (b) 6 % GOS (c) 6 % MD	14 days	GOS: (a) ↑ fat and glucose intake** (b) ↓ ileal glucose digestibility** (c) ↓ total tract OM, CP, carbohydrate and glucose digestibility** (d) ↑ wet fecal volume and fecal score (softer feces)** MD: (a) ↑ fat and glucose intake** (b) ↓ ileal glucose digestibility** (c) ↓ total tract CP digestibility** (d) ↑ wet and dry fecal volume and fecal score (softer feces)**

Table 10.1 *(Continued)*

Reference	Outcome variables quantified	Animals/ treatment	Dietary information	Daily prebiotic dose; source	Time on treatment	Major findings
Howard *et al.*, 2000	(a) Fecal characteristics (b) Small and large intestinal microbial ecology (c) N metabolism (d) Total tract nutrient digestibility	7 F adults	(a) Diet type: dry (b) Primary ingredients: pregelatinised cornstarch, PBM and poultry fat (c) Chemical composition: not provided	(a) 6 % cellulose (b) 6 % beet pulp (c) 1.5 % scFOS (d) Fiber blend (6 % beet pulp, 2 % gum talha, and 1.5 % scFOS)	35 days	scFOS: (a) ↑ total tract DM digestibility* (b) ↓ DM intake (as % BW)** (c) ↑ fecal microbial N excretion (as % N intake)** (d) ↑ fecal microbial N excretion (as % fecal N output and as % of total N output)* (e) ↑ aerobic bacteria in distal colon* (f) ↑ ileal *Clostridium* spp.*** Fiber blend: (a) ↑ N output (g day^{-1})** (b) ↑ fecal microbial N excretion (g day^{-1} and as % N intake)** (c) ↑ fecal microbial N excretion (as % total N output)*
Strickling *et al.*, 2000	(a) Fecal characteristics (b) Fecal metabolites (c) Fecal microbial ecology (d) Ileal metabolites (e) Ileal microbial ecology (f) Ileal nutrient digestibility (g) N metabolism (h) Total tract nutrient digestibility	7 F ileal cannulated adults	(a) Diet type: dry (b) Primary ingredients: corn, PM, SBM, cornstarch, white grease and corn gluten (c) Chemical composition: 30 % CP, 15 % fat, 48 % NFE and 1.5 % CF	(a) Control (no prebiotic) (b) 0.5 % OF (c) 0.5 % MOS (d) 0.5 % XOS	21 days	FOS, MOS, and XOS: (a) ↑ fecal DM %* (b) ↓ ileal butyrate concentrations MOS: (a) ↑ ileal propionate and ↓ ileal butyrate vs. XOS and FOS (b) ↓ fecal *C. perfringens* vs. XOS and FOS

Willard *et al.*, 2000	(a) Fecal microbial ecology	6 F adults	(a) Diet type: dry (b) Primary ingredients: cornstarch, poultry meat, and PBM (c) Chemical composition: 90–93 % DM, 28–29 % CP, 7–8 % fat, 47–49 % NFE, 1–2 % CF and 6 % ash	(a) Control (no prebiotic) (b) 1 % FOS	28–56 days	No significant biological effects
Beynen *et al.*, 2002	(a) Fecal characteristics (b) Fecal microbial ecology (c) Mineral absorption (d) N metabolism	5 adults (4 M and 1 F)	(a) Diet type: dry (b) Primary ingredients: cornstarch, PBM, poultry fat, rice, corn and beet pulp (c) Chemical composition: 88–93 % DM, 28–29 % CP, 16–17 % fat, 2 % CF and 7–8 % ash	(a) Control (no prebiotic) (b) 1 % OF	3 weeks	OF: (a) ↑ fecal total anaerobic and aerobic bacteria, streptococci, clostridia and bifidobacteria** (b) ↑ fecal lactobacilli* (c) ↑ apparent Ca and Mg absorption**
Swanson *et al.*, 2002a	Study 1 (a) Fecal characteristics (b) Fecal metabolites (c) Fecal microbial ecology (d) Total tract nutrient digestibility	Study 1 5 adults (M and F)	Study 1 (a) Diet type: dry (b) Primary ingredients: pregelatinised cornstarch, MBM, PBM, poultry fat, and beet pulp (c) Chemical composition: 94 % DM, 87 % OM, 24 % CP, 18 % fat, 6 % TDF and 13 % ash	Study 1 (a) Control (no prebiotic or probiotic) (b) 4 g scFOS (c) 2×10^9 CFU *Lactobacillus acidophilus* probiotic (d) 4 g scFOS + 2×10^9 CFU *L. acidophilus*	Study 1 28 days	Study 1 scFOS: (a) ↑ fecal propionate concentrations*** (b) ↑ fecal butyrate and lactate concentrations** (c) ↓ fecal *C. perfringens** (d) ↓ fecal indole, phenols and total indole and phenol concentrations*
	Study 2 (a) Fecal characteristics (b) Fecal metabolites (c) Fecal microbial ecology (d) Total tract nutrient digestibility	Study 2 5 adults (M and F)	Study 2 (a) Diet type: dry (b) Primary ingredients: pregelatinised cornstarch, MBM, PBM, poultry fat and beet pulp (c) Chemical composition: 94 % DM, 87 % OM,	Study 2 (a) Control (no prebiotic or probiotic) (b) 4 g scFOS (c) 2×10^9 CFU *L. acidophilus* (d) 4 g scFOS + 2×10^9 CFU *L. acidophilus*	Study #2 28 days	Study 2 scFOS: (a) ↑ fecal butyrate concentrations** (b) ↑ lactate concentrations* (c) ↑ fecal total aerobic bacteria and bifidobacteria** (d) ↑ fecal lactobacilli*

Table 10.1 *(Continued)*

Reference	Outcome variables quantified	Animals/ treatment	Dietary information	Daily prebiotic dose; source	Time on treatment	Major findings
			24 % CP, 18 % fat, 6 % TDF and 13 % ash			(e) ↓ fecal isobutyrate, isovalerate and total BCFA concentrations*** (f) ↓ fecal indole concentrations*
Swanson *et al.*, 2002b	(a) Fecal characteristics (b) Fecal metabolites (c) Fecal microbial ecology (d) Ileal nutrient digestibility (e) Immune characteristics (f) Total tract nutrient digestibility	4 F ileal cannulated adults	(a) Diet type: dry (b) Primary ingredients: PBM, rice, poultry fat, and beet pulp (c) Chemical composition: 92 % DM, 87 % OM, 37 % CP, 21 % fat and 5 % TDF	(a) Control (no prebiotic) (b) 2 g scFOS (c) 2 g MOS (d) 2 g scFOS +2 g MOS	14 days	scFOS: (a) ↓ fecal indole concentrations* (b) ↓ total phenol and indole concentrations** MOS: (a) ↑ fecal pH* (b) ↓ fecal total aerobic bacteria* (c) ↑ blood lymphocyte %** scFOS and MOS: (a) ↓ fecal total anaerobic bacteria* (b) ↑ ileal IgA concentrations (mg g^{-1} DM or mg g^{-1} CP)* (c) ↓ fecal indole concentrations* (d) ↓ total phenol and indole concentrations**
Swanson *et al.*, 2002c	(a) Ileal microbial ecology (b) Immune characteristics (c) Fecal microbial ecology	8 F ileal cannulated adults	(a) Diet type: dry (b) Primary ingredients: PBM, rice, poultry fat and beet pulp (c) Chemical composition: 92 % DM, 87 % OM, 37 % CP, 21 % fat and 5 % TDF	(a) Control (no prebiotic) (b) 4 g scFOS +2 g MOS	14 days	scFOS and MOS: (a) ↑ ileal lactobacilli** (b) ↑ fecal total aerobic bacteria, lactobacilli, and bifidobacteria** (c) ↑ blood lymphocyte % and ↓ blood neutrophil concentration and %*

Zentek *et al.*, 2002	(a) Fecal characteristics (b) Fecal metabolites (c) Total tract nutrient digestibility	4 F adults	(a) Diet type: dry (b) Primary ingredients: GM, rice, soy oil and fish meal (c) Chemical composition: 95 % DM, 37 % CP, 24 % fat, 5 % CF and 4 % ash	(a) Control (no prebiotic) (b) 1 g kg^{-1} BW MOS (c) 1 g kg^{-1} BW TGOS (d) 1 g kg^{-1} BW lactose (e) 1 g kg^{-1} BW lactulose	15 days	MOS: (a) ↓ total tract DM, CP and NFE digestibilities** (b) ↑ total tract CF digestibility** (c) ↓ fecal DM %, unbound water and pH** (d) ↓ fecal ammonia concentrations** TGOS: (a) ↑ total tract CF digestibility** Lactulose: (b) ↑ total tract CF digestibility**
Flickinger *et al.*, 2003	Study 1 (a) Fecal characteristics (b) Fecal metabolites (c) Fecal microbial ecology (d) Total tract nutrient digestibility (e) Urine metabolites	Study 1 4 M adults	Study 1 (a) Diet type: dry. (b) Primary ingredients: corn, wheat, MBM, SBM, beef tallow and CGM (c) Chemical composition: 96 % DM, 91 % OM, 25 % CP and 7 % fat	Study 1 (a) Control (no prebiotic) (b) 0.3 % OF (c) 0.6 % OF (d) 0.9 % OF	Study 1 22 days	Study 1 OF: (a) ↓ total tract fat digestibility*** (b) ↓ total tract DM and OM digestibilities** (c) ↓ total tract CP digestibility* (d) ↓ fecal ammonia concentrations* (e) ↑ fecal propionate concentrations** (f) ↑ fecal total SCFA concentrations* (g) ↓ fecal total aerobic bacteria*
	Study 2 (a) Fecal characteristics (b) Fecal metabolites (c) Fecal microbial ecology (d) Ileal nutrient digestibility (e) Total tract nutrient digestibility	Study 2 4 F ileal cannulated adults	Study 2 (a) Diet type: dry (b) Primary ingredients: PBM, corn, rice, poultry fat, fish oil, soy oil, beef tallow and beet pulp (c) Chemical composition: 93 % DM, 93 % OM, 34 % CP and 23 % fat	Study 2 (a) Control (no prebiotic) (b) 1 g scFOS (c) 2 g scFOS (d) 3 g scFOS	Study 2 14 days	Study 2 Linear trends of scFOS: (a) ↓ wet and dry fecal output (as % of DMI)* (b) ↓ fecal score (harder faeces)* (c) ↑ fecal total aerobic bacteria** (d) ↓ fecal *C. perfringens***

Table 10.1 *(Continued)*

Reference	Outcome variables quantified	Animals/ treatment	Dietary information	Daily prebiotic dose; source	Time on treatment	Major findings
Hesta *et al.*, 2003	(a) Fecal characteristics (b) N metabolism (c) Total tract nutrient digestibility	8 adults (gender not defined)	(a) Diet type: dry (b) Primary ingredients: MBM, GM, or PM (c) Chemical composition: 91–93 % DM, 27–50 % CP, 6–10 % fat and 4–16 % ash	(a) MBM control (b) MBM + 3 % OF or 3 % IMO (c) GM control (d) GM + 3 % OF or 3 % IMO (e) PM control (f) PM + 3 % OF or 3 % IMO	14 days	OF or IMO: (a) ↑ wet and dry fecal volume** (b) ↓ total tract DM, fat, and CP digestibilities** (c) ↑ fecal bacterial N (as % of fecal DM, % of fecal N and N intake)** (d) ↑ fecal ammonia production (g day^{-1})**
Propst *et al.*, 2003	(a) Fecal characteristics (b) Fecal metabolites (c) Ileal nutrient digestibility (d) Total tract nutrient digestibility	7 F ileal cannulated adults	(a) Diet type: dry (b) Primary ingredients: PBM, rice, poultry fat and beet pulp (c) Chemical composition: 93 % DM, 93 % OM, 33 % CP and 24 % fat	(a) Control (no prebiotic) (b) 0.3 % OF (c) 0.3 % inulin (d) 0.6 % OF (e) 0.6 % inulin (f) 0.9 % OF (g) 0.9 % inulin	14 days	OF: (a) ↓ total tract DM, OM and CP digestibilities** (b) ↑ wet fecal volume* (c) ↑ fecal ammonia concentrations*** (d) ↑ fecal acetate, propionate, butyrate and total SCFA concentrations*** (e) ↓ fecal isovalerate concentrations*** Inulin: (a) ↓ total tract DM, OM and CP digestibilities** (b) ↑ fecal ammonia concentrations*** (c) ↑ fecal propionate, butyrate and total SCFA concentrations*** (d) ↑ fecal acetate concentrations* (e) ↑ fecal valerate concentrations* (f) ↑ isovalerate concentrations***

						(g) ↑ fecal phenylethylamine concentrations* (h) ↓ fecal phenol concentrations* (i) ↓ fecal total phenol concentrations**
Twomey *et al.*, 2003	(a) Fecal characteristics (b) Fecal metabolites (c) Total tract nutrient digestibility	6 adult dogs (M and F)	(a) Diet type: dry (b) Primary ingredients: wheat, barley, wheat bran and pollard and PBM (c) Chemical composition: 93–94 % DM, 13–14 % CP, 9–10 % fat and 41–48 % NFE	(a) Control (no prebiotic or enzyme) (b) Control + inulinase enzyme (c) 3 % OF (d) 3 % OF + inulinase (e) 6 % OF (f) 6 % OF + inulinase	13 days	OF: (a) ↓ total tract fat and GE digestibilities** (b) ↑ fecal score (softer feces)** (c) ↓ fecal pH and fecal DM concentration*** (d) ↑ fecal lactate concentrations** (e) ↑ total SCFA and lactate concentrations*
Zentek *et al.*, 2003	(a) Fecal characteristics (b) Fecal microbial ecology	9 adults (2 M and 7 F)	(a) Diet type: dry (b) Primary ingredients: GM, beef lung, corn, soy oil and cellulose (c) Chemical composition: 74–94 % DM, 25–73 % CP, 10–12 % fat and 2–4 % CF	(a) Control (no prebiotic) (b) 3 % inulin (c) 3 % glucose	21 days	Inulin: (a) ↑ fecal score (harder faeces)*** (b) ↑ fecal DM %*** (c) ↓ fecal pH*** (d) ↓ fecal *C. perfringens**** (e) ↑ fecal bifidobacteria*
Karr-Lilienthal *et al.*, 2004	(a) Fecal N and DAPA concentrations	Control: 14 adults (M and F) Treatments: 12 adults (M and F)	(a) Diet type: dry (b) Primary ingredients: corn, rice, SBM, MBM, poultry meal and CGM (c) Chemical composition: 91–92 % DM, 92–93 % OM, 23–25 % CP and 11–13 % fat	(a) Control (no prebiotic) (b) 1 % chicory (c) 1 % MOS (d) 1 % chicory + 1 % MOS	28 days	Chicory: (a) ↑ fecal N concentrations**

Table 10.1 *(Continued)*

Reference	Outcome variables quantified	Animals/ treatment	Dietary information	Daily prebiotic dose; source	Time on treatment	Major findings
Grieshop *et al.*, 2004	(a) Fecal characteristics (b) Fecal microbial ecology (c) Total tract nutrient digestibility (d) Immune characteristics	34 adults (20 M and 14 F)	(a) Diet type: dry (b) Primary ingredients: corn, rice, SBM, MBM, poultry meal and CGM (c) Chemical composition: 91–92 % DM, 92–93 % OM, 23–25 % CP and 11–13 % fat	(a) Control (no prebiotic) (b) 1 % chicory (c) 1 % MOS (d) 1 % chicory + 1 % MOS	28 days	Chicory: (a) ↑ total tract fat digestibility* (b) ↑ bifidobacteria** (c) ↑ blood neutrophil concentrations* MOS: (a) ↑ food intake* (b) ↑ wet fecal volume** (c) ↑ bifidobacteria** (d) ↓ *E. coli*** (e) ↓ blood lymphocyte concentrations* Chicory and MOS: (a) ↑ food intake* (b) ↑ wet fecal volume** (c) ↑ fecal score (softer faeces)** (d) ↓ blood lymphocyte concentrations** (e) ↑ blood neutrophil concentrations*

BCFA, branched-chain fatty acids; BW, body weight; CF, crude fiber; CFU, colony forming units; CGM, corn gluten meal; DAPA, diaminopimelic acid; DM, dry matter; DMI, dry matter intake; F, female; FOS, fructooligosaccharide; GE, gross energy; GM, greaves meal; GOS, α-glucooligosaccharide; IgA, immunoglobulin A; IMO, isomaltooligosaccharide; M, male; MBM, meat and bone meal; MD, maltodextrin-like oligosaccharide; MOS, mannanoligosaccharide-containing yeast cell wall; N, nitrogen; NFE, nitrogen free extract; OF, oligofructose; OM, organic matter; PBM, poultry by-product meal; SBM, soybean meal; SCFA, short-chain fatty acids; scFOS, short-chain fructooligosaccharide; TDF, total dietary fiber; PM, poultry meal; TGOS, transgalactooligosaccharide; XOS, xylooligosaccharide.

*$P < 0.10$; **$P < 0.05$; ***$P < 0.01$.

(5) isomaltooligosaccharides (IMO); (6) lactose; (7) laevan; (8) maltose; (9) melezitose; (10) melibiose; (11) palatinose; (12) Panorich®, a high panose syrup consisting of 30 % panose, 23 % glucose, 17 % maltose, 16 % branched oligosaccharides (dp $\geq$4), 9 % isomaltose, 3 % maltotriose and 2 % isomaltotriose; (13) Raffinose; (14) scFOS; (15) stachyose; (16) sucrose; (17) tagatose; (18) xylooligosaccharides (XOS); and (19) xylan. Initial screening was performed using bacteria/carbohydrate combinations to identify those that inhibited pathogenic growth in 24-h batch co-culture experiments. Each of the batch culture fermenters was inoculated with pure cultures of one of the *Lactobacillus* strains and one of the pathogens. Cell-free supernatants of co-cultures found to decrease pathogenic growth in the initial screening were tested further for pathogen inhibition in pure culture at neutral or acidic pH. Substrates reported to induce an inhibitory effect on pathogenic growth included: (1) Panorich® and maltose for *L. mucosae*; (2) Biotose® for *L. acidophilus*; and (3) Biotiose®, glucose, and maltose for *L. reuteri*. Further analyses carried out using supernatants from these compounds resulted in pathogenic inhibition when incubated in combination with all three strains of *Lactobacillus*. Overall, these results demonstrate the ability of some carbohydrates to aid in pathogen resistance. Because the results indicated that acidity was not the sole inhibitory agent, the authors suggested that the induction of bacteriocins is another possible mechanism of action. Inclusion of more carbohydrates having known 'prebiotic' activity in future experiments would be valuable.

These same researchers evaluated *in vitro* fermentation properties of prebiotics using canine faecal inocula (Tzortzis *et al*., 2004b). The main objective of this experiment was to evaluate *in vitro* fermentability of novel α-galactooligosaccharides that were synthesised using an enzyme extract from a canine *L. reuteri* strain. In this experiment, canine faecal inoculum was used to measure *in vitro* fermentability of a synthesised oligosaccharide (galactosyl melibiose mixture) vs. three reference carbohydrates [oligofructose (OF), melibiose, and raffinose]. Primary outcomes included bacterial growth (bifidobacteria, bacteroides, lactobacilli, clostridia, *E. coli* and eubacteria) as measured by fluorescent *in situ* hybridisation (FISH) and SCFA production after 10 and 24 h of fermentation. Compared with baseline values, all substrates increased total bacterial counts, bifidobacteria, and lactobacilli. In addition, OF fermentation decreased ($P < 0.05$) *E. coli* and increased ($P < 0.05$) clostridia concentrations. As expected, all substrates increased SCFA and lactate concentrations when compared with the baseline.

The primary objectives of a fourth *in vitro* experiment were to: (1) evaluate the ability of lactic acid bacteria in canine and feline faeces to degrade oxalate *in vitro*; and (2) evaluate the effects of various prebiotics on *in vitro* oxalate degradation by selected oxalate-degrading lactate-producing bacteria in canine faeces (Weese *et al*., 2004). Although several contributory factors have been identified, hyperoxaluria is believed to play the most important role in urinary calcium oxalate saturation, leading towards the formation of calcium oxalate uroliths (Ogawa *et al*., 2000). Urinary oxalate excretion is dependent on several factors including intestinal absorption. The role of intestinal microbes on oxalate degradation in dogs and cats has not hitherto been studied; however, experiments performed in other species suggest that microbial populations significantly impact upon intestinal oxalate concentrations. To attain the first objective, 20 canine and 19 feline faecal samples were collected and incubated with oxalate at 37 °C for 48 h to identify bacterial strains capable of degrading oxalate. From the initial experiment, five lactic acid bacterial strains were identified (two strains of *Leuconostoc lactis*, two strains of *Lactobacillus acidophilus*, and one strain of *Lactobacillus plantarum*). In a second experiment, various carbohydrates were incubated with these selected lactic acid bacteria and oxalate. Carbohydrates evaluated included: (1) arabinogalactan; (2) gum Arabic; (3) guar gum; (4) two sources of inulin; (5) lactitol; and (6) maltodextrin. Significant interspecies variation occurred in response to the inclusion of prebiotics, reinforcing the notion of testing numerous strains when evaluating prebiotic compounds. Nonetheless, inclusion of prebiotics increased ($P < 0.05$) oxalate degradation, suggesting a potential role in reducing calcium oxalate urolithiasis.

10.3 Canine Prebiotic Experiments

Twenty-three canine prebiotic publications may be found in the literature (Table 10.1). In addition to the various dosages used across experiments, the number of prebiotic types and/or forms tested in these experiments stands at 12. Common trends observed from these experiments are highlighted below. In contrast to many of the prebiotic experiments using rodents, most canine experiments have been designed to test prebiotics at low dietary concentrations (≤ 1 %). Although the responses observed in the canine literature are oftentimes less dramatic than those reported in rodents, concentrations greater than 1 % are not applicable to canine nutrition and health from the perspectives of digestive physiology and economics.

Four fructans have so far been tested in canines, including: (1) chicory (a natural source of long chain fructans); (2) inulin (up to 60 dp); (3) OF (dp 8–10); and (4) scFOS (dp 3–5). Fructans have been the most often tested prebiotics in dogs, with at least one being evaluated in 20 of the 23 published experiments. Other than YCW, that has been used in six canine experiments, the following have been evaluated only once: (1) α-galactooligosaccharides (GOS); (2) IMO; (3) lactosucrose; (4) lactulose; (5) maltodextrin-like oligosaccharides (MD); (6) transgalactooligosaccharides (TGOS); and (7) XOS. Thus, this summary is largely descriptive of fructan action in general. Even though these nondigestible carbohydrates have similar physicochemical properties, differentiating the types and forms from one another is extremely important in assessing mode of action.

Stool consistency and volume produced are important criteria used by consumers when selecting a dog food. Thus, food intake and faecal characteristics are important outcomes of prebiotic supplementation and have been reported in most experiments. When providing prebiotics at concentrations in the diet of 1–2 %, as occurs in most of the canine prebiotic experiments, food intake is generally unaffected. However, even at constant food intake across treatments, prebiotic supplementation may result in greater wet faecal volume, decreased faecal dry matter (DM) percentage, and reduced faecal pH. Although prebiotic consumption may result in the production of softer faeces, canine experiments have generated mixed results. Prebiotics have been reported to increase stool frequency and moisture content in humans, thereby effectively preventing and treating constipation (Hidaka *et al.*, 1986; Kleesen *et al.*, 1997). Similar benefits apply to dogs when supplemented at low concentrations (1–2 %).

Macronutrient digestibility is a primary contributor to faecal characteristics and was a primary outcome measured in 13 of the canine prebiotic experiments. While all 13 experiments measured total tract digestibility, five also measured nutrient digestibility at the terminal ileum. While the results appear to be dose-dependent, the literature suggests prebiotic consumption decreases total tract macronutrient [e.g. OM, crude protein (CP)] digestibility by dogs. Similar to other fermentable fibre sources, prebiotic supplementation often increases faecal nitrogen (bacterial mass) concentrations (Cummings *et al.*, 1979; Cummings and Bingham, 1987). Ileal CP digestibility, a more accurate measure of protein availability to the host, has not been affected in the canine experiments published thus far. Because fermentable substrates are known to influence N metabolism, it was the focus of five canine prebiotic experiments. In three of these experiments, prebiotic consumption resulted in increased faecal N excretion in the form of bacterial N. The increased use of colonic N for bacterial protein synthesis, resulting in greater faecal N and decreased total tract N digestibility, is not a negative outcome of prebiotic consumption. As will be discussed below, these outcomes may, in fact, prove to be beneficial as regards overall health of the colon.

Rodent studies have reported improved mineral absorption in animals fed prebiotics. This effect is believed to be due to an increased production of SCFA, which is said to improve colonic absorption of Na^+, Ca^{2+}, and Cl^- (Binder and Mehta, 1989; Lutz and Scharrer, 1991). Increased absorption is thought to be due to exchange mechanisms (e.g. Na^+—H^+ exchange) present in the colon. The work of Beynen *et al.* (2002) has been the only canine experiment to evaluate prebiotic effects on mineral absorption. Similar to rodent studies, increased Ca and Mg absorption was reported in dogs consuming 1 % (w/v) OF. As mineral absorption efficiency is important in several canine life-stages, physiological states, and/or breeds, more research on this topic is needed.

By definition, a nondigestible carbohydrate is considered to be a 'prebiotic' only if it stimulates the activity or number of one or a select number of microorganisms. Therefore, intestinal microbial populations have been one of the most common outcome variables measured in canine experiments (14 experiments). Lactate-producing bacteria such as *Lactobacillus* and *Bifidobacterium* are considered beneficial members of the colonic microbiota. Even though *Clostridium spp.* and *E. coli* are part of the commensal bacteria and may reside in healthy animals without disease for years, they are considered to be potential pathogens and often are viewed as examples of detrimental bacteria. Similar to the human literature (Gibson *et al.*, 1995; Bouhnik *et al.*, 1999), canine experiments often have reported beneficial outcomes of prebiotic supplementation on microbial populations. Approximately 50 % of the canine experiments reported greater faecal bifidobacteria and lower faecal clostridia concentrations in dogs consuming prebiotics (Table 10.1). Although not measured in all experiments, lactobacilli concentrations also were reported to increase in faeces of prebiotic-fed dogs. In one experiment, lactobacilli concentrations were reported to increase in both ileal and faecal samples (Swanson *et al.*, 2002c). The response of other microbial species to prebiotic supplementation has been inconsistent. Given these results, it appears that prebiotics beneficially manipulate intestinal microbial populations in dogs. With the increased use of molecular techniques to identify bacterial species and evaluate dietary responses on colonic microbiota, this line of research will continue to increase our understanding of microbe–host interactions and the associated influences of dietary manipulation.

While bacterial numbers alone may indicate some degree of intestinal health status, fermentation end-product concentrations also are important criteria when evaluating prebiotic potential. Acetate, propionate, and butyrate are the primary SCFA produced from carbohydrate fermentation, and butyrate is the main energy source for colonocytes (Roediger, 1980). Along with lactate, an intermediary product of bacterial fermentation, these organic acids decrease luminal pH and assist in pathogen resistance. *In vitro* experiments suggest that SCFA increase the expression of intestinal heat shock proteins, a highly conserved family of stress proteins crucial to the integrity of the gastrointestinal tract (Ren *et al.*, 2001). Because of their demonstrated importance, a considerable number of the canine prebiotic experiments (9) measured faecal SCFA and/or lactate concentrations. Of these experiments, about 50 % reported increased faecal concentrations in dogs consuming prebiotics. Faecal acetate, propionate, total SCFA, and lactate concentrations followed a common trend, increasing as a result of prebiotic consumption. Fecal butyrate concentration results were equivocal, with prebiotics leading to increased or decreased concentrations depending on the experiment.

Taken together, these experiments suggest a common trend for increased faecal SCFA concentrations, a potentially beneficial outcome of prebiotic supplementation. Because SCFA are rapidly absorbed by the host (von Englehardt *et al.*, 1989), it is unknown how well faecal SCFA concentrations correlate with luminal SCFA concentrations. Although the measurement of SCFA concentrations in the proximal colon would be most useful, sample collection in this part of the gastrointestinal tract is often not possible in canine experiments. Although the dataset is small (two experiments), it appears that prebiotic supplementation increases intestinal length, weight, and surface area, colonic blood flow, and small intestinal carrier-mediated glucose uptake in dogs. These effects on intestinal morphology may be expected from prebiotic supplementation, as increased carbohydrate fermentation and consequent SCFA production are known to result in intestinal hypertrophy.

In contrast to the benefits that occur from carbohydrate fermentation, the outcomes of microbial breakdown of proteins in the colon are usually detrimental. Increasing protein flow to the colon provides more substrates for potential pathogens such as *Clostridium perfringens*, which is known for its ability to degrade amino acids and produce faecal odour. In addition to being responsible for faecal odour, protein catabolites are harmful to the intestinal epithelia. For example, high ammonia concentrations are suspected to contribute to colon carcinogenesis (Visek, 1978), while phenols have been positively associated with intestinal disease (Ramakrishna *et al.*, 1991). Fermentable carbohydrates, including prebiotics, may decrease colonic protein catabolite concentrations by providing gut microbiota with an additional energy supply. In the colon, bacteria act as N sinks, utilising undigested proteins and their metabolites in the presence of energy for their protein synthesis (Cummings *et al.*, 1979). Bacteria use ammonia as a major source of N, and other forms of protein or amino acids are deaminated to ammonia before being used metabolically (Jackson, 1995). Thus, by providing sufficient energy in the form of fermentable carbohydrates, bacteria are able to use available amino acids for their own protein synthesis rather than using them for energy, a process that results in the production of putrefactive compounds.

Seven canine experiments measured faecal protein catabolite concentrations as a means of evaluating prebiotic efficacy. Faecal ammonia concentrations were highly variable across experiments and did not indicate any clear response to prebiotic supplementation. Branched-chain fatty acid (BCFA) concentrations followed a similar pattern, resulting in contradictory data. The most convincing evidence was the effects of prebiotics on phenol and indole concentrations, which were substantially decreased in four experiments. No changes in phenol and indole concentrations were noted in the other three experiments. Given the recent popularity of high protein diets use by humans as well as canines, a need to identify ingredients capable of decreasing the putrefactive compounds produced from excess protein intake exists. The results of the experiments performed thus far suggest a potential role for prebiotics in this regard.

The intestine is the largest immune organ of the body (Jalkanen, 1990), with the gut-associated lymphoid tissues (GALT) possessing approximately 80 % of the body's immunological-secreting cells (Brandtzaeg *et al.*, 1989) and more than 50 % of the immune effector cells (McKay and Perdue, 1993). It has been well established that the development of GALT is highly dependent on the colonisation of bacterial populations in the gut. Even after the gastrointestinal immune system is fully developed, immune cells are on constant guard against pathogen invasion. Given the role of the microbiota on GALT and the potential for prebiotics to manipulate gut microbial ecology, it is logical to hypothesise enhanced gut immunity in response to prebiotic supplementation. To date, three canine experiments have evaluated the effects of prebiotics on immune indices in the blood [white blood cell concentrations, serum immunoglobulin A (IgA)] and/or gastrointestinal tract (IgA concentrations in ileal digesta). Immunoglobulin A is important in mucosal immunity as it inhibits the attachment and penetration of bacteria in the lumen, increases mucus secretion (McKay and Perdue, 1993), and prevents inflammatory reactions that potentially would result in damage to the epithelial tissues (Russell *et al.*, 1989). The presence of normal IgA concentrations may play a role in some intestinal diseases, as reduced IgA concentrations have been associated with Crohn's disease in humans (MacDermott *et al.*, 1986) and with small intestinal bacterial overgrowth in dogs (Batt *et al.*, 1991). Each of the three experiments reported significant effects of prebiotics on immune cell populations. Results across experiments have been conflicting and no clear trends can be identified at this time. Future experiments should not only measure immune cell number, but also test their functional capacity (e.g. lymphocyte blastogenesis; phagocytic activity of neutrophils) to more accurately evaluate prebiotic effects on immune capacity.

10.4 Feline Prebiotic Experiments

A literature search on prebiotic use in cats identified only four peer-reviewed publications (Table 10.2). Although cats are strict carnivores and are metabolically different than dogs in many ways, the potential benefits from prebiotic supplementation also exist in this species. Given the small number of experiments performed to date, it is evident that more research is needed in this area. Of the feline prebiotics experiments that have been published, inulin, lactosucrose, OF and FOS have been tested.

Because outcome variables were different in each of the four experiments, few trends can be drawn from this small dataset. However, a few effects are similar to those observed in canine experiments and demonstrate potential in feline diets. First, prebiotic supplementation may lead to an increased number of defaecations per day and wet faecal volume, decreased faecal DM percentage and softer faeces. Secondly, prebiotics appear to decrease total tract nutrient digestibility. As in dogs, decreased CP digestibility

Table 10.2 *In vivo experiments, listed in chronological order, evaluating the effects of prebiotics in cats*

Reference	Outcome variables quantified	Animals/ treatment	Dietary information	Daily prebiotic dose; source	Time on treatment	Major findings
Terada *et al.*, 1993	(a) Fecal characteristics (b) Fecal metabolites (c) Fecal microbial ecology	8 adults (3 M and 5 F)	(a) Diet type: dry breakfast; moist dinner (b) Primary ingredients: not provided (c) Chemical composition: breakfast: 90 % DM, 30 % CP, 10 % fat, 22 % CF and 6 % ash; dinner: 23 % DM, 19 % CP, 1 % fat, 0.5 % CF, and 2 % ash	(a) Control (b) 175 mg lactosucrose	2 weeks	Lactosucrose: (a) ↑ fecal bifidobacteria and lactobacilli*** (b) ↓ fecal fusobacteria, lecithinase-positive clostridia, enterobacteria and staphylococci*** (c) ↓ fecal lecithinase-negative clostridia** (d) ↓ fecal ammonia, ethylphenol and indole concentrations** (e) ↓ urinary ammonia concentrations**
Sparkes *et al.*, 1998	(a) Fecal microbial ecology	12 adults (6 M and 6 F)	(a) Diet type: dry (b) Primary ingredients: not provided (c) Chemical composition: 93 % DM, 39 % CP, 19 % fat and 29 % NFE	(a) Control (no prebiotic) (b) 0.75 % FOS	12 weeks	FOS: (a) ↓ fecal *E. coli* and ↑ fecal lactobacilli and bacteroides** (b) ↓ fecal *C. perfringens**
Groeneveld *et al.*, 2001	(a) N metabolism	8 adults (3 M and 5 F)	(a) Diet type: moist (b) Primary ingredients: not provided (c) Chemical composition: 18 % DM, 8 % CP, 4 % fat and 2 % ash	(a) Control (no prebiotic) (b) 0.25 g inulin $kg^{-1} BW^{0.75}$	19–22 days	Inulin: (a) ↓ total tract N digestibility** (b) ↑ fecal N excretion**

Hesta *et al.*, 2001	Study 1 (a) Fecal characteristics	Study 1 8 adults (gender not defined)	Study 1 (a) Diet type: dry (b) Primary ingredients: not provided (c) Chemical composition: 37 % CP, 23 % fat, 3 % CF and 6 % ash	Study 1 (a) Control (no prebiotic) (b) 3 % OF (c) 6 % OF (d) 9 % OF	Study 1 12 days	Study 1 6 % and 9 % OF: (a) ↓ fecal DM %, fecal pH and fecal score (softer faeces)*** (b) ↑ no. of defecations day^{-1} and wet fecal volume***
	Study 2 (a) Fecal characteristics (b) Fecal metabolites (c) Total tract nutrient digestibility	Study 2 8 adults (gender not defined)	Study 2 (a) Diet type: dry (b) Primary ingredients: not provided (c) Chemical composition: 37 % CP, 23 % fat, 3 % CF, and 6 % ash	Study 2 (a) Control (no prebiotic) (b) 3 % OF (c) 3 % inulin (d) 6 % inulin	Study 2 17 days	Study 2 3 % inulin: (a) ↓ urine production** (b) ↓ total tract CP and fat digestibilities*** (c) ↑ fecal bacterial N %*** 6 % inulin: (a) ↓ urine production** (b) ↓ fecal DM %** (c) ↓ total tract CP and fat digestibilities*** (d) ↑ fecal bacterial N %*** (e) ↓ fecal acetate (as % of total SCFA)*** (f) ↑ fecal total SCFA and valerate concentrations** 3 % OF: (a) ↓ urine production** (b) ↓ total tract CP and fat digestibilities*** (c) ↑ fecal bacterial N %***

BW, body weight; CF, crude fiber; CP, crude protein; DM, dry matter; F, female; FOS, fructooligosaccharide; M, male; N, nitrogen; NFE, nitrogen free extract; OF, oligofructose; SCFA, short-chain fatty acids.
*$P < 0.10$; **$P < 0.05$; ***$P < 0.01$.

corresponds with increased faecal N concentrations. Third, prebiotics manipulate microbial populations in a beneficial manner, resulting in increased bifidobacteria and lactobacilli and decreased *C. perfringens* concentrations. Lastly, supplementation appears to result in greater faecal SCFA and lower faecal protein catabolites in cats fed prebiotics vs. cats fed the control diet.

10.5 Prebiotic Assessment

The authors have attempted to briefly summarise all published data from canine and feline prebiotic experiments to date. Because so many outcome variables have been measured, it often is difficult to effectively compare prebiotics against each other. Because qualitative outcomes are often difficult to compare, Palframan *et al.* (2003) made the first attempt to develop a quantitative method of comparing prebiotics. The proposed prebiotic index (PI) equation was based on changes in key bacterial groups during fermentation. Bacterial groups included in the analyses were bifidobacteria, lactobacilli, clostridia and bacteroides. When applied to actual experimental data, output from the PI equation was reported to agree with qualitative conclusions drawn from experiments (Palframan *et al.*, 2003). These authors recently expanded this equation to include SCFA and lactate data, resulting in a more accurate measure of prebiotic effect (MPE) (Vulevic *et al.*, 2004). Given the importance of SCFA and lactate production, incorporating such data into the equation was well justified.

As more researchers study the impacts of protein catabolites on intestinal health, it is apparent that these compounds are highly toxic and may increase susceptibility to cancer and/or intestinal disease. To date, the family of protein catabolites having the most consistent response to prebiotic supplementation has been the phenols and indoles. Phenols and indoles, which originate from breakdown of aromatic amino acids, have been consistently and dramatically decreased in dogs and cats consuming prebiotics. Therefore, in addition to the bacteria, SCFA and lactate variables already included in the current MPE equation, we propose the addition of a 'phenols/indoles' variable. Because this revised equation would also take into consideration some of the putrefactive compounds associated with hindgut fermentation, its addition to the equation would be expected to further increase the accuracy of prebiotic assessment.

10.6 Conclusions and Future Direction

The increasing number of published canine and feline experiments over the past decade portrays the interest surrounding the inclusion of prebiotics in companion animal diets. From the limited number of experiments published in this area, it appears that prebiotic supplementation has several beneficial effects in the gastrointestinal tract of dogs and cats (i.e. positive shifts in microbial populations, decreases in faecal protein catabolites, changes in immune status). However, more research is required to identify optimal doses, life-stages most likely to benefit, and disease states likely to be avoided or treated with prebiotic supplementation. Most of the experiments performed to date have used healthy adult dogs and cats. In the future, experiments also must test prebiotic supplementation on animals of different life-stages (e.g. weanlings, gestation/lactation, geriatric animals) and suffering various disease states.

Although several hypotheses have been postulated, more research also is needed to understand mechanisms by which prebiotics function. To completely understand these mechanisms, scientists must first accurately identify all microbial species inhabiting the gut and determine how they interact with one another and the epithelial cells of the gut. Although this task is far from complete, researchers have begun to make some progress with the use of molecular biological techniques. In fact, several microbial genomes already have been sequenced.

In addition to classical methodology, the use of molecular biological techniques will advance knowledge in several respects. First, scientists will be able to more accurately identify and quantify microbial species present in the gut, especially those that are unable to be grown in culture. Second, gene mapping will identify and locate genes of importance/interest. Third, by using microarray technology, which measures hundreds to thousands of genes simultaneously, gene expression profiles may be generated of microbial species or gut epithelium. This technology generates a global view of gene expression, enabling scientists to see the 'big picture' as opposed to only a few genes of interest. The use of laser capture microdissection and RNA amplification will allow researchers to generate gene expression profiles from minute quantities of gut tissue. This technology will enable researchers to measure responses of specific gut epithelial cells to gut microbes and metabolites. Finally, bioinformatic modeling and statistical analyses may be used to make sense of the enormous datasets generated by microarray technology. The completion of additional experiments using dogs and cats of different life-stages and physiological states, in combination with molecular biological techniques, will greatly enhance our understanding of prebiotic function, perhaps enhancing the health and well being of these important animal species.

References

Batt, R. M., Barnes, A., Rutgers, H. C. and Carter, S. C. 1991. Relative IgA deficiency and small intestinal bacterial overgrowth in German shepherd dogs. *Res. Vet. Sci.* **50**: 106–111.

Beynen, A. C., Baas, J. C., Hoekemeijer, P. E., Kappert, H. J., Bakker, M. H., Koopman, J. P. and Lemmens, A. G. 2002. Faecal bacterial profile, nitrogen excretion and mineral absorption in healthy dogs fed supplemental oligofructose. *J. Anim. Physiol. Anim. Nutr.* **86**: 298–305.

Binder, H. J. and Mehta, P. 1989. Short-chain fatty acids stimulate active sodium and chloride absorption *in vitro* in the rat distal colon. Gastroenterol. **96**: 989–996.

Bouhnik, Y., Vahedi, K., Achour, L., Attar, A., Salfati, J., Pochart, P., Marteau, P., Flourié, B., Bornet, F. and Rambaud. J.-C. 1999. Short-chain fructo-oligosaccharide administration dose-dependently increases fecal bifidobacteria in healthy humans. *J. Nutr.* **129**: 113–116.

Brandtzaeg, P., Halstensen, T. S. and Kett, K. 1989. Immunobiology and immunopathology of human gut mucosa: humoral immunity and intraepithelial lymphocytes. *Gastroenterol.* **97**: 1562–1584.

Buddington, R. K., Buddington, K. K. and Sunvold, G. D. 1999. Influence of fermentable fiber on small intestinal dimensions and transport of glucose and proline in dogs. *Am. J. Vet. Res.* **60**: 354–358.

Cummings, J. H. and Bingham, S. A. 1987. Dietary fiber, fermentation and large bowel cancer. *Cancer Surv.* **6**: 601–621.

Cummings, J. H., Hill, M. J., Bones, E. S., Branch, W. J. and Jenkins, D. J. A. 1979. The effect of meat protein and dietary fiber on colonic function and metabolism. II. Bacterial metabolites in feces and urine. *Am. J. Clin. Nutr.* **32**: 2094–2101.

Diez, M., Hornick, J.-L., Baldwin, P. and Istasse, L. 1997. Influence of a blend of fructo-oligosaccharides and sugar beet fiber on nutrient digestibility and plasma metabolite concentrations in healthy Beagles. *Am. J. Vet. Res.* **58**: 1238–1242.

Diez, M., Hornick, J. L., Baldwin, P., Van Eenaeme, C. and Istasse, L. 1998. The influence of sugar-beet fibre, guar gum and inulin on nutrient digestibility, water consumption and plasma metabolites in healthy Beagle dogs. *Res. Vet. Sci.* **64**: 91–96.

Flickinger, E. A., Schreijen, E. M. W. C., Patil, A. R., Hussein, H. S., Grieshop, C. M., Merchen, N. R. and Fahey Jr, G. C. 2003. Nutrient digestibilities, microbial populations, and protein catabolites as affected by fructan supplementation of dog diets. *J. Anim. Sci.* **81**: 2008–2018.

Flickinger, E. A., Wolf, B. W., Garleb, K. A., Chow, J., Leyer, G. J., Johns, P. W. and Fahey Jr, G. C. 2000. Glucose-based oligosaccharides exhibit different *in vitro* fermentation patterns and affect *in vivo* apparent nutrient digestibility and microbial populations in dogs. *J. Nutr.* **130**: 1267–1273.

Gibson, G. R., Beatty, E. R., Wang, X. and Cummings, J. H. 1995. Selective stimulation of bifidobacteria in the human colon by oligofructose and inulin. *Gastroenterol.* **108**: 975–982.

Gibson, G. R. and Roberfroid, M. B. 1995. Dietary modulation of the human colonic microbiota: introducing the concept of prebiotics. *J. Nutr.* **125**: 1401–1412.

Grieshop, C. M., Flickinger, E. A., Bruce, K. J., Patil, A. R., Czarnecki-Maulden, G. L. and Fahey Jr, G. C. 2004. Gastrointestinal and immunological responses of senior dogs to chicory and mannan-oligosaccharides. *Arch. Anim. Nutr.* **58**: 483–493.

Groeneveld, E. A., Kappert, H. J., Van der Kuilen, J. and Beynen, A. C. 2001. Consumption of fructooligosaccharides and nitrogen excretion in cats. *Int. J. Vitam. Nutr. Res.* **71**: 254–256.

Hesta, M., Janssens, G. P. J., Debraekeleer, J. and De Wilde, R. 2001. The effect of oligofructose and inulin on faecal characteristics and nutrient digestibility in healthy cats. *J. Anim. Physiol. Anim. Nutr.* **85**: 135–141.

Hesta, M., Roosen, W., Janssens, G. P. J., Millet, S. and De Wilde, R. 2003. Prebiotics affect nutrient digestibility but not faecal ammonia in dogs fed increased dietary protein levels. *Britt. J. Nutr.* **90**: 1007–1014.

Hidaka, H., Eida, T., Takizawa, T., Tokunaga, T. and Tashiro, Y. 1986. Effects of fructooligosaccharides on intestinal flora and human health. *Bifido. Microflora* **5**: 37–50.

Hooper, L. V., Wong, M. H., Thelin, A., Hansson, L., Falk, P. G. and Gordon, J. I. 2001. Molecular analysis of commensal host–microbial relationships in the intestine. *Science* **291**: 881–884.

Howard, M. D., Kerley, M. S., Mann, F. A., Sunvold, G. D. and Reinhart, G. A. 1999. Blood flow and epithelial cell proliferation of the canine colon are altered by source of dietary fiber. *Vet. Clin. Nutr.* **6**: 8–15.

Howard, M. D., Kerley, M. S., Sunvold, G. D., and Reinhart, G. A. 2000. Source of dietary fiber fed to dogs affects nitrogen and energy metabolism and intestinal microflora populations. *Nutr. Res.* **20**: 1473–1484.

Jackson, A. A. 1995. Salvage of urea-nitrogen and protein requirements. *Proc. Nutr. Soc.* **54**: 535–547.

Jalkanen, S. 1990. Lymphocyte homing into the gut. *Immunopathol.* **12**: 153–164.

Karr-Lilienthal, L. K., Grieshop, C. M., Spears, J. K., Patil, A. R., Czarnecki-Maulden, G. L., Merchen, N. R. and Fahey Jr, G. C. 2004. Estimation of the proportion of bacterial nitrogen in canine feces using diaminopimelic acid as an internal bacterial marker. *J. Anim. Sci.* **82**: 1707–1712.

Kleesen, B., Sykura, B., Zunft, H. J. and Blaut, M. 1997. Effects of inulin and lactose on fecal microflora, microbial activity, and bowel habit in elderly constipated persons. *Am. J. Clin. Nutr.* **65**: 1397–1402.

Lutz, T. and Scharrer, E. 1991. Effect of short-chain fatty acids on calcium absorption by the rat colon. *Exp. Physiol.* **76**: 615–618.

MacDermott, R. P., Nash, G. S., Bertovich, M. J., Mohrman, R. F., Kodner, I. J., Delacroix, D. L. and Vaerman, J. P. 1986. Altered patterns of secretion of monomeric IgA and IgA subclass 1 by intestinal mononuclear cells in inflammatory bowel disease. *Gastroenterol.* **91**: 379–385.

McKay, D. M. and Perdue, M. H. 1993. Intestinal epithelial function: the case for immunophysiological regulation. *Dig. Dis. Sci.* **38**: 1377–1387.

Ogawa, Y., Miyazato, T. and Hatano, T. 2000. Oxalate and urinary stones. *World J. Surg*. **24**: 1154–1159.

O'Neill, D. and Phillips, V. 1992. A review of the control of odour nuisance from livestock buildings: Part 3. Properties of the odorous substances which have been identified in livestock wastes or in air around them. *J. Agric. Eng. Res.* **53**: 23–50.

Palframan, R., Gibson, G. R. and Rastall, R. A. 2003. Development of a quantitative tool for the comparison of the prebiotic effect of dietary oligosaccharides. *Lett. Appl. Microbiol.* **37**: 281–284.

Propst, E. L., Flickinger, E. A., Bauer, L. L., Merchen, N. R. and Fahey Jr, G. C. 2003. A dose–response experiment evaluating the effects of oligofructose and inulin on nutrient digestibility, stool quality, and fecal protein catabolites in healthy adult dogs. *J. Anim. Sci.* **81**: 3057–3066.

Ramakrishna, B. S., Roberts-Thomas, I. C., Pannall, P. R. and Roediger, W. E. W. 1991. Impaired sulphation of phenol by the colonic mucosa in quiescent and active colitis. *Gut* **32**: 46–49.

Ren, H., Musch, M. W., Kojima, K., Boone, D., Ma, A. and Chang, E. B. 2001. Short-chain fatty acids induce intestinal epithelial heat shock protein 25 expression in rats and IEC 18 cells. *Gastroenterol.* **121**: 631–639.

Roediger, W. E. W. 1980. Role of anaerobic bacteria in the metabolic welfare of the colonic mucosa in man. *Gut* **21**: 793–798.

Russell, T. J. 1998. The effect of natural source of non-digestible oligosaccharides on the fecal microflora of the dog and effects on digestion. Friskies R & D Center/Missouri, Copyright © Friskies – Europe 1998.

Russell, M. W., Reinholdt, J. and Kilian, M. 1989. Anti-inflammatory activity of human IgA antibodies and their Fab α fragments: inhibition of IgG-mediated complement activation. *Eur. J. Immunol.* **19**: 2243–2249.

Sparkes, A. H., Papasouliotis, K., Sunvold, G., Werrett, G., Gruffydd-Jones, E. A., Egan, K., Gruffydd-Jones, T. J. and Reinhart, G. 1998. Effect of dietary supplementation with fructo-oligosaccharides on fecal flora of healthy cats. *Am. J. Vet. Res.* **59**: 436–440.

Spoelstra, S. 1980. Origin of objectionable odorous components in piggery wastes and the possibility of applying indicator components for studying odour development. *Agric. Environ.* **5**: 241–260.

Strickling, J. A., Harmon, D. L., Dawson, K. A. and Gross, K. L. 2000. Evaluation of oligosaccharide addition to dog diets: Influences on nutrient digestion and microbial populations. *Anim. Feed Sci. Technol.* **86**: 205–219.

Swanson, K. S., Grieshop, C. M., Flickinger, E. A., Bauer, L. L., Chow, J., Wolf, B. W., Garleb, K. A. and Fahey Jr, G. C. 2002a. Fructooligosaccharides and *Lactobacillus acidophilus* modify gut microbial populations, total tract nutrient digestibilities, and fecal protein catabolite concentrations in healthy adult dogs. *J. Nutr.* **132**: 3721–3731.

Swanson, K. S., Grieshop, C. M., Flickinger, E. A., Bauer, L. L., Healy, H.-P., Dawson, K. A., Merchen, N. R. and Fahey Jr, G. C. 2002b. Supplemental fructooligosaccharides and mannanoligosaccharides influence immune function, ileal and total tract nutrient digestibilities, microbial populations and concentrations of protein catabolites in the large bowel of dogs. *J. Nutr.* **132**: 980–989.

Swanson, K. S., Grieshop, C. M., Flickinger, E. A., Healy, H.-P., Dawson, K. A., Merchen, N. R. and Fahey Jr, G. C. 2002c. Effects of supplemental fructooligosaccharides plus mannanoligosaccharides on immune function and ileal and fecal microbial populations in adult dogs. *Arch. Anim. Nutr.* **56**: 309–318.

Terada, A., Hara, H., Kato, S., Kimura, T., Fujimori, I., Hara, K., Maruyama, T. and Mitsuoka, T. 1993. Effect of lactosucrose (4^G-β-D-galactosylsucrose) on fecal flora and fecal putrefactive products of cats. *J. Vet. Med. Sci.* **55**: 291–295.

Terada, A., Hara, H., Oishi, T., Matsui, S., Mitsuoka, T., Nakajyo, S., Fujimori, I. and Hara, K. 1992. Effect of dietary lactosucrose on faecal flora and faecal metabolites of dogs. *Microb. Ecol. Health Dis.* **5**: 87–92.

Twomey, L. N., Pluske, J. R., Rowe, J. B., Choct, M., Brown, W. and Pethick, D. W. 2003. The effects of added fructooligosaccharide (Raftilose® P95) and inulinase on faecal quality and digestibility in dogs. *Anim. Feed Sci. Technol.* **108**: 83–93.

Tzortzis, G., Baillon, M.-L. A., Gibson, G. R. and Rastall, R. A. 2004a. Modulation of anti-pathogenic activity in canine-derived *Lactobacillus* species by carbohydrate growth substrate. *J. Appl. Microbiol.* **96**: 552–559.

Tzortzis, G., Goulas, A. K., Baillon, M.-L. A., Gibson, G. R. and Rastall, R. A. 2004b. *in vitro* evaluation of the fermentation properties of galactooligosaccharides synthesised by α-galactosidase from *Lactobacillus reuteri*. *Appl. Microbiol. Biotechnol.* **64**: 106–111.

Vickers, R. J., Sunvold, G. D., Kelley, R. L. and Reinhart, G. A. 2001. Comparison of fermentation of selected fructooligosaccharides and other fiber substrates by canine colonic microflora. *Am. J. Vet. Res.* **62**: 609–615.

Visek, W. J. 1978. Diet and cell growth modulation by ammonia. *Am. J. Clin. Nutr.* **31** (Suppl. 1): S216–S220.

von Englehardt, W., Ronnau, K., Rechkemmer, G. and Sakata, T. 1989. Absorption of short-chain fatty acids and their role in the hindgut of monogastric animals. *Anim. Feed Sci. Tech.* **23**: 43–53.

Vulevic, J., Rastall, R. A. and Gibson, G. R. 2004. Developing a quantitative approach for determining the in vitro prebiotic potential of dietary oligosaccharides. *FEMS Microbiol. Lett.* **236**: 153–159.

Weese, J. S., Weese, H. E., Yuricek, L. and Rousseau, J. 2004. Oxalate degradation by intestinal lactic acid bacteria in dogs and cats. *Vet. Microbiol.* **101**: 161–166.

Willard, M. D., Simpson, R. B., Cohen, N. D. and Clancy, J. S. 2000. Effects of dietary fructooligosaccharide on selected bacterial populations in feces of dogs. *Am. J. Vet. Res.* **61**: 820–825.

Willard, M. D., Simpson, R. B., Delles, E. K., Cohen, N. D., Fossum, T. W., Kolp, D. and Reinhart, G. 1994. Effects of dietary supplementation of fructo-oligosaccharides on small intestinal bacterial overgrowth in dogs. *Am. J. Vet. Res.* **55**: 654–659.

Zentek, J., Marquart, B. and Pietrzak, T. 2002. Intestinal effects of mannanoligosaccharides, transgalactooligosaccharides, lactose and lactulose in dogs. *J. Nutr.* **132**: 1682S–1684S.

Zentek, J., Marquart, B., Pietrzak, T., Ballèvre, O. and Rochat, F. 2003. Dietary effects on bifidobacteria and *Clostridium perfringens* in the canine intestinal tract. *J. Anim. Physiol. Anim. Nutr.* **87**: 397–407.

11

Prebiotics: Past, Present and Future

Jeff D. Leach, Robert A. Rastall and Glenn R. Gibson

In this concluding chapter we examine the history and evolution of prebiotics in the human diet and discuss current and future issues.

11.1 Prebiotics – the Past

The occurrence of fructans in a significant portion of the world's flora all but guaranteed that prebiotic inulin and oligofructose were consumed by our Pliocene and Pleistocene ancestors millions of years ago. As our early ancestors moved from the rainforest to the parched savanna-woodlands of subtropical Africa, subsurface tubers, rhizomes, corms, and perennial bulbs, many rich in prebiotics, would have been a familiar and important source of energy. Many of these same resources served as staples for the modern hunter-gatherer and farming groups still inhabiting subtropical environs. However, digestion-inhibiting compounds and plant toxins present in many below-ground food sources would have limited their role as staples in early diet until technological adaptations, such as fire, were introduced (Ragir, 2000). Nevertheless, as early members of the genus *Homo* began their evolutionary march to mammalian dominance, the inclusion of prebiotics in the diet would have no doubt conferred a selective advantage for the consuming population.

The antiquity and evolutionary role of prebiotics in the human diet is very poorly understood. Continued advances in our understanding of the health benefits of prebiotics through modern *in vitro* and *in vivo* studies would benefit from the time depth provided by archaeology. There is a growing awareness, in the nutritional and medical communities, that aspects of prehistoric or hunter-gatherer diets may serve as models for developing therapeutic diets that reduce risk to diseases of modern civilisation (Eaton and Konner,

Prebiotics: Development and Application Edited by G.R. Gibson and R.A. Rastall.

1985; Cordain *et al.*, 2002). As the archaeological evidence reveals, prebiotics have long been part of the human diet and in quantities for some areas and time periods that far exceed those currently consumed by modern populations (Van Loo *et al.*, 1995).

The physical evidence for plant consumption by our early ancestors is virtually nonexistent, owing to poor preservation of organic plant parts in the archaeological records. We must travel millions of years forward to the Upper Palaeolithic (~40 000–12 000 years ago) of Western Europe and the Mediterranean Basin and to the Early Holocene (~10 000 years ago) of North America before significant direct and indirect evidence of prebiotic food consumption is evident.

Decades of large-scale archaeological research in North America has documented extensive exploitation of prebiotic rich plants such as agave (*Agave* spp.), sotol (*Dasylirion* spp), camas (e.g. *Camasia quamash*, *C. leichtlinii*) and wild onion (*Allium* spp.). While a great number of inulin-bearing plants were known as food sources among the prehistoric and historic groups of North America (Wandsnider 1997), these particular plants by far provide the oldest evidence of prebiotic consumption in North America, dating back over 9000 years.

In the Lower Pecos Region of the Chihuahuan Desert in west Texas along the US–Mexican border, deeply stratified cave deposits document the use of agave, sotol and onion that date back nearly 9200 years. Kept dry and preserved by the large overhangs that characterise many of the caves and shelters of the region, an extraordinary collection of human coprolites and preserved macro botanical plant remains suggest that pit-baked prebiotic foods (e.g. agave, sotol, onion) were a mainstay of this desert economy (Sobolik, 1990, 1994).

East of the Lower Pecos on western edge of the Edwards Plateau in central Texas, the deeply buried Wilson-Leonard site has produced a 2 m diameter rock-lined earth oven (Figure 11.1) used to cook the nutritious onion-like bulbs of camas (*Camassia* spp.). Charred camas bulbs recovered during excavation of the oven produced a date of ca. 8200 years before present (Bousman *et al.*, 2002). Though no charred bulbs of camas were recovered from deeper excavations, 'stone-lined hearths' underlying the camas oven were dated to ca. 9410–9990 years before present, hinting at possible earlier evidence of prebiotic use.

At the Stigewalt site in southeastern Kansas, remains of large (>2 m diameter), rock-filled earth ovens with charred onion (*Allium* spp.) bulbs dated ca. 8810–7910 years before present (Thies, 1990). As with the large oven at the Wilson-Leonard site in Central Texas, the occurrence of hand-excavated pits lined with pre-heated stones, seem to be consistently associated with the cooking of prebiotic foods. This same pattern continues throughout the American Southwest, where thousands of agave roasting pits (also known as *mescal* pits) are scattered about the landscape (Leach, 2005). Similarly, in the American northwest, large, rock-lined ovens (Figure 11.2) were used to cook as much as 1500 kg of edible camas bulbs in a single firing event (Thoms, 2003).

In the American Southwest, ideal surface conditions and slow rates of soil accumulation, accompanied by repeated use of oven facilities and subsequent accumulation of oven debris (discarded cooking stones) over multiple seasons, has made it possible to map thousands of cooking facilities, which often reach over 1 m in height and cover areas tens of metres in diameter. Synthesis of hundreds of radiocarbon dates from cook-stone facilities across extensive areas of southern North America (Leach, 2005) has revealed a

Figure 11.1 *Deeply buried earth-oven at the Wilson-Leonard site in Central Texas. Charred bulbs of camas (Camassia spp.) recovered from the feature date to ca. 8,200 years before present. (Photograph courtesy of Alston V. Thoms, Texas A&M University)*

Figure 11.2 *An example of a camas oven from the Calispell Valley, northeastern Washington, showing an intact heating element of high density cook stones. The oven shown is approximately 3 meters in diameter and could have easily held 1,000 kgs of nutritious camas bulbs (Camassia spp). Once the stones were heated in the bottom of the pit with fast burning wood charcoal, a thick layer of green leaves was placed on the hot rocks first and then the bulbs were added, covered with another layer of green leaves and pine bark (for added insulation). All of this was then sealed with a layer of earth to a height of 1.5 to 2.0 m and allowed to cook between 48 and 72 hours. The oven shown is ca. 3,500 years old. (Photograph courtesy of Alston V. Thoms, Texas A&M University)*

steady increase in prebiotic food consumption beginning over 9000 years ago, culminating in substantial intensification around 1250 years ago. The intensification of prebiotic foods in southern North America (specifically the American Southwest) coincides with increased reliance on cultivated crops such as corn (*Zea mays*), squash (*Cucurbita* sp.) and beans (*Phaseolus* sp.) and large-scale growth in human population. Therefore, while populations were making the transition to a diet heavily dependent on starchy cultivars, prebiotic foods continued to play an important and often increasing regional role in the nutritional economy.

As we see in North America, the occurrence of cook-stone technology, in the absence of recoverable plant remains, may be used as a proxy indicator to the exploitation of prebiotic foods in the archaeological record. While a great number of foods are known to have been processed with cook-stone (Wandsnider, 1997), the occurrence of large (>1 m diameter), ovens are consistently associated with many prebioitc foods.

Throughout Western Europe, similar remains of massive cooking facilities are known to occur in Wales, England, Scotland, Ireland and Scandinavia. Referred to locally as *fulacht fiadh*, recent urban development has led to the excavation of a number of these mounds, which can reach over 1 m in height and several metres in diameter, representing dozens, if not hundreds, of individual oven events. While moist ground conditions have all but destroyed any evidence of the plants that *may* have been processed in these features, radiocarbon dates on small amounts of carbonised wood charcoal from initial heating of cook-stone indicate the majority of mounds were constructed within the last 6000 years. Similar cook-stone mounds of varying sizes, dating roughly within the same time period, are known in southern parts of Australia (Holdaway *et al.*, 2002; Simon Holdaway, personal communication, 2005).

By far the oldest known evidence of cook-stone technology in Europe comes from the cave site of Abri Pataud in the Dordogne region of southern France. In excavations by a joint American–French team between 1958 and 1964, a series of cook-stone features, some >1 m in diameter, were dated to ca. 33 000–18 000 years ago (Movius Jr, 1966). While it is impossible to know if prebiotic plant tissue was processed in these ancient features, as no direct evidence in the form of plant material was reported, their use in cooking vegetal material is inferred from the overwhelming evidence of similar features recorded throughout the world. Similar stoned-lined cooking features have been dated to about 30 500 years before present on the Japanese island of Tanegashima (Dogome, 2000).

It is worth mentioning, that the earliest examples of extensive exploitation of select prebiotic grasses and cereals has recently been uncovered by receding waters at the site of Ohala II, along the margins of the Sea of Galilee in Israel. Dated at ca. 23 000 years before present, a remarkable collection of >90 000 plant parts from 142 taxa have been recovered thus far (Weiss *et al.*, 2004). Among the many small-grained grasses recovered (e.g. *Alopecurus utriculatus/arundinaceus*, *Bromus pseudobrachystachys*, *Hordeum glaucum*, *Hordeum marinum*, *Puccinella* cf. *convolute*) were the cereals barley (*Hordeum spontaneum*) and emmer wheat (*Triticum dicoccoides*), progenitors of the domesticates. Interestingly, at this same site, a small, 'oven-like hearth' in a shallow basin with burned rocks in a circular pattern was interpreted as a bread-baking oven due to its spatial association with ground stone tools that yielded starch grains of grasses, barley and wheat (Piperno *et al.*, 2004).

From the current discussion it is clear that our distant ancestors consumed, in varying quantities, plants containing prebiotics. The interesting association between cook-stone technology and prebiotics offers some proxy of initial intensification, in the absence of direct recovery of prebiotic plant tissue. We suspect, that while our ancestors have always included amounts of prebiotic plants in their diet through daily foraging activities and that some evidence for use of cook-stone is present during the Middle Palaeolithic (Mellars, 1996), its not until the onset of the Upper Palaeolithic (~40 000 years ago), with its ornaments, decorated tools, deliberate storage facilities, crudely tailored clothing, art and clear demographic pulses (Steiner, 2002), that prebiotic plant foods *may* begun to play an increasing role in the dietary evolution of our species.

Increased demographic pressure resulted in shrinking territories, making access to preferred plants and high-return animal and aquatic resources, less reliable. It is under this cultural pressure that initial intensification (increased diet breadth) of under utilised below-ground resources, many rich in prebiotics, possibly took place. This form of land-use intensification (Thoms, 2003) was the beginning of a long-term, albeit punctuated, prebiotic revolution made possible by the adaptation of cook-stone technology. The evolutionary implications of prebiotic consumption on the development and relative success of our species is unknown, and thus requires further research. However, it is safe to say, that advances in processing technology, brought about during the industrial revolution, in conjunction with the increase in 'westernised diets' and its accompanying medical maladies, have forever altered the delicate evolutionary-induced balance between food and human health, thereby resetting our metabolic clocks once again.

11.2 Prebiotics – the Present

As described, prebiotics have been an integral part of the human diet for many centuries. However, it is far more recently that their nutritional properties were recognised and this is outlined elsewhere in this book. The use of oligosaccharides to effect increases in gut bacteria seen as beneficial has been in existence in Japan for many years, however the term 'prebiotic' was only first coined in the mid-1990s (Gibson and Roberfroid, 1995). This clearly built upon the success of probiotics for microflora management approaches. The approach targets indigenous beneficial bacteria in the gut and uses nonviable food ingredients to selectively promote them. They are finding increasing use in functional foods, since there are few stability issues and they are resistant to heat. As described elsewhere, some prebiotics can also improve food quality characteristics such as mouthfeel and other textural aspects. They have also been exploited as low calorie fat replacers.

Human milk can be considered as the original prebiotic for gut microflora management in that its constituents have a very powerful stimulatory effect upon bifidobacteria. These in turn operate several mechanisms that can inhibit common pathogens (infectious agents of the gut). As a result, the gut flora of a breast-fed infant is high in bifidobacteria and these children experience fewer gastrointestinal problems than those who are bottle fed. This has led to the worldwide formula feed industry incorporating prebiotics into their products.

Criteria for classifying a food ingredient as a prebiotic include:

(1) resistance to gastric acidity, hydrolysis by mammalian enzymes and gastrointestinal absorption;
(2) fermentation by the indigenous intestinal microflora;
(3) selective stimulation of the growth and/or activity of intestinal bacteria associated with health and well-being (such as bifidobacteria, lactobacilli).

Dietary carbohydrates, such as fibres are candidate prebiotics, but oligosaccharides (short-chain sugar molecules) have proven the most effective, through their ability to confer selectivity to the target bacterial species. A range of oligosaccharides have been tested using various *in vitro* methods, animal models and human clinical trials.

11.2.1 Lactulose

Lactulose is a synthetic disaccharide in the form Gal β1-4 Fru. Lactulose was originally used as a laxative as it is not hydrolysed or absorbed in the small intestine. Lactulose has also received attention as a bifidogenic factor and has been administered as such (Tamura, 1983). Tuohy *et al.* (2002) reported the prebiotic effect of lactulose in a controlled human study.

11.2.2 Inulin and Fructo-oligosaccharides

Inulin is a saccharide of the form Glu α1-2[β Fru 1-2]$_n$ where $n > 10$ (Crittenden and Playne, 1996). The structural relatives of inulin, fructo-oligosaccharides (FOS, a lower molecular weight version) have been the best documented oligosaccharides for their effect on intestinal bifidobacteria and are considered important prebiotic substrates. They are produced in large quantities in several countries and are added to various products such as biscuits, drinks, yoghurts, breakfast cereals, table spreads and sweeteners. Human trials with FOS and inulin, include those with a controlled diet (Gibson *et al.*, 1995; Buddington *et al.*, 1996; Kleesen *et al.*, 1997; Tuohy *et al.*, 2001a,b), and demonstrate prebiotic activity of the substrates. Bifidobacteria are able to breakdown and utilise FOS due to their possession of an appropriate enzyme (a highly competitive β-fructofuranosidase).

11.2.3 Galacto-oligosaccharides

Galacto-oligosaccharides are galactose-containing oligosaccharides of the form Glu α 1-4[β Gal 1-6]$_n$ where $n = 2$–5, and are produced from lactose syrup using the transgalactosylase activity of the enzyme β-galactosidase (Crittenden, 1999). In humans, 10 g day^{-1} transgalacto-oligosaccharides significantly reduced breath hydrogen (Bouhnik *et al.*, 1997) whereas this increased in human flora associated rats fed 5 % or 10 % (w/v) TOS (Andrieux and Szylit, 1992).

The above-mentioned carbohydrates seem to have the most convincing evidence for prebiotic effects. However, others are accumulating in importance (particularly in the Japanese market) and include the following.

11.2.4 Soybean Oligosaccharides

Main oligosaccharides contained in soybeans are raffinose and stachyose. Human trials have been carried out to assess the prebiotic activity of soybean oligosaccharides (Gibson *et al.*, 2000). Variation was observed between volunteers but overall, they showed some prebiotic activity.

11.2.5 Lactosucrose

Lactosucrose is produced from a mixture of lactose and sucrose using the enzyme β-fructofuranosidase (Playne and Crittenden, 1996) and was found to be bifidogenic in pure culture studies (Tamura, 1983; Fujita *et al.*, 1991).

11.2.6 Isomalto-oligosaccharides

Isomalto-oligosaccharides (IMO) are composed of glucose monomers linked by α1-6 glucosidic linkages. Studies with a three-stage continuous culture model of the gut have shown that IMO fermentation maintained a lactic acid flora whilst also allowing the generation of butyrate (Olano-Martin *et al.*, 2000). As this is thought to be a desirable metabolite of colonic function, it may be that IMOs are effective prebiotics.

11.2.7 Gluco-oligosaccharides

An oligosaccharide preparation has been enzymatically synthesised, using a glucosyltransferase from *Leuconostoc mesenteroides,* to transfer glucose molecules from a sucrose donor to an acceptor, namely maltose (Valette *et al.*, 1993). *In vitro* data look promising, however controlled human studies are needed to confirm the prebiotic effects.

11.2.8 Xylo-oligosaccharides

Xylo-oligosaccharides (XOS) are chains of xylose molecules linked by β1-4 bonds and mainly consist of xylobiose, xylotriose and xylo-tertraose (Hopkins *et al.*, 1998). Few studies have been conducted on XOS fermentation by gut bacteria.

For the above-named oligosaccharides to be fully considered as prebiotics, then more evidence on their bacterial fermentation, and other nutritional properties, is required.

11.3 Prebiotics – the Future

As the field of prebiotics has developed, so has the methodology for investigating functionality, in particular microflora compositional changes as a response to the fermentation. Much of the early (and some of the current) literature describes studies performed on pure cultures. Typically, this involves the selection of a range of strains of *Bifidobacterium* spp., *Lactobacillus* spp. and other representative bacteria such as *Bacteroides* spp., *Clostridium* spp. and *Escherichia coli*. A carbohydrate is usually judged to be prebiotic if species of bifidobacteria (for example) metabolise the oligosaccharides more efficiently than other bacteria. The problem with this approach is, of course, that the strains selected cannot truly be considered as representative of the colonic microbiota. Such studies cannot establish that the test substrate is selectively metabolised and should be used for initial screening purposes only.

A more meaningful *in vitro* method for studying prebiotic oligosaccharides is the use of mixed culture (faecal inocula). Study of the changes in populations of selected genera or species can then establish whether the fermentation is selective. The use of faecal inocula probably gives a representation of events in the distal colon. However, more proximal areas are likely to have a more saccharolytic nature and both the composition and activities of the microbiota indigenous to the colon is variable dependent upon the region sampled.

One problem with the use of faeces, is identification of the genera and species present. Traditionally, this has been accomplished by culturing on a range of purportedly selective agars followed by morphological and biochemical tests designed to confirm culture identities. This approach is adequate to establish that a prebiotic selectively enriches defined 'desirable' organisms and depletes 'undesirable' organisms but does not give a true picture of the population changes occurring. A much more reliable approach involves the use of molecular-based methods of bacteria identification. These have advantages over culture-based technologies in that they have improved reliability and encompass more of the microflora diversity. More information is needed on the fine structure of the changes brought about by regular intake of prebiotics. With the new generation of molecular microbiological techniques now becoming available, it will be possible to gain definitive information on the species rather than genera that are influenced by the test carbohydrate. If comparative information is to be gathered on structure–function relationships in prebiotic oligosaccharides, a rigorous approach to the evaluation of these molecules will be required. Such thorough comparative studies will allow informed choices when incorporating prebiotics into foods and should increase confidence amongst consumers and regulatory authorities. As these procedures are scientifically validated and become more frequently used, the effects of prebiotics on the full gut flora diversity will also become apparent.

A recent review updated the concept of prebiotics and suggested that other components suitable for inclusion in the diet, may exert specific effects upon gut bacteria (Gibson *et al.*, 2004). These were germinated barley foodstuffs, oligodextrans, gluconic acid, gentio-oligosaccharides, pectic-oligosaccharides, mannan oligosaccharides, lactose, glutamine and hemicellulose rich substrates, resistant starch and its derivatives, oligosaccharides from melibiose, lactoferrin-derived peptides and *N*-acetylchito-oligosaccharides. As the use of more refined and reliable technologies are applied to prebiotic research then the list of candidate materials for food use is likely to grow.

Moreover, it may well be that an expansion of activities into microbial ecosystems other than the intestinal may be likely (e.g. skin applications, urinary tract, nonhuman applications). The current market approach for prebiotics is dominated by human use, particularly those susceptible to gut infections (e.g. infants, elderly, hospitalised persons). However, they are also finding increased use for animal application. In terms of companion pets, this may be related to improved gastrointestinal health as well as reduced faecal odour. For agricultural aspects, the following may be relevant:

- greater resistance to infection;
- increased growth yield and feed conversion;

- improved digestion and absorption;
- better milk/egg quality and yield;
- improved carcass quality;
- reduced contamination.

In particular, as the European Union will not allowed antibiotics for routine agricultural use with effect from 2006, the above may become more important.

Prebiotic foods, either on the market or under development, include:

- dairy products;
- beverages and health drinks;
- spreads;
- infant formulae and weaning foods;
- cereals;
- bakery products;
- confectionery chocolates, chewing gum;
- savoury product, soups;
- sauces and dressings;
- meat products;
- dried instant foods;
- canned foods;
- food supplements;
- animal feeds;
- pet foods.

Again, as the proof of efficacy increases then further applications may become clear and the above list expanded upon.

Clinical trials to determine the value of prebiotics in managing specific gut mediated disorders are ongoing, as is the use of a quantifiable index to compare efficacy (Vulevic *et al.*, 2004). More human trials are required to prove effect and identify definitive health promoting activities and mechanisms behind them. These should be hypothesis driven and well controlled. Trials in patients suffering from, and/or at risk of, clinical disorders are currently sparse. Varying expertise and techniques now exist however. Exploitation of the latest technologies and collaboration from various disciplines will help to identify outcomes.

There are several avenues of research that can be further exploited for prebiotic use. These include:

- Increased functionality. For example, the incorporation of anti-adhesive capacities against gut pathogens and their toxins.
- Preferred use in food products and perhaps defined products for particular target groups (e.g. infants, elderly, different countries, frequent travellers, institutionalised persons, those at particular risk of infection).
- Differential, species level, effects – should individual species of bifidobacteria/lactobacilli resident in the gut be seen to be more beneficial than others.
- Distal colon delivery – as most chronic disease of the colon arise distally, it would be of value to target this region of the bowel.
- Defined health outcomes and mechanisms – a collaborative effort has produced reliable research tools to determine prebiotic efficacy. With further moves into the

post-genomic era, more mechanistically driven studies in humans are feasible. These will be hypothesis driven and exploit new approaches such as microarray technology, metabolomics, proteomics. Good biomarkers of effect are already evident (microflora changes, metabolic end product formation).

The activities of the European funded cluster of projects, Food, GI-tract Functionality and Human Health (PRO-EU-HEALTH), has both probiotics and prebiotics as a central theme (www.vtt.fi/virtual/proeuhealth). There is ongoing research into:

- diet and ageing (CROWNALIFE);
- host–microbe interactions (EU AND MICROFUNCTION);
- molecular biology to detect gut bacteria (MICROBE DIAGNOSTICS);
- new processing technologies for pro/prebiotics (PROTECH);
- safety aspects (PROSAFE);
- pathogen inhibition mechanisms (PROPATH);
- chronic gut disease (PROGID);
- second generation products (DEPROHEALTH).

There are 64 research partners in 16 European countries, with a research budget of around 20 million euros. Such a significant investment has made a major impact into probiotic and prebiotic mechanisms. The project website contains appropriate updates. Important inroads into the prophylactic management of improved gut related health have been made, new research tools developed that exploit modern technologies and hypotheses are being tested in various laboratory, animal and human trials.

A further important initiative for this area has been the development of an international society dedicated to prebiotic and probiotic research (International Scientific Association for Probiotics and Prebiotics, ISAPP). This organisation has produced several research articles and summary workshop reports, with input from scientists all over the world (www.isapp.net). Its intention is to develop the principal areas for gut related dietary intervention, mechanisms of action, preferred food use, standards, research collaboration and dissemination (Rastall *et al.*, 2005; Sanders *et al.*, 2005). Examples of research reported and ongoing with prebiotics are human studies into ulcerative colitis, irritable bowel syndrome, colon cancer, gastroenteritis, immune response, mineral bioavailability, coronary heart disease, necrotising enterocolitis, autism, vaginal thrush and obesity. As with probiotics, the trials should be mechanistically driven, well controlled and use the best methodologies available (Abbott, 2004). Such careful planning is required to optimally exploit this important part of public health related science.

References

Abbott, A. 2004. Gut reaction. *Nature* **427**, 284–286.

Andrieux, C. and Szylit, O. 1992. Effects of galacto-oligosaccharides (TOS) on bacterial enzyme activities and metabolite production in rats associated with a human faecal flora. *Proceedings of the Nutrition Society* **51**, 7A.

Bouhnik, Y., Flourie, B., D' Agay-Abensour, L., Pochart, P., Gramet, G., Durand, M. and Rambaud, J. C. 1997. Administration of *trans*-galactooligosaccharides increases fecal bifidobacteria and modifies colonic fermentation metabolism in healthy humans. *Journal of Nutrition* **127**, 444–448.

Bousman, C. B., Collins, M. B., Golberg, P., Stafford, T., Guy, J., Baker, B. W., Steele, D. G., Kay, M., Kerr, A., Fredlund, G., Dering, P., Holliday, V., Wilson, D., Gose, W., Dial, S., Takac, P., Balinsky, R., Masson, M. and Powell, J. F. 2002. The Paleoindian – Archaic transition in North America: new evidence from Texas. *Antiquity* **76**, 980–990.

Buddington, R. K., Williams, C. H., Chen, S. and Witherly, S. A. 1996. Dietary supplement of neosugar alters the fecal flora and decreases activities of some reductive enzymes in human subjects. *American Journal of Clinical Nutrition* **63**, 709–716.

Cordain, L., Eaton, S. B., Brand-Miller, J., Mann, N. and Hill, K. 2002. The paradoxical nature of hunter-gatherer diets: meat-based, yet non-atherogenic. *European Journal of Clinical Nutrition* **56** (Suppl. 1), S42–S52.

Crittenden, R. G. 1999. Prebiotics. In: *Probiotics: A Critical Review* (G.Tannock, ed.). Horizon Scientific Press, Wymondham, pp. 141–156.

Crittenden, R. G. and Playne, M. J. 1996. Production, properties and applications of food-grade oligosaccharides *Trends in Food Science and Technology* **7**, 353–361.

Dogome, H. 2000. Summary (English). In: *The Yokomine C Site* (in Japanese). Town Board of Education, Minami Tane, Kagoshima, pp. 1–2.

Eaton, S. B. and Konner, M. 1985. Paleolithic nutrition. A consideration of its nature and current implications. *New England Journal of Medicine* **312**, 283–289.

Fujita, K., Hara, K., Sakai, S., Miyake, T., Yamashita, M., Tsunetomi, Y. and Mitsuoka, T. 1991. Effect of 4^{G}-β-D-galactosylsucrose (lactosucrose) on intestinal flora and its digestibility in human. *Denpun Kagaku* **38**, 249–255.

Gibson, G. R., Beatty, E. R., Wang, X. and Cummings, J. H. 1995. Selective stimulation of bifidobacteria in the human colon by oligofructose and inulin. *Gastroenterology* **108**, 975–982.

Gibson, G. R. and Roberfroid, M. B. 1995. Dietary modulation of the human colonic microbiota: introducing the concept of prebiotics. *Journal of Nutrition* **125**, 1401–1412.

Gibson, G. R., Berry Ottaway, P. and Rastall, R. A. 2000. *Prebiotics: New Developments in Functional Foods.* Chandos Publishing Limited, Oxford.

Gibson, G. R., Probert, H. M., Van Loo, J. A. E., Rastall, R. A. and Roberfroid, M. B. 2004. Dietary modulation of the human colonic microbiota: updating the concept of prebiotics. *Nutrition Research Reviews* **17**, 259–275.

Holdaway, S. J., Fanning, P. C., Jones, M., Shiner, J., Witter, D. and Nicholls, G. 2002. Variability in the chronology of late holocene aboriginal occupation on the arid margin of Southeastern Australia. *Journal of Archaeological Science* **29**, 351–363.

Hopkins, M. J., Cummings, J. H. and Macfarlane, G. T. 1998. Inter-species differences in maximum specific growth rates and cell yields of bifidobacteria cultured on oligosaccharides and other simple carbohydrate sources. *Journal of Applied Microbiology* **85**, 381–386.

Kleessen, B., Sykura, B., Zunft, H.-J. and Blaut, M. 1997. Effects of inulin and lactose on fecal microflora, microbial activity and bowel habit in elderly constipated persons. *American Journal of Clinical Nutrition* **65**, 1397–1402.

Leach, J. D. (ed.) 2005. *Learning from Once Hot Rocks.* British Archaeological Reports: International Series, London.

Mellars, P. 1996. *The Neanderthal Legacy: An Archeological Perspective from Western Europe.* Princeton University Press, Princeton, pp. 296–301.

Movius Jr, H. L. 1966. The hearths of the Upper Périgordian and Aurignacian Horizons at the Abri Pataud, Les Eyzies (Dordogne), and their possible significance. *American Anthropologist* **2**, 296–325.

Olano-Martin, E., Mountzouris, K. C., Gibson, G. R. and Rastall, R. A. 2000. *In vitro* fermentability of dextran, oligodextran and maltodextran by human gut bacteria. *British Journal of Nutrition* **83**, 247–255.

Piperno, D. R., Weiss, E., Holst, I. and Nadel, D. 2004. Processing of wild cereal grains in the Upper Paleolithic revealed by starch grain analysis. *Nature* **430**, 670–673.

Playne, M. J. and Crittenden, R. 1996. Commercially available oligosaccharides. *Bulletin of the International Dairy Foundation* **313**, 10–22.

Ragir, S. 2000. Diet and food preparation: rethinking early hominid behavior. *Evolutionary Anthropology* **9**, 153–155.

Rastall, R., Gibson, G., Gill, H., Guarner, F., Klaenhammer, T., Pot, B., Reid, G., Rowland, I. and Sanders, M. E. 2005. Modulation of the microbial ecology of the human colon by probiotics, prebiotics and synbiotics to enhance human health: an overview of enabling science and potential applications. *FEMS Microbiology Ecology* **52**, 145–152.

Sanders, M. E., Guarner, F., Mills, D., Pot, B., Rafter, J., Rastall, R., Reid, G., Ringel, Y., Rowland, I., Saarela, M. and Tuohy, K. 2005. Selected topics in probiotics and prebiotics: meeting report for the 2004 International Scientific Association for Probiotics and Prebiotics. *Current Issues in Intestinal Microbiology* **6**, 55–68.

Sobolik, K. D. 1994. Paleonutrition of the Lower Pecos region of the Chihuahuan Desert. In: *Paleonutrition: The Diet and Health of Prehistoric Americans* (K. D. Sobolik, ed.). Southern Illinois University: Center for Archaeological Investigations, Carbondale, Occasional Paper No. 22, pp. 247–264.

Sobolik, K. D. 1990. A nutritional analysis of diet as revealed in prehistoric human coprolites. *The Texas Journal of Science* **42**, 23–36.

Steiner, M. C. 2002. Carnivory, coevolution, and the geographic spread of the genus Homo. *Journal of Archaeological Research* **10**(1), 1–63.

Tamura, Z. 1983. Nutriology of bifidobacteria. *Bifidobacteria Microflora* **2**, 3–16.

Thies, R. M.1990. *The Archeology of the Stigewalt Site, 14LT351.* Kansas State Historical Society, Contract Archeology Series, Publication 7. Kansas State Historical Society, Archeology Department, Lawrence.

Thoms, A. V. 2003. Cook-stone technology in North America: evolutionary changes in domestic fire structures during the Holocene. In: *Colloque et Experimention: Le Feu Domestique et Ses Structures au Neolithic aux Auges des Metaux* (M.-C. Frere-Sautot, ed.) Collection Prehistories No. 9, Editions Monique Mergoil, pp. 87–96.

Tuohy, K. M., Finlay, R. K., Wynne, A. G. and Gibson, G. R. 2001a. A human volunteer study on the prebiotic effects of HP-Inulin – gut bacteria enumerated using fluorescent *in situ* hybridisation (FISH). *Anaerobe* **7**, 113–118.

Tuohy, K. M., Kolida, S., Lustenberger, A. and Gibson, G. R. 2001b. The prebiotic effects of biscuits containing partially hydrolyzed guar gum and fructooligosaccharides – a human volunteer study. *British Journal of Nutrition* **86**, 341–348.

Tuohy, K. M., Ziemer, C. J., Klinder, A., Knobel, Y., Pool-Zobel, B. L. and Gibson, G. R. 2002. A human volunteer study to determine the prebiotic effects of lactulose powder on human colonic bacteria. *Microbial Ecology in Health and Disease* **14**, 165–173.

Valette, P., Pelenc, V., Djouzi, Z., Andrieux, C., Paul, F., Monsan, P. and Szylit, O. 1993. Bioavailability of new synthesised glucooligosaccharides in the intestinal tract of gnotobiotic rats. *Journal of Science of Food Agriculture* **62**, 121.

Van Loo, J., Coussement, P., De Leenheer, L., Hoebregs, H. and Smits, G. 1995. On the presence of inulin and oligofructose as natural ingredients in the Western diet. *Critical Reviews in Food Science and Nutrition* **35**, 525–552.

Vulevic, J., Rastall, R. A. and Gibson, G. R. 2004. Developing a quantitative approach for determining the *in vitro* prebiotic potential of dietary oligosaccharides. *FEMS Microbiology Letters* **236**, 153–159.

Wandsnider, L. 1997. The roasted and the boiled: food composition and heat treatment with special emphasis on pit-hearth cooking. *Journal of Anthropological Archaeology* **16**, 1–48.

Weiss, E., Wetterstrom, W., Nadel, D. and Bar-Yosef, O. 2004. The broad spectrum revisited: evidence from the plant remains. *Proceedings of the National Academy of Sciences* **101**(26), 9551–9555.

Index

Prebiotics: Development and Application Edited by G.R. Gibson and R.A. Rastall.